African Smallholders
Food Crops, Markets and Policy

FSC
www.fsc.org
MIX
Paper from responsible sources
FSC® C013604

The book is in fond remembrance of the late Dr Gasper Ashimogo

AFRICAN SMALLHOLDERS
Food Crops, Markets and Policy

Edited by

Göran Djurfeldt,

Department of Sociology, Lund University, Lund, Sweden

Ernest Aryeetey

The Brookings Institution, Washington, DC, USA

and

Aida C. Isinika

Institute of Continuing Education, Sokoine University of Agriculture, Morogoro, Tanzania

www.cabi.org

CABI is a trading name of CAB International

CABI Head Office
Nosworthy Way
Wallingford
Oxfordshire OX10 8DE
UK

Tel: +44 (0)1491 832111
Fax: +44 (0)1491 833508
E-mail: cabi@cabi.org
Website: www.cabi.org

CABI North American Office
875 Massachusetts Avenue
7th Floor
Cambridge, MA 02139
USA

Tel: +1 617 395 4056
Fax: +1 617 354 6875
E-mail: cabi-nao@cabi.org

A catalogue record for this book is available from the British Library, London, UK.

Library of Congress Cataloging-in-Publication Data

African smallholders : 'food crops, markets and policy' / edited by Goran Djurfeldt, Ernest Aryeetey and Aida C. Isinika.
p. cm.
Includes bibliographical references and index.
ISBN 978-1-84593-716-4 (alk. paper)
1. Agriculture--Economic aspects--Africa, Sub-Saharan. 2. Agriculture and state--Africa, Sub-Saharan. 3. Agricultural development--Africa, Sub-Saharan. 4. Food supply--Africa, Sub-Saharan. 5. Africa, Sub-Saharan--Economic conditions. I. Djurfeldt, Göran, 1945- II. Aryeetey, Ernest, 1955- III. Isinika, Aida C., 1951- IV. Title.

HD2117.A3447 2011
338.10967--dc22

2010037677

ISBN-13: 978 1 84593 716 4

Commissioning editor: Sarah Hulbert
Production editor: Shankari Wilford

Typeset by SPi, Pondicherry, India.
Printed and bound in the UK by CPI Antony Rowe, Chippenham, UK.

Contents

Contributors vii

Acknowledgements ix

1 Introduction 1
Göran Djurfeldt, Ernest Aryeetey and Aida C. Isinika

2 African Agriculture: from Crisis to Development? 23
Hans Holmén and Göran Hydén

3 The Millennium Goals, the State and Macro-level Performance – an Overview 45
Hans Holmén

4 Smallholders Caught in Poverty – Flickering Signs of Agricultural Dynamism 74
Magnus Jirström, Agnes Andersson and Göran Djurfeldt

5 A New Era for Sub-Saharan African Agriculture? Changing Drivers of Maize Production 107
Agnes Andersson, Göran Djurfeldt, Björn Holmquist, Magnus Jirström and Sultana Nasrin

6 Maize Remittances, Market Participation and Consumption among Smallholders in Africa 138
Agnes Andersson

7 Meeting the Financial Needs of Smallholder Farmers in Ethiopia 156
Wolday Amha

8 **Agricultural Diversification, Food Self-sufficiency and Food Security in Ghana – the Role of Infrastructure and Institutions** 189
Fred M. Dzanku and Daniel Sarpong

9 **Conditions for Achieving Sustained Agricultural Intensification in Africa: Evidence from Kenya** 214
Stephen K. Wambugu, Joseph T. Karugia and Willis Oluoch-Kosura

10 **The Fertilizer Support Programme and the Millennium Development Challenge in Zambia: Is Government a Problem Solution?** 237
Hyde Haantuba, Mukata Wamulume and Richard Bwalya

11 **Has the Nigerian Green Revolution Veered Off Track?** 257
Tunji Akande, Agnes Andersson, Göran Djurfeldt and Femi Ogundele

12 **Addressing Food Self-sufficiency in Tanzania: a Balancing Act of Policy Coordination** 281
Aida C. Isinika and Elibariki E. Msuya

13 **Focusing on the Majority – Rethinking Agricultural Development in Mozambique** 316
Peter E. Coughlin

14 **Conclusions: What Direction for the Future of African Agriculture?** 354
Ernest Aryeetey, Göran Djurfeldt and Aida C. Isinika

Index 373

Contributors

Tunji Akande, Nigerian Institute of Social and Economic Research (NISER), PMB 05, UIPO, Ibadan, Nigeria.

Wolday Amha, Association of Ethiopian Microfinance Institutions (AEMFI), Africa Avenue, Kirkos Sub city, Kebele 01, House no. 227, P.O. Box 338 code 1110, Addis Ababa, Ethiopia.

Agnes Andersson, Department of Human Geography, Lund University, Sölveg. 10, SE-223 62 Lund, Sweden.

Ernest Aryeetey, The Brookings Institution, Washington, DC 20036, USA.

Richard Bwalya, Institute of Economic and Social Research, University of Zambia, P.O. Box 30900, Lusaka 10101, Zambia.

Peter E. Coughlin, EconPolicy Research Group Ltd, Av. Valentim Siti 218, 1° andar, Maputo, Mozambique; Caixa Postal 3296, Maputo 2, Mozambique.

Göran Djurfeldt, Department of Sociology, Lund University, P.O. Box 114, SE-221 00 Lund, Sweden.

Fred M. Dzanku, Institute of Statistical, Social and Economic Research, University of Ghana, P.O. Box LG 74, Legon, Accra, Ghana

Hyde Haantuba, Agricultural Consultative Forum, 30G Sable Road, Lusaka 10101, Zambia.

Hans Holmén, Department of Geography, Linköping University, SE-581 83 Linköping, Sweden.

Björn Holmquist, Department of Statistics, Lund University, P.O. Box 743, SE-220 07 Lund, Sweden.

Göran Hydén, Department of Political Science, University of Florida, 234 Anderson Hall, Gainesville, FL 32611-7325, USA.

Aida C. Isinika, Institute of Continuing Education, Sokoine University of Agriculture, P.O. Box 3044, Morogoro, Tanzania.

Magnus Jirström, Department of Human Geography, Lund University, Sölveg. 10, SE-223 62 Lund, Sweden.

Joseph T. Karugia, Department of Agricultural Economics, University of Nairobi, P.O. Box 29053–00625 Nairobi, Kenya.

Elibariki E. Msuya, Department of Agricultural Economics and Agribusiness, Sokoine University of Agriculture, P.O. Box 3007, Morogoro, Tanzania.

Sultana Nasrin, Department of Statistics, Lund University, P.O. Box 743, SE-220 07 Lund, Sweden.

Femi Ogundele, Nigerian Institute of Social and Economic Research (NISER), PMB 05, UIPO, Ibadan, Nigeria.

Willis Oluoch-Kosura, Department of Agricultural Economics, University of Nairobi, P.O. Box 29053– 00625 Nairobi, Kenya.

Daniel Sarpong, Department of Agricultural Economics and Agribusiness, College of Agriculture and Consumer Sciences, University of Ghana, P.O. Box LG 68, Legon, Accra, Ghana.

Stephen K. Wambugu, Department of Agribusiness Management and Trade, Kenyatta University, P.O. Box 43844-00100, Nairobi, Kenya.

Mukata Wamulume, Institute of Economic and Social Research, University of Zambia, P.O. Box 30900, Lusaka 10101, Zambia.

Acknowledgements

The *Afrint* team and the editors want to acknowledge the support of the Swedish Research Council and Swedish International Development Cooperation Agency (Sida), which, together with Lund University, financed the research. The work builds upon cooperation between researchers at Lund and Linköping Universities and the Ethiopian Economic Association; Addis Ababa University; African Economic Research Consortium (AERC); Department of Geography, Kenyatta University; Department of Agricultural Economics and Agribusiness, Makerere University, Kampala; Institute of Continuing Education, Sokoine University of Agriculture; Centre for Social Research and Faculty of Social Science, University of Malawi; Institute of Economic and Social Research (INESOR) and Development Studies Department, University of Zambia; EconPolicy Research Group Ltd, Maputo; Nigerian Institute for Social and Economic Research (NISER), Ibadan; Institute of Statistical, Social and Economic Research (ISSER) and Department of Agricultural Economics and Agribusiness, University of Ghana, Legon–Accra. Finally, we thank our advisors: Göran Hydén, Oliver Saasa and Richard Mkandawire.

1 Introduction

Göran Djurfeldt,[1] Ernest Aryeetey[2] and Aida C. Isinika[3]

[1]*Department of Sociology, Lund University, Lund, Sweden;* [2]*The Brookings Institution, Washington, DC, USA;* [3]*Institute of Continuing Education, Sokoine University of Agriculture, Morogoro, Tanzania*

Impending apocalypse? Or a lost Arcadia to be recreated with organic agriculture and traditional knowledge? In popular accounts the image of food production and consumption in sub-Saharan Africa (SSA) varies from impending apocalypse to a newly lost but recoverable Arcadia.

Images of catastrophe are obviously the more ubiquitous of the two scenarios and they are premised on the poor performance of agriculture over the last decades. Sub-Saharan Africa is widely known as the only major region in the world which has failed to progress in terms of food security, with more or less stagnant levels of production per capita (although with positive growth in overall production).

Visions of Arcadia may be less common but can be found in discourses on ecotourism and in some, mostly non-governmental-sponsored efforts to propagate organic agriculture to African farmers, both as a way to increase food security and, by implication, as a way of recreating a lost Arcadia.

In this book we try to place ourselves somewhere between apocalypse and Arcadia. The middle ground between these extremes may be less street-smart but closer to the reality of African smallholders and truer to their chances of a better life.

The editors of this volume do not subscribe to dogmatism, be it Malthusian or à la Rousseau. We are inspired by the late Noble Laureate Norman Borlaug's dictum: 'I personally cannot live comfortably in the midst of abject hunger and poverty and human misery', he once said.[1] Abolishment of hunger, as part of a process of human development, is a supreme value we subscribe to.

This, again, is a fundamental part of, and requirement for, human development, as defined, for example, by Amartya Sen (Sen, 2001, 2009). Development

[1] According to Thomas Lumpkin: Lumpkin, T.A., 14 September 2009, Farewell to Norman Borlaug: the world loses its leading spokesman for the fight against hunger. CIMMYT, Centro Internacional de Mejoramiento de Maíz y Trigo, El Batan, Texcoco, Mexico.

is to create the conditions for people to develop their potential as human beings. To do away with hunger is an essential requirement for this.

It is important for researchers not to be dogmatic about the means to achieve the abolishment of hunger, be it by organic agriculture or high-tech biotechnology, by large-scale plantations or smallholder farming, by development of markets or by state intervention.

This book focuses on smallholders in sub-Saharan Africa. Although the subcontinent is soon projected to have more than 50% of its population in urban areas, the near majority of Africans will, for some time yet, inhabit its rural areas, from the arid or semi-arid savannahs to the humid forests, from coastal plains to mountainous highlands. Given solidly agrarian societies, the majority of rural dwellers are occupied in agriculture or in animal husbandry. They may have non-farm and non-agrarian sources of income but they continue to depend on food crops, not only for selling but also for feeding themselves and their children.

Poverty in sub-Saharan Africa is a predominantly rural and agricultural phenomenon. The large majority of all poor are farmers and herders. Given the character of poverty, and as long as the poor remain smallholders, alleviation of poverty remains an agricultural task. This volume is dedicated to the task of alleviating poverty among African smallholders.

The subtitle of the volume signals three themes: *Food Crops, Markets and Policy*. Avoiding the risk of overloading it, we could have added a fourth one: *Technology*.

We concentrate on the staples and on four major crops: maize, sorghum, rice and cassava (known as manioc and tapioca in other parts of the world). We have collected data on other staples and other food crops, as well as non-food crops, but in less detail than for the major crops mentioned. A core finding, repeating what was found in earlier work by the same team (Larsson, 2005), is that African smallholders are producing far below their potential (see Jirström *et al.*, Chapter 4, this volume). We define the production frontier somewhat differently than is customarily done. Ordinary farmers are judged by comparing their yields not with those obtained for a certain crop in research stations but with those obtained by their peers. Comparing yields reached by the top 5% of farmers in a village with those reached by the remaining 95%, we find yield gaps of 50% and above.

The gap may be partly explained by differences in soils and their nutrient contents, by differing water regimes, drainage and other factors, which are possible to manipulate by technology in the broadest sense of the word: by soil and water management, by crop selection and breeding, pest management, etc. We are interested in the state of these technologies and in the possibilities of extending improved technologies to smallholders.

Even more core to our concern is the effect of such technologies on yields of food crops and, by implication, on food security. Mostly we assume that improved harvests of food crops are beneficial to the food security of the smallholders. That assumption is tested and found to be sound in the chapter by Dzanku and Sarpong (Chapter 8, this volume).

We are further interested in what enables and constrains the linking of smallholders to markets. To a large extent this is a question of infrastructure, but equally important are the institutional prerequisites and the potential of institutional reforms in the development of smallholder agriculture. Markets and commercialization do indeed explain much of the dynamism that we observe (Andersson *et al.*, Chapter 5, this volume), but the recent history of African agriculture hints that markets cannot do the job on their own.

The period since the Structural Adjustment Programmes (SAPs), which in many of our case study countries were launched in the early 1980s, can be seen as one of laissez-faire agricultural policies, with low levels of investment on the part of governments and donors. It thus became a test of how much markets on their own can do to propel the growth of agriculture and in alleviating the poverty suffered by agricultural producers. The record of this laissez-faire period in the history of agricultural policy is both clear and tragic: agricultural development stalled and poverty, if anything, increased.

Markets cannot, on their own, do the job of alleviating poverty and reaching the Millennium Development Goals (MDGs). It was in implicit recognition of this that the United Nations took the lead in formulating the MDGs and in getting massive backup by both governments and donors in trying to achieve them. To this government- and donor-backed initiative, in the African setting we can add the New Partnership for Africa's Development (NEPAD), which is an economic development programme of the African Union (AU), adopted by African heads of state in 2001. NEPAD includes the Comprehensive Africa Agriculture Development Program (CAADP), which was adopted 2 years later (Comprehensive Africa Agriculture Development Programme, n/d). This programme has an ambitious, well-structured and informed agenda for African agriculture. Yet more substance was added at the second summit of the heads of states and governments of the AU in Maputo in 2003, when, in a document at a concurrent meeting with the ministers of agriculture, it was stated that: 'To this end, we agree to adopt sound policies for agricultural and rural development, and commit ourselves to allocating at least 10% of national budgetary resources for their implementation within five years' (Conference of Ministers of Agriculture of the African Union, 2004).

A 10% budget allocation to agriculture is high by recent historical standards; on average, it was only a fraction of that in sub-Saharan Africa from the 1980s to the early years of the 'noughties'. According to Fan *et al.* (2008), investments in agriculture in Asia today constitute 8–14% of total government budgets. During the Asian Green Revolution, from the late 1960s, according to the same source, investments were from 15% and upwards. All accounts made up, total investments were at least 50% higher than the goal set by NEPAD.

After this short introduction to the key themes of this volume and before coming to a more detailed description of its contents, we need to go somewhat more into its background.

The Earlier Book and Research

The present volume is a sequel to an earlier work (Djurfeldt *et al.*, 2005b), where one of us (Djurfeldt) was among the editors. A key theme for *Afrint I*, as we call it, was the Asian Green Revolution and its relevance to Africa. The book was timely and part of a series of intellectual and academic inputs that led to a growing acceptance for its major tenet, i.e. that an African Green Revolution must be state-driven, market-mediated and smallholder-based. Although it was not put in exactly those terms, the thrust of the *World Development Report 2008: Agriculture for Development* (World Bank, 2007) was in the same direction. Thus it is fair to say that there was a reversal of the tide in agricultural policies in the early noughties.

In the work just mentioned, we were able, in the macro-level policy studies, to document the beginning of the reversal from the laissez-faire of the 1980s and 1990s to a more interventionist policy. In the micro-level studies, on the other hand, we did not find much in terms of policy-induced dynamism, except perhaps in Nigeria (Akande, 2005). The Ethiopian government of Meles Zenawi had enthusiastically accepted the policy advice of Norman Borlaug, broadcast by the Sasakawa Foundation and the Carter Foundation, but our attempts to trace the effects of this policy on the ground led to largely negative conclusions (Djurfeldt *et al.*, 2008). Growth of staple food production seemed largely to build on area-extensive growth, both in Ethiopia and in the other countries studied. While the potential of scientific industrial inputs are easy to document, also on the ground, they contributed little to the dynamism observed. This, in turn, seems to be possible to attribute to ineffective or counter-effective agricultural policies (Djurfeldt *et al.*, 2008).

Events since data collection for the first book in 2002 include not only the reversal in the intellectual and policy debate but also the global food price crisis starting in 2008 and the global financial crisis erupting the year after. While the long-term effects of these crises are not possible to discern at the moment of writing (March 2010), it is widely believed that there will be an increased level of prices in the world markets for food crops, including those we are studying here. In part, increased prices are due to the competition of biofuel crops for scarce agricultural land. Irrespective of the soundness of the prognosis about higher price levels, a clear signal has been sent to African policy makers, and many appear to have got the message, that the easy way out that many opted for in the 1980s and 1990s, i.e. to import grain sold at throwaway prices by Organisation for Economic Co-operation and Development (OECD) countries, will be permanently closed. Many observers expected that this tendency would be reinforced by a dismantling of producer and export subsidies in the OECD, which so far has not really happened.

Anyhow, African governments are more reluctant than they were before the crises to rely on laissez-faire in feeding their citizens with subsidized grain from the West. As the Maputo declaration suggests, governments envisage having to go down the much more narrow road of trying to get their own farmers to produce not only what they need to feed to themselves but also to feed the growing population of urban areas.

The shift in intellectual and business cycles that we have discussed came in handy for the sequel to the first study, i.e. the *Afrint II* project, from which the current book is an output and for which field work was conducted in 2008.

The Current Book and the Underlying Research

The *Afrint II* project set out to investigate how the changed agricultural policy climate affected government policies in the nine countries studied already as part of the preceding project: Ethiopia, Ghana, Kenya, Malawi, Mozambique, Nigeria, Tanzania, Uganda and Zambia. By repeating the cross-sectional survey made in over 100 villages in 2002 and converting it into a panel, it is possible to trace village- and household-level effects of agricultural policies and other macro-level processes. Parts of this research are reported in the current volume.

In Chapter 2, Hans Holmén and Göran Hydén point out that sub-Saharan Africa's inability to improve its food security by domestic means is a matter of recurring concern. Being initially seen as caused by negligent domestic policies, it was dramatically worsened by imposed structural adjustment policies in the 1980s and 1990s. Structural Adjustment Programmes reduced the role of the state and reduced or eliminated governmental support systems and were accompanied by reduced levels of aid, especially for agriculture. Recently, the African food crisis has been aggravated by increasing world market prices for food and the global financial crisis, which both affect the region's ability to rely on food imports. This has led to food riots and social unrest but also to renewed self-assertiveness among African political elites and a new recognition of the importance of agriculture for food security and development, a renewed emphasis on African food crop research and on the necessity to enhance the productivity of small farmers. Impressive accomplishments have been made in a short time but will the 'soft' and neo-patrimonial governments in sub-Saharan Africa have the capacity in the long run to manage a development process based on smallholders?

In Chapter 3, Hans Holmén writes about 'The Millennium Goals, the State and Macro-level Performance' and shows that poverty reduction and achievement of the MDGs, hereunder poverty reduction and food security, are professed priorities for governments in sub-Saharan Africa. In order to attain them, governments have declared an ambition to facilitate private sector participation and enhance the role of farmers' organizations. At the same time, and post structural adjustment, the role of the state is growing, not only as a facilitator but also by direct involvement. The chapter examines trends in food crop production, productivity, input use and market development in the nine case study countries and in relation to smallholder agriculture. To reach beyond official declarations and policy prose, and in order to explain observed trends, it highlights the actual roles of the state in relation to food crop intensification (supportive or hands off, small or large farm priority), market development, budget allocations and investment priorities (research, extension, infrastructure), gender dimensions of reform programmes, capacities, degree and scope of private sector involvement and, finally, the role of farmers' associations. The analysis reveals the sincerity of the above commitment and whether – and to what extent – contemporary reforms are state-driven, market-mediated and smallholder-based.

Chapter 4 on 'Smallholders Caught in Poverty – Flickering Signs of Agricultural Dynamism' by Magnus Jirström, Agnes Andersson and Göran Djurfeldt draws on the two cross-section samples generated by *Afrint I* and *Afrint II*. The data offer an opportunity to overview any major changes in smallholder production during a period of both relatively rapid overall economic growth on the continent and a dramatic increase in global food prices affecting farm households. In addition, it contributes empirically to debates which have been running on a number of themes, for example on farm size, the role of staple crop production and of non-farm household incomes.

'A New Era for Sub-Saharan African Agriculture? Changing Drivers of Maize Production' is the title of Chapter 5, written by Agnes Andersson, Göran Djurfeldt, Björn Holmquist, Magnus Jirström and Sultana Nasrin. The purpose of this chapter is to analyse and discuss the drivers behind changes in staple food production, focusing on maize for the period 2002 to 2008, re-evaluating and discussing the role of the three key processes identified in 2002, namely the role of commercial drivers, farm technology and the agrarian policies of the state. This is done on the basis of a reduced form model of production, which draws on data from a panel of 1805 maize-growing smallholder households in the nine African countries interviewed in 2002 and 2008.

Chapter 6 by Agnes Andersson deals with a neglected subject: 'Maize Remittances, Market Participation and Consumption among Smallholders in Africa'. In-kind remittances of foodstuffs among family members constitute a tangible but hitherto unexplored reflection of spatial linkages in the developing world. This chapter sheds some light on in-kind remittances of staple foodstuffs and assesses the wider reciprocal and livelihood implications for the remitters. Findings are substantiated through data on remittances of Africa's main staple crop, maize, collected in 2008 as part of the *Afrint II* resurvey of the maize and cassava belt. The paper draws on cross-sectional data covering 2900 maize producers in 91 villages and discusses the relationship between in-kind remittances and market participation among the remitting households, as well as the food security implications of remittances. The study concludes that in-kind remittances of maize and other staple foodstuffs constitute crucial sources of supplementary food and may act as a source of food security for both rural and urban recipients. This is especially so where markets cannot be trusted to deliver foodstuffs due to various infrastructural, institutional and policy constraints. Meanwhile, the subsistence obligations of remitting smallholder households and the extent of urban dependence on family-produced food may be underestimated since the remittances are informal and therefore invisible.

After these more general chapters, follow a number of chapters based on country-level data. Chapter 7 is on Ethiopia: 'Meeting the Financial Needs of Smallholder Farmers in Ethiopia' by Wolday Amha. Improving financial access to smallholder farmers has been one of the most prominent instruments in the development programmes and strategies used by the Ethiopian government and its development partners. Despite the efforts of finance providers, governments, donors, and others to expand outreach in delivering financial services to smallholder farmers, there is still a huge unmet demand for such services. Thus, there is a need to revisit the entire approach of the delivery system in order to satisfy this unmet demand. Drawing on the *Afrint I* and *II* surveys, the problems discussed in this chapter include assessment of the policies, strategies and regulatory framework and the meso-level players that affect the delivery of financial services to smallholder farmers in Ethiopia.

The following Chapter 8 is on Ghana and is written by Fred Dzanku and Daniel Sarpong: 'Agricultural Diversification, Food Self-sufficiency and Food Security in Ghana – the Role of Infrastructure and Institutions'. Staple crops are cultivated by all farm households in the eight villages studied in 2002 and 2008. There has been little change in the production of these staples over the period of the observed data. Indeed, in many of the villages there have been declines in both cultivated area and yield of staples. Very little change in production practices has been observed on staple crop farms across regions and villages. At the same time, there appear to be changes in the production of non-staples, particularly vegetables, in most of the study villages. These non-staples, apart from contributing immensely to household food consumption, contribute substantially to cash income and thus have the potential of reducing poverty and improving household welfare. The relatively more intensive application of improved farming techniques to non-staple crop farms suggests a diversion of scarce resources for the production of these crops. This raises important questions, among which are: What have been the intra- and inter-household changes in resource allocation between staple and non-staple crops over the period 2002 to 2008? What accounts for these changes? Are there any significant food security implications? Have any observed changes in resources allocation led to significant changes in household welfare? And what has been the contribution of the meso- and macro-environment to observed changes in farm household resource allocation between crops over time? The study notes important variations and differences between districts in the same region and villages within the same district, which suggests that specific decentralized polices and tailor-made programmes may be necessary if any significant economic welfare transformation is to occur within farm households and farming villages.

Chapter 9, 'Conditions for Achieving Sustained Agricultural Intensification in Africa: Evidence from Kenya', is co-authored by Stephen Wambugu, Joseph Karugia and Willis Olouch-Kosura. Increased agricultural productivity and competitiveness is critical for national growth and development in almost all African countries and for the achievement of the MGDs, as well as for tackling the current and emerging food, fuel and financial crises. This chapter examines the conditions for achieving sustained agricultural intensification using evidence from micro- and macro-data from Kenya, as well as the six 'I's that represent significant proximate variables influencing agricultural performance, namely Incentives, Inputs, Infrastructure, Institutions, Initiatives and Innovations. The chapter further demonstrates how a change in these 'I's affects agricultural productivity. Furthermore, the authors discuss agricultural intensification and a number of public interventions to promote it, and spell out their implications for the realization of Millennium Development Goal of halving, by 2015, the share of people suffering from extreme poverty and hunger. Emphasis is laid on maize production, since the lack of maize signals famine and poverty in Kenya, even when other food crops may be available. The chapter examines the conditions that led to a revitalization of increased agricultural productivity in the period 2003 to 2007, after an enabling policy environment that favoured the six 'I's was put in place. The authors also present scenarios likely to emerge after the skirmishes that rocked the country soon after the December 2007 general elections.

Chapter 10 on Zambia is written by Hyde Haantuba, Mukata Wamulume and Richard Bwalya. Despite liberalization of the input and output markets, the authors contend that the Zambian government has still continued being active in these markets on the basis that the private sector does not have the capacity yet. They have done this through operating programmes such as the Fertilizer Support Program (FSP) in the input markets and the Food Reserve Agency (FRA) in the output market. However, of late, there have been concerns raised on the impacts of these interventions on both the market and the government's ability to meet its other commitments in the sector, such as promotion of extension, infrastructure development and research. This is because almost 50% of the agricultural budget is currently being spent on fertilizer support, leaving little for other activities. Furthermore, concerns have been raised on the crowding-out effects on private sector participation as well as crowding-out effects on other alternative crops as these support programmes are only targeted at maize. This study used *Afrint* data from a sample of 423 households drawn from selected districts in the southern and central provinces of Zambia to determine the impacts of these fertilizer support programmes on smallholder production patterns as well as the government's ability to meet the Millennium Development Goal 1 on food security. A review of the literature shows that although the FSP has resulted in increased fertilizer use as well as increased yields among smallholder farmers, it has also resulted in crowding-out of private sector participation in the fertilizer markets. Similarly, the programme is impacting negatively on crop diversification efforts. Analysis of factors that influenced productivity among the sampled households in 2007 showed that, apart from fertilizer usage (expenditure on artificial fertilizers), other factors, such as market access, amount of labour and ownership of productive assets such as oxen, influenced the quantities of maize produced.

Chapter 11, by Tunji Akande, Agnes Andersson, Göran Djurfeldt and Femi Ogundele, is titled 'Has the Nigerian Green Revolution Veered Off Track?' Macroeconomic and sectoral policies and programmes initiated by the government between 2002 and 2007 were aimed at rapid growth in the agricultural sector and progress in reducing poverty. There have also been product-specific programmes, like the presidential initiatives on cassava, rice, maize and other crops, livestock and fisheries products. There have been encouraging pay-offs as agriculture's growth rose from 3.5% per annum in 1990–1999 to 5.9% per annum in 2000–2007, close to the CAADP target of 6% growth, which is necessary for African countries to meet the Millennium Development Goal 1 targets by 2015. The Nigerian government itself set a higher, and possibly unrealistic, growth rate of 10% for the country to meet the MDG 1 targets. It is within this context that this paper examines critically the response of the agricultural sector to the various policies and programmes, with specific reference to farm-level productivity, exemplified primarily by the major food crop in the country (maize). Available statistics from the various analyses show that, even though recent growth trends reveal some modest increases in productivity over time, yield levels are generally below potential and in some cases declining. This reflects the fact that much of the growth or increase in output has come from expansion in the land area under cultivation.

'Addressing Food Self-sufficiency in Tanzania: a Balancing Act of Policy Coordination', written by Aida Isinika and Elibariki Msuya, is Chapter 12. This chapter reviews aspirations by the government of Tanzania to transform agriculture and thereby to reduce income poverty while also achieving food security, especially for the 80% of the population who mainly depend on primary agricultural production for their livelihoods. This ambition is articulated in various national, regional and global development goals (e.g. the MDGs) to which Tanzania is a signatory. More recently (2009), another policy statement 'Kilimo Kwanza' expressed the ambition to foster a Tanzanian Green Revolution. However, analysis of data at macro and micro levels shows that set targets are not being met. Use of inputs and productivity has declined since 1999. Production and productivity for maize and rice, the main food staples, has been declining or rising slower than the population growth rate. The agricultural sector has been growing at about 4.2%, which is lower than the target of 6% set under CAADP. At the household level, farmers face challenges of access to inputs, lack of essential services, poor marketing infrastructure, dysfunctional markets, poor organization and land tenure.

The penultimate Chapter 13 is on Mozambique and is written by Peter Coughlin. Capital-poor and rarely receiving advice from extension workers, Mozambique's small farmers are ensnared in a low-technology, low-output trap. Except in rice irrigation and concessionaire schemes, most farmers in Mozambique use traditional methods with no chemicals, no improved seeds, no animal traction and, typically, no improved farming techniques beyond crop rotation and intercropping. For them, the Green Revolution is far away. Without capital to back them up, the extension workers' messages – even when good – are hard to implement. Worse yet, improved techniques are often suboptimal or even make a loss unless the farmers have improved storage, enabling them to wait to sell their crops only after prices recover from harvest-time lows. By contrast, when extension agents work inside the context of a project or concessionaire scheme that furnishes inputs on credit and perhaps invests in infrastructure, then farmers implement their messages much more readily. What is the lesson? It seems, Coughlin argues, that adoption of improved practices must occur together with a steady, programmed improvement in the farmers' investment capacity (capital). Without that, when a project ends, impoverished farmers necessarily revert to traditional, low-input, low-technology farming systems. Can this change? Can the huge majority of small farmers be effectively reached by and benefit from policies and efforts to enhance their productivity and market access? This is the overall question addressed in this chapter. With the recent large increases in international prices for a wide gamut of crops, agricultural investments will be more profitable and less risky, especially for organic fertilizers, animal traction, small-scale irrigation, and improved seeds and storage. This bodes a big change in farmers' receptivity to advice by extension officers if backed by project credit to promote irrigation, animal traction, improved storage and cooperative marketing for farmers organized in associations. Receptivity alone is, however, insufficient. Escape from the low-level production trap requires large, synchronized infrastructural and industrial investments to facilitate commerce and create value chains. Villages also require capital investment: focused, moderately sized, short term and preferably rotational, so that the funds move on to other farmers and villages. The vision and the effort must be big, also at the level of the village. If not, the majority will be ignored and impoverished – for generations.

Chapter 14 is the concluding one, written by the editors, and attempts to draw overall conclusions from the current study as well as the preceding one and to point out directions for future research as well as policy.

Methodology of the *Afrint* Projects

Data collection for the first round of the *Afrint* project was made in 2002 and for the second one in late 2007 or early 2008. The data collected as part of the second round are referred to as 2008 data. In the first round, preparations were made for making the survey into a panel, so that the households interviewed could be traced.

From the outset we selected five case study countries: Ghana, Kenya, Malawi, Nigeria and Tanzania. The selection was made on theoretical grounds and the research design for *Afrint I*. Our initial perspective was very much inspired by Boserup (1965), which implied that we were looking for signs of area intensification in countries and areas with higher population density and favourable agricultural potential. Outside francophone Africa, these five countries were ideally suited, in our view, to charting progress in intensification, induced from below by farmers themselves, or state induced, as in the Asian Green Revolution.

The first round of the *Afrint* project also had an Asian leg, where we carried through comparative historical case studies in order to develop one or more models of Asian Green Revolutions, to be used as an analytical framework for the African case studies. These studies have been reported in several works (Djurfeldt, 2005; Djurfeldt and Jirström, 2005; Djurfeldt *et al.*, 2005a).

Basic finance for *Afrint I* was eroded with the sudden appreciation of the US dollar in 2000–2001. We were forced to seek additional finance from the Swedish International Development Cooperation Agency (Sida), which was granted under the condition that we expanded our country sample. Thus we ended up with a sample which, admittedly, was too heterogeneous from our theoretical point of view. To the original five countries, four more were added: Ethiopia, Mozambique,[2] Uganda and Zambia. From a Boserupian point of view,

[2] Local counterparts in Mozambique were found only after long searching, which delayed the survey by 1 year and prevented the country from being part of the sample discussed in Djurfeldt, G., Holmén, H., Jirström, M. and Larsson, R. (eds) (2005b) *The African Food Crisis: Lessons from the Asian Green Revolution*, CAB International, Wallingford.

the three last-mentioned countries are of questionable relevance, since they are land-abundant and not likely candidates for a population-driven intensification. Ethiopia,on the other hand,is peculiar in an African context, with its long history of plough agriculture and feudal-like social formation. As a result of the model development in the Asian leg of the project, however, the theoretical focus shifted, as it often does in the social sciences, from a Boserupian one stressing demographic factors to one which places more emphasis on economic and political drivers. In this process, our heterogeneous sample of countries has proved less cumbersome to work with than one might have expected.

Formally, the *Afrint* sample was drawn in four stages, of which the country selection described above was the first one. The next stage was regions within countries, followed by selection of villages within countries, with selection of farm households as the last stage. All stages except the final one have been based on purposive sampling. Data collection was sought to be made at all four levels.

At country or macro level, the team agreed on a data collection and analytical format in a methodology workshop held in 2002. Macro-studies would be mainly desk studies, complemented with interviews with key persons. The key problem was 'to explore the political and economic preconditions for intensification' (quoted from a methodology paper written for the 2002 workshop). Although the key problem was expanded upon in various documents, including the terms of reference stated in the contracts with the country teams, the latter were given fairly loose reigns to pursue the macro-level studies in a manner adapted to the local context. These studies were reported separately (Afrint, 2010) and included two books (Akande, 2006; Coughlin, 2006). The comparative analysis of these studies was reported in two papers by Hans Holmén (2005a,b).

In retrospect, the 2002 terms of reference might have provided too much liberty in the design of the macro-level studies, which led to difficulties in making a comparative study of the cases, although Holmén (2005a,b) made a good job of it. In the second round, therefore, it was decided to ask the country teams to fill out comparable data sheets on a long number of indicators, to facilitate the comparative analysis. Thus the comparative analysis made after the second round may have allowed for a deeper and more penetrating analysis (see Holmén, Chapter 3, this volume). While the data facilitated comparative analysis, the various country teams have drawn on the specific characteristics of their respective cases in making country-level analyses (see Chapters 7 to 13, this volume).

We reproduce below the instructions for the second to fourth level of sampling, which were proposed to the 2002 methodology workshop.

Sampling procedure

Our objective is to study the performance of smallholders in areas of sub-Saharan Africa that have the potential for substantial improvements in production and yields of staple food crops. The project departs from the assumption that such a potential most likely can be found in areas that meet certain minimum requirements in terms of average annual rainfall and other ecological conditions, as well as in the access to markets (infrastructure), implying that investments aimed at raising productivity would be relatively lower in such areas. This means that areas that are too arid or too remote in terms of infrastructure are less likely to respond to market incentives and show the agricultural dynamism we hope to capture in this study. For this reason we have excluded most of the Sahelian countries from the country sampling frame, limiting the selection of country cases from the group of countries located in what we may depict as the 'maize and cassava belt' (see Byerlee and Eicher, 1997:14; and Nweke *et al.*, 2002).

Also, within this large area of the continent, our choice of countries has not been random but purposive, so as to ensure sufficient variation in the reception of Asian models in the form of national agricultural policies (where the large-scale adoption of the Sasakawa Foundation–Global 2000 model in Ethiopia and the market-inspired development in Uganda can be said to represent the extremes of the variation supplied by the African cases).

We also propose that the households to be sampled within these countries be selected with respect to the agricultural potential of the areas in which they reside. This is illustrated by Fig. 1.1, showing agricultural dynamism as a continuum, where 'low' depicts low productivity potential following aridity and/or remoteness to markets. At the other extreme, 'high' refers to cases where ecological endowments and marketing infrastructure have combined to create some of the most dynamic and productive environments in Africa (examples are Mt Kilimajaro in Tanzania, parts of the Kenyan highlands, areas surrounding the main cities, etc.). Albeit interesting, we consider the latter type of areas as extreme cases or 'outliers'.

The intention is thus to capture the dynamism in the areas that are 'above average' in terms of ecological and market (infrastructure) endowments but excluding the most extreme cases in this regard. We thus propose elevating this selection criterion to *a recurring methodological principle* in this project, guiding the sampling of the regions, villages and households. In this way we believe we will be most likely to find the agricultural dynamism we are looking for.

We believe that the geographical areas encircled in Fig. 1.1 will provide sufficient variation as to the factors we assume to be crucial for improved

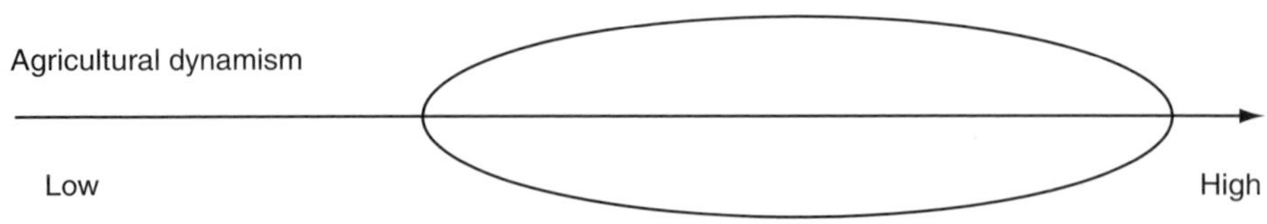

Fig. 1.1. Sampling frame.

performance by African smallholders. In addition, these areas contain the majority of the population in the African sub-Saharan region. Thus, the agricultural development of these areas most likely holds the key to Africa's future food provisioning.

It is the performance of the smallholder farming population within these areas that is our subject of study and which constitute our sampling frame. In the eight countries where we have planned for in-depth studies, the total sample size drawn from this smallholder population will be between 2000 and 2400 households. The methodological question is: how do we best draw this sample so as to ensure sufficient variation in the causal factors and so that we are able to identify the driving forces of improved productivity among smallholders?

Obviously, the sampling strategy will have to be a multi-stage one, where the first stage (country selection) has already been discussed above.

- Stage 1. Countries (purposive sample), eight units;[3]
- Stage 2. Regions (sites) within countries (purposive), two or more units;
- Stage 3. Villages/sub-villages (purposive, stratified), two to ten units;
- Stage 4. Households (stratified/random sample), 300–400 units per country.

Statistically the problem of sampling can be formulated as one of getting enough variance both in the dependent variables and in the independent variables, at household level and above. Getting enough variance in these factors will allow us to explore hypotheses about causal mechanisms underlying correlations that can be established in our data.

Ideally, a multi-stage sampling design should be self-weighting and produce unbiased survey estimates through optimal sampling methods at the various stages in the design (such as sampling the various units with a probability proportional to size, e.g. PPS). Such a randomized design would also allow estimation of sampling errors and thus computation of statistical tests and confidence intervals.

However, since for logistical reasons we are constrained to select a fairly small number of units at the various stages of the sampling process, any probability-based sampling technique would run the risk of not capturing the entire range of variance in the causal factors. Thus a purposive design is preferable in this case. Thus, we cannot aim for a sample that is representative in a statistical sense. Instead we aim at a sample which is illustrative of conditions in the maize–cassava belt, excluding both low-potential, dry and remote areas and extreme outliers at the other end of the scale, i.e. privileged, high-potential areas of the kind already exampled.

Sampling regions/sites

A purposive sampling at stage 2 (region/site) is thus more feasible. In this respect, regions must be sufficiently large so as to contain the prescribed variation of villages/sub-villages along the 'agricultural dynamism' continuum

[3] Mozambique was added later.

presented above, yet, at the same time, be sufficiently small not to present overwhelming difficulties when it comes to survey logistics, costs and time frames. With only two sites randomly selected we would run the risk of facing large sampling errors associated with cluster-specific variables correlating with the dependent variables, thus making it difficult to differentiate between 'real' and 'cluster-dependent' effects. With the method of purposive sampling of regions/sites we are suggesting here, however, we may secure the required variation between the sampling units at the next stage in the design, i.e. villages or sub-villages. In this respect, utmost care must be taken by the teams to select regions that are sufficiently heterogeneous with respect to the villages they contain.

Villages/sub-villages

At stage 3, how should we sample villages/sub-villages? Also in this case, we should be guided by the range of variation defined in Fig. 1.1. We don't need to include the least-endowed villages, distant from the market, with poor infrastructure, poor soils, etc., since in these villages we are not likely to find the agricultural dynamism that we would like to capture. Ideally, both the units at stage 2 (regions) and at stage 3 (villages) should overlap with administrative units, in which case stratification of villages reflecting their positions along the 'market dynamism' axis can be made more easily through interviews with key informants (e.g. agricultural staff, government officials) having a good overview of their neighbourhood agricultural area.

While the number of units in stage 2 (regions/sites) is limited, efforts should be taken so as to increase the number of villages/sub-villages. It should be possible to distribute the interviews in each region over a number of villages substantially above two. However, it would also seem that having more than, say, ten units would not be feasible and would intolerably increase the costs of surveying.

Sampling households

Having thus worked through the sampling design, it is in the fourth and last stage that we opt to select households for interviews, with 300–400 respondents distributed over a number of villages in each of the regions. By which method should these households be selected? Here we foresee some kind of conventional probability sampling, with the option of first stratifying the sampling frame on a crucial core variable, such as 'technology adoption rate'. Depending on the local situation, it may also be desirable to over-sample small strata, such as commercially oriented farmers, women-headed households and perhaps ethnic minorities. If this is at all possible depends on the information contained in the list of households from which households have to be sampled.

Access to an updated list of all the households residing in the sampled units of stage 3, or the possibility of creating such a list, is therefore crucial for household sampling. It then follows that the stage 3 units must not be too big in terms of population size, in order not to cause insurmountable problems in the creation of sampling frames (household lists). In this respect, villages may prove too big a unit for this kind of exercise. Sub-village, or even an administrative unit below this one, may have to be considered.

Stratification of the household sampling frame can be done through the kind of ranking techniques described by, for example, Grandin (1988), in relation to wealth ranking. This kind of participatory rural appraisal (PRA) technique provides a rapid and reliable classification of households vis-à-vis the ranking criteria. In respect of the relatively small household sample size per village, stratification secures variance in the core (stratification) variable(s) and thus reduces sampling errors.

To summarize, we foresee a four-stage sample design, with purposive sampling at all stages, except the last one, where we propose sampling of households after having made up household lists and stratification criteria by means of PRA ranking techniques. We will now discuss the design of the survey questionnaire (quoted from internal working paper on methodology).

With one exception, the above sampling strategy was followed. No country team went for stratification at village level, however, which was wise, since ranking exercises as proposed by Grandin (1988), if they are not very carefully done, have proved to be quite unreliable.[4]

Although the agreed-upon strategy was described in detail, individual country teams sometimes made idiosyncratic interpretations of the strategy. So, for example, in the first round only four villages within four regions were chosen in Ethiopia, which made for too low a variance on the variable distance to market. This was corrected for in the second round, when another four villages were chosen. Similarly, in Nigeria too many villages (about 50) were chosen, making for small samples within each village and very high within-village standard errors, making it difficult to test hypotheses on village-level factors. In the second round about half of these villages were dropped and additional households added to the remaining village samples. This makes the Nigerian panel smaller than it otherwise would have been (see Akande *et al.*, Chapter 11, this volume).

Similarly, the Ghana team described its village-level strategy as 'convenience sampling', which conjures an image of villagers queuing up to be interviewed, with the local elite first in the row. In the second round, possible bias was checked through comparison with an additional simple random sample. It was found that there was no traceable bias in the first round sample.

In Uganda the local team chose four villages and delivered data that were difficult to use. This led to contracting a new team for Uganda. When the new team tried to trace the 2002 households in the 2002 villages, this led to too high rates of attrition, which is why it was decided to go for entirely new villages. Thus there are no panel data for Uganda.

[4] Even if very carefully done, this method has its own difficulty in avoiding pitfalls, as demonstrated by Larsson, R. (2001) *Between Crisis and Opportunity: Livelihoods, Diversification, and Inequality among the Meru of Tanzania*. Department of Sociology, Lund Dissertations in Sociology 41, Lund.

In Mozambique, finally, the lead investigator detected that, in order to simplify their work, a local team had divided a big village into several segments, thus artificially blowing up the number of villages to the required number. This was rectified in 2008 and the country sample now contains fewer villages than it appeared to do in 2005, when the first round was made.

In the remaining countries, sampling was done in accordance with the commonly accepted guidelines. When we compare point estimates from the sample with those from other sources, for example yields for the various crops with FAO statistics, no apparent sample bias has been detected (see Jirström *et al.*, Chapter 4, this volume).

The *Afrint I* questionnaire

After having established basic facts about the household, such as age of respondent and age of farm,[5] gender of farm manager, etc., interviewers in the 2002 round sought to describe the crop pattern, i.e. if in the last season or in the two previous seasons the farmers were growing any of our four main crops: maize, cassava, sorghum or rice. Other crops were classified as either 'other food crops' or 'non-food cash crops' (i.e. typically export crops). The same information was ascertained about the crop pattern in the reference year. For each food crop, data was collected on area, irrigation (if any) and amount produced in the current season and the two previous seasons. The interviewer also asked if yields in the reference year were typically higher or lower than currently. Then followed a series of questions on uses of the output for own consumption, for seed, in-kind wages, etc.

The following section dealt with methods of cultivation and technology and covered use of both scientific–industrial inputs and 'pre-industrial' technologies. This was done both for the most recent season and for the reference year, making it possible to ascertain changes over the life course of the farm. Marketing was covered in terms of volumes sold, prices received and details about the forms of marketing. In *Afrint I* we also included two attitudinal questions, covering the constraints to increased production. These questions yielded little useful information, for reasons discussed below.

Questions about other food crops and non-food crops were much less detailed. We covered the crops grown and sold, the total area under such crops and the methods of cultivation, marketing outlets and changes in these respects since the reference year.

Another section in the questionnaire dealt with land resources, such as total area of the farm, the existence of set-asides and, again, changes since the reference year. Livestock was similarly covered, but not in much detail, before coming to the demographic particulars: household composition, non-resident members, labour force, hiring of wage labour, etc. Yet another section dealt with institutional conditions relating to markets and to land access.

Incomes and expenditures were dealt with, but not in detail. Income sources, including off-farm and non-farm sources of income, were ranked in

[5] This is what we call the reference year, i.e. the year in which the current manager of the farm started on his/her own in farming.

terms of importance. Ownership of consumer durables and housing standard was also documented. At the end of the questionnaire, the interviewer was asked to rank the household in terms of its wealth.

The household questionnaire was designed keeping in mind that the overall interview time should not exceed 2 hours. This is in order to avoid respondent fatigue and to avoid asking for precise answers to questions when the respondent is unable to be precise. The first point implied concentration on core issues and avoiding longish excursions into peripheral details. The second one meant going down in scale, from ratio to nominal or ordinal ones. For example, we avoided asking retrospective questions that are over-precise, such as 'How many bags of maize did you ordinarily get from a hectare of maize when you were a young farmer?' Instead we asked: 'Did you get more or less maize from a given piece of land when you were a young farmer than you do today?'

We thus adhered to 'Patton's rule' (Patton, 1980), according to which it is easier to get precise answers to questions that deal with the present and with behavioural or factual matters than to questions which deal with a distant past and with attitudes and knowledge rather than behaviour. Thus we asked for more precision when dealing, for example, with the most recent (or impending) harvest: 'How many bags of maize did you (or do you expect) to harvest this year?' Metric conversion of local measurement, such as bags, was made in the field by the interviewers.

An illustration of the danger of being over-precise is when we asked for quantities of fertilizer used (in 2002). We got very imprecise answers and thus high variance and large standard errors. It is preferable to reduce the scale of the variable to a dummy: 'Did you use chemical fertilizer on maize last season?' This gives more reliable answers, although it has consequences for the statistical methods that can be used.

Not unexpectedly, special difficulties were encountered when trying to establish the basic facts of cassava cultivation. Volumes of production are exceedingly difficult to establish in a survey like ours. This is essentially because cassava is harvested continuously during the year. Moreover, cassava is generally weighed by and sold as roots, the prices of which differ depending upon their size and quality. The 2002 data on cassava production proved more or less useless, and in 2008 no attempt was made to ascertain production for this crop.

Overcoming language barriers is also much easier when following Patton's rule and keeping to mundane, down-to-earth issues like amounts of food crops harvested. So, for example, we got much less precise data when trying to chart the usage of 'pre-industrial inputs' such as crop rotation, manuring, fallowing, intercropping etc. When trying to use these data it is evident that they contain much noise, because expected correlations are weak or non-existent. This is probably because we relied on on-the-spot translation from an English questionnaire to the local language.[6] To cover such issues more efficiently requires translation to local languages and, moreover, to rustic versions of these rather than to the urban, middle-class language. It is probably wise also

[6] Translation to other languages was only made in Ethiopia (to Amhara) and in Mozambique (to Portuguese). This does not eliminate the language problems, because many respondents would not be proficient in these languages.

to re-translate, i.e. to translate back from the vernacular to English in order to check that the intended meaning is rendered.

Cursory coverage of inputs, including inputs of labour, implies that we are not in a position to estimate profitability, labour productivity and other variables that an economist would ask for.

Village questionnaire

The section headings of the 2002 village questionnaire gives a fair impression of what is covered: population size and land use, agro-ecology, infrastructure and markets, state interventions (food relief, non-governmental organization or donor presence, agricultural extension), markets (for both output and inputs), farmer organizations, land and land tenure, credit, history of agricultural intensification (introduction of high-yielding varieties, fertilizer, etc.), and labour and gender aspects.

The village questionnaire also contained a number of open-ended questions, where the country teams were asked to analyse and characterize the village in terms of agricultural potential, level of agricultural development, constraints to development, etc. These questions proved so multi-dimensional that they were virtually useless for a comparative analysis. In the 2008 village questionnaire, therefore, a much more detailed coverage of the same matters was attempted by means of series of detailed questions on the same dimensions as mentioned above.

Respondents to village interviews were key persons, such as village leaders and extension agents. Investigators were also instructed to conduct focus group interviews with representatives of various segments of the village population, including women farmers.

The *Afrint II* questionnaire and the panel

When going for a second round and a panel in 2008, we went for a balanced panel design, i.e. constructing the 2008 sample so that, in itself, it would be representative of village populations in 2008. This also involved sampling descendants when a household had been partitioned since 2002. In case of sizeable in-migration to a village, we also provided for sampling from the newly arrived households.

The 2002–2008 panel thus is a subset of the two cross-sectional samples. In itself this subset is not statistically representative of the village population in any of the two years. Since this is the case, one should be wary of making point estimates from the panel. Such estimates should instead be made from the two cross sections.

Establishing a panel implies that questions should be repeated and thus calls for small changes to the 2002 questionnaire. In principle, then, the 2008 questionnaire is identical to that used in 2002. In practice we made some changes, such as adding a few questions on household income, making it possible to estimate total income and shares of income, for example from food cropping and non-farm sources. We also added a food security indicator: number of meals usually had per day. This indicator will come in handy when a third-round panel is carried through.

The questionnaires used at village and household levels in the two rounds are available on the internet.[7]

Attrition

The overall attrition rate in the 2008 resurvey is 20.6 and varies considerably between countries, as Table 1.1 makes clear.

Ethiopia is exceptionally low in terms of attrition. Besides good survey organization, this stems from the fact that we drew our sample from the memberships lists of the peasant associations. The moderate attrition rates in Ghana, Kenya and Nigeria are the reflection of excellent survey organizations set up by the country teams. The high rate in Mozambique is probably due to the high mobility among the rural population, which in turn was due to the post-conflict situation that country was still caught in when the survey was made (in 2005). Malawi, Tanzania and Zambia, finally, had problems with their survey organization, which, unfortunately, resulted in higher attrition rates.

A quick analysis of the distribution of attrition on some key variables shows that heads of households who could not be retrieved or re-interviewed in 2008 tended to be:

1. Older (66 years on average, compared to 47 years for those re-interviewed).
2. Women-headed (21% compared to 17%).
3. Higher-educated (6.6 compared to 4.8 years).

Table 1.1. Attrition rates in 2008 resurvey, per cent. (From: own survey data.)

Country	Per cent
Ethiopia	0.6
Ghana	14.1
Kenya	11.3
Malawi	24.0
Nigeria	12.9
Tanzania	34.7
Zambia	27.5
Mozambique	28.9
Total	20.6

[7] See: http://gem.sam.lu.se/soc/socgdjweb/Questionnaires/Questionnaires.htm.

4. Have smaller households (3.3 compared 3.7 members) and belong to the poorest wealth group (35% versus 24%).
5. Finally, 13% of the attrition households said that they did not have full control of the land they cultivated in 2002.

On the other hand, there were no differences between attrition households and others in terms of:

1. Size of land cultivated in 2002.
2. Micro-business or large-business source of income.[8]

The above gives a picture not only of those who fail to reproduce their business but, conversely, also of those who succeed in doing so and who, as a result, are over-represented in the panel. First of all, the poor and those with smaller households are less likely to survive in farming. Secondly, women and widows are more prone to drop out, often because their security of tenure is lower, so, if they don't remarry, they tend to return to their native villages. In both cases they disappear from the sample. That older farmers tend to drop out more often is, of course, no surprise, especially if they have no descendants to take over the farm.

That higher-educated heads of households have higher rates of attrition than others is probably a reflection of higher mobility chances. When opportunities improve in urban areas, as they did in the period 2002 to 2008, when growth rates were high, one would expect the better-educated to be quicker to grasp the opportunities created. On the other hand, it is also interesting to note that micro- or large-scale business is not associated with higher rates of dropout. This would go against any hypothesis saying that such business is a platform for leaving agriculture altogether.

Acknowledgements

The *Afrint II* project features collaboration between researchers in nine African countries.[9] The team was led by Göran Djurfeldt, Lund University,

[8] All these findings are statistically significant at the 1% level of significance or lower.
[9] The country teams were: for Ethiopia, Dr Wolday Amha, Ethiopian Economic Association; Dr Teketel Abebe, Addis Ababa University; Dr Mulat Demeke, Addis Ababa University; for Ghana, Professor Ernest Aryeetey, Institute of Statistical, Social and Economic Research (ISSER), Legon-Accra; Dr Daniel Bruce Sarpong, Department of Agricultural Economics and Agribusiness, University of Ghana; Mr Fred Danku, Institute of Statistical, Social and Economic Research (ISSER), Legon-Accra; for Kenya, Professor Willis Oluoch-Kosura, African Economic Research Consortium (AERC); Dr Stephen K. Wambugu, Department of Geography, Kenyatta University; Dr Joseph Karugia, the same department; for Malawi, Mr John Kadzandira, Centre for Social Research, University of Malawi, Zomba and Dr Wapulumuka O. Mulwafu, Faculty of Social Science, University of Malawi, Zomba; for Mozambique, Dr Peter Coughlin, EconPolicy Research Group Ltd, Maputo; for Nigeria, Professor Olatunji Akande, Nigerian Institute for Social and Economic Research (NISER), Ibadan and Dr Olorunfemi Oladapo Ogujndele, the same institute; for Tanzania, Professor Aida Isinika, Institute of Continuing Education, Sokoine Agricultural University; for Uganda, Dr Bernard Bashaasha, Department of Agricultural Economics and Agribusiness, Makerere University, Kampala; and for Zambia, Mr Mukata Wamulume, Institute of Economic and Social Research (INESOR) and Ms Charlotte Wonani, Development Studies Department, University of Zambia.

Sweden, and involved a team of researchers from Lund and Linköping Universities.

References

Afrint (2010) Publications Afrint I. Available at: http://blog.sam.lu.se/afrint/?page_id=35 (accessed 4 April 2010).

Akande, T. (2005) The role of state in the Nigerian Green Revolution. In: Djurfeldt, G., Holmén, H., Jirström, M. and Larsson, R. (eds) *The African Food Crisis: Lessons from the Asian Green Revolution*. CAB International, Wallingford, UK, pp.161–179.

Akande, T. (2006) *Food Policy in Nigeria: an Analytical Chronicle*. New World Press, Ibadan.

Boserup, E. (1965) *Conditions of Agricultural Growth*. George Allen and Unwin, London.

Byerlee, D. and Eicher, C.K. (eds) (1997) *Africa's Emerging Maize Revolution*. Lynne Rennier, Colorado.

Comprehensive Africa Agriculture Development Programme (CAADP) (n/d) Available at: http://www.nepad-caadp.net/about-caadp.php (accessed 4 April 2010).

Conference of Ministers of Agriculture of the African Union (2004) Report of the Ministers of Agriculture (accessed 4 April 2010).

Coughlin, P.E. (2006) *Agricultural Intensification in Mozambique Infrastructure, Policy and Institutional Framework – When Do Problems Signal Opportunities?* EconPolicy Research Group, Maputo.

Djurfeldt, G. (2005) Global perspectives on agricultural development. In: Djurfeldt, G., Holmén, H., Jirström, M. and Larsson, R. (eds) *The African Food Crisis: Lessons from the Asian Green Revolution*. CAB International, Wallingford, UK, pp. 9–23.

Djurfeldt, G. and Jirström, M. (2005) The puzzle of the policy shift – the early green revolution in India, Indonesia and the Philippines. In: Djurfeldt, G., Holmén, H., Jirström, M. and Larsson, R. (eds) *The African Food Crisis: Lessons from the Asian Green Revolution*. CAB International, Wallingford, UK, pp.43–63.

Djurfeldt, G., Holmén, H., Jirström, M. and Larsson, R. (2005a) African Food Crisis – the Relevance of Asian Experiences. In: Djurfeldt, G., Holmén, H., Jirström, M. and Larsson, R. (eds) *The African Food Crisis: Lessons from the Asian Green Revolution*. CAB International, Wallingford, UK, pp. 1–8.

Djurfeldt, G., Holmén, H., Jirström, M. and Larsson, R. (eds) (2005b) *The African Food Crisis: Lessons from the Asian Green Revolution*. CAB International, Wallingford, UK.

Djurfeldt, G., Larsson, R., Holmquist, B., Jirström, M. and Andersson, A. (2008) African farm dynamics and the sub-continental food crisis – the case of maize. *Food Economics – Acta Agriculturae Scandinavica, Section C* 5, 75–91.

Fan, S., Johnson, M., Saurkar, A. and Makombe, T. (2008) *Investing in African Agriculture to Halve Poverty by 2015*. IFPRI Discussion Paper 00751, February 2008. Development Strategy and Governance Division, IFPRI, Washington, DC.

Grandin, B. (1988) *Wealth Ranking in Smallholder Communities: a Field Manual*. Intermediate Technology Publications, London.

Holmén, H. (2005a) Spurts in production – Africa's limping Green Revolution. In: Djurfeldt, G., Holmén, H., Jirström, M. and Larsson, R. (eds) *The African Food Crisis: Lessons from the Asian Green Revolution*. CAB International, Wallingford, UK, pp. 65–85.

Holmén, H. (2005b) The state and agricultural intensification in sub-Saharan Africa. In: Djurfeldt, G., Holmén, H., Jirström, M. and Larsson, R. (eds) *The African Food Crisis: Lessons from the Asian Green Revolution*. CAB International, Wallingford, UK, pp. 87–112.

Larsson, R. (2001) *Between Crisis and Opportunity: Livelihoods, Diversification, and Inequality among the Meru of Tanzania*. Department of Sociology, Lund Dissertations in Sociology 41, Lund, Sweden.

Larsson, R. (2005) Crisis and potential in smallholder food production – evidence from micro level. In: Djurfeldt, G., Holmén, H., Jirström, M. and Larsson, R. (eds) *The African Food Crisis: Lessons from the Asian Green Revolution*. CAB International, Wallingford, UK, pp. 113–138.

Lumpkin, T.A. (2009) Farewell to Norman Borlaug: the world loses its leading spokesman for the fight against hunger. CIMMYT, Centro Internacional de Mejoramiento de Maíz y Trigo, El Batan, Texcoco, Mexico.

Nweke, F.I., Spencer, D.S.C. and Lynam, J.K. (2002) *The Cassava Transformation – Africa's Best-kept Secret*. Michigan State University Press, East Lansing, Michigan.

Patton, M.Q. (1980) *Qualitative Evaluation Methods*. Sage, Beverly Hills, California.

Sen, A. (2001) *Development as Freedom*. Oxford University Press, Oxford, UK.

Sen, A. (2009) *The Idea of Justice*. Penguin, London.

World Bank (2007) *World Development Report 2008: Agriculture for Development*. The World Bank, Washington, DC.

2 African Agriculture: from Crisis to Development?

Hans Holmén[1] and Göran Hydén[2]

[1]*Department of Geography, Linköping University, Linköping, Sweden;*
[2]*University of Florida, Gainesville, USA*

In the first decade of the third millennium, the African food crisis is real. Sub-Saharan Africa's persistent inability to feed its growing population adequately has become a matter of widespread concern. The subcontinent (henceforth SSA) has been deemed the most food-insecure major region in the world. In the early third millennium, per capita food production in SSA is at the same level as it was in 1961 (Godfray *et al.*, 2010). Not only is the African food crisis real, 10 years into the third millennium it has been dramatically accentuated. A number of factors, endogenous as well as exogenous to Africa (capacity constraints, faulty domestic policies, structural adjustment, recent world market price hikes for food and fuel and the current global financial crisis), have combined to turn the African food problem into 'a full-blown development crisis' (ERD, 2009:12). The irony of the matter is that 'Africa has the capacity to feed itself' (Ejeta, 2010:831). Larsson (2005) has shown that sub-Saharan Africa has a considerable untapped potential for agricultural productivity increases. What concerns us here is if and how this potential is being, or can be, made use of.

Small-scale family farming is the economic backbone in sub-Saharan Africa, where smallholders, to a considerable degree, are oriented towards food production, primarily for own consumption. In most cases technology use is rudimentary and yields are low. At the same time the region's population is expected to more than double within the next 50 years (UN, 2003). This has called for sometimes drastic solutions to solve the dilemma of how to feed Africa, proposals that range from rapid modernization and 'big-push' agricultural investments to almost abandonment of agriculture. In the perception of academics and donors, the role and importance of agriculture has shifted over the years between periods of high expectations and those of not so high expectations. These fluctuating emphases to some extent reflect advances in understanding Africa but also – perhaps more so – fads, ideologies and donors' strategic considerations. Advisors and favoured standpoints are legion, but progress is frustrated as 'much policy advice on the agricultural economy in African countries remains based on unrealistic analysis and assumptions' (Omamo and Farrington, 2004:1).

This chapter reviews the potential and constraints for agricultural improvements in the context of where African agriculture is perceived to be in the beginning of the second decade of the new millennium. The point is that not only are the global context and SSA's links to this context different from what they used to be, in so far as previous Green Revolutions (e.g. those in Asia) provide any lessons, their options and contexts then differ to some extent from SSA's options and contexts today. As such, the chapter will also help set the stage for the analyses in subsequent chapters. The first section traces the reasons for the marginalization of African agriculture. The second looks at the attempts that have been made to resuscitate agriculture on the continent. The third and final section focuses on the remaining hurdles facing current efforts to improve agriculture in Africa.

How the Crisis Arose

During the last 30 years or so, SSA's declining capacity to feed its growing population has commonly been explained as being caused by bad governance and neglect of food crop agriculture. Resource-strapped governments have prioritized cash and export crops at the expense of a worsening food situation. Not infrequently, therefore, one finds in development literature comments that there are places in the world – notably in SSA – where governments fail to perform their 'core tasks' (ERD, 2009). This is a problematic conclusion because it is not at all self-evident what these core tasks are. Whereas in a modern welfare state the core tasks of governments are, to a high degree, redistributive and to guarantee basic amenities to the citizens, in Africa after independence (and even today) the most urgent task was to establish the state and to broadcast political power over sparsely populated and non-integrated territories inherited from their former colonial masters. In other words, political consolidation often became more important than development.

Lacking all kinds of resources (financial, institutional, administrative, etc.) and hampered by low population densities and severely limited transport and communication infrastructures, this has proved to be a slower and much more cumbersome process than first believed. Herbst (2000:55f) stressed that 'the cost of extending formal authority in Africa was very high'.[1] In fact, this is still often the case. African states were, and often still are, weak, with limited influence in rural areas. Many African governments responded to spatially uneven distribution of 'political transaction costs' by directing attention more towards urban rather than rural segments of populations (Kydd and Dorward, 2001). Bates (1981, quoted in Herbst, 2000:18) noted that

[1] The problems posed by vast peripheries with low population densities and inadequate transportation infrastructure have not only hampered nation-building but have also constrained the expansion of markets in large parts of SSA.

'African politicians equated their political survival with appeasing their urban populations via subsidies even if the much larger, and poorer, rural populations had to be taxed.'

This could be done because, at the time, food was not perceived as a problem. In fact, SSA was a net exporter of food well into the 1970s (Larsson *et al.*, 2002). For governments there seemed to be little need to worry about the food situation. Moreover, since urban populations were not so numerous (SSA is still the least urbanized major region in the world), they could be fed by way of imports and/or through state farms and some commercialized agriculture near major towns. This also meant that the vast majority of farmers remained predominantly subsistence-oriented. Agricultural investments instead went to export crops (tea, coffee, cocoa) that could bring much-needed hard currencies to the treasury. In this reorientation, donors and international financial institutions (IFIs) gladly lent a helping hand.

This, however, is not to say that African governments totally neglected smallholders or food crop agriculture. To the contrary, there were many efforts, and some successes, to support food crop production, which, however, for various reasons, could not be sustained (Holmén, 2005). Most commonly, governments tried to substitute for missing rural markets by public interventions and the creation of state-led cooperatives and supplying subsidized inputs and monopolistic marketing of produce through parastatals and marketing boards. Systems of pan-territorial and pan-seasonal pricing were applied in order to provide predictability to producers unacquainted with production for a 'market'. Over time, however, these systems became less a supportive measure and more a means for taxation, and they became increasingly burdened by malpractices, inefficiencies and high operational costs. Small farmers often sought to circumvent them.

It may be an open question whether the above sketched development should be interpreted primarily as policy failures or capacity constraints. In whichever case, in the 1980s donors and IFIs increasingly came to regard the African state as an obstacle to development. Development aid declined and much of what remained was bypassing the state, instead being directed to non-governmental organizations (NGOs). Beginning in the 1980s, Structural Adjustment Programmes (SAPs) and the 'rolling back of the state' were implemented all over the subcontinent. For agriculture this meant an end (gradually) to subsidies and dismantling of state-led cooperatives and marketing boards. Instead, input supply and provision of extension services, as well as marketing of produce, were to be handled by private traders. This actually made things worse. Markets were largely missing; there were too few traders around and those who were to be found suffered from all kinds of capacity problems. The share of Western aid going to agriculture fell by around three-quarters between 1980 and 2006 (The Economist, 2009). Donors seemed to lose interest in African food crop agriculture and instead tended to recommend prioritization of export crops to pay for (at the time) cheap imports (e.g. World Bank, 2003). From a food security perspective, the result was disastrous. Food crises 'tripled in sub-Saharan Africa between 1980 and early 2000s' (ERD, 2009:7). As a consequence, 'in the past 20 years, the number of Africans who live below the global poverty line ($1 per day) has increased by more than fifty percent' (Ejeta, 2010:831).

In the second half of the first decade of the third millennium a series of shocks – all of external origin – threw salt into already open wounds. To many, this came as a surprise. As the ERD (2009:11) points out, 'many experts and pundits thought [the global financial crisis] would pass sub-Saharan Africa by, because of the tiny size of its financial sector and its low integration into the global financial system.' Africa, however, was not spared. But the global financial crisis (from 2008 onwards) was only the tip of the iceberg. Since 2003, international prices of a wide range of commodities, notably food and fuel, 'have surged upward in dramatic fashion, in many cases more than doubling within a few years' (Heady and Fan, 2008:1). Within only a few years, SSA felt the impact of soaring oil prices, rapidly increasing food prices and, in various ways, the effects of a global financial crisis. High oil prices meant that already high import bills for fuel became even more expensive, leaving less of governments budgets for other expenses. High oil prices also result in high fertilizer prices, making fertilizer even less accessible for African smallholders than before. Also, in other ways, increasing fuel prices push food prices upwards through their effects on both input prices and transport costs.

Godfray *et al.* (2010:812) note that 'in mid-2008 there was an unprecedented rapid increase in food prices, the cause of which is still being debated.'[2] According to the Food and Agricultural Organization (FAO), in 2008 the world food price index was more than twice its level in the year 2000. For cereals, it was almost three times as high as in 2000 (FAO, 2010a). The FAO further reports that, even if food prices had declined somewhat from their peak in 2008, 'in southern and eastern Africa ... maize prices are 25 to 75 percent higher than in the pre-food crisis level of two years ago' (FAO, 2010b).[3] In theory, this could provide an incentive for small farmers to enhance production and to finally embrace the market. In practice, however, the majority of smallholders in SSA are net buyers of food. Hence, they tend to withdraw from the (emerging) market rather than engaging with it. Consequently, in SSA, 'increasing numbers of farmers are growing only food crops for home consumption and storage, and reducing levels of purchased inputs applied' (IFAD, 2008:7).

[2] Among the reasons for food price hikes are often mentioned (or assumed) increased demand for food (and animal feed) from populous countries with high growth rates (China and India). Heady and Fan (2008) find this less convincing, pointing out that these countries are self-sufficient regarding food and tend to export food rather than import it. Instead they point at increasing oil prices, depreciation of the US$, and biofuels as triggers of escalating food prices in 2008 (see also ADB, 2008).

[3] It needs to be underlined that it is not the magnitude of the price hike that is the major problem but rather its abruptness, which gave farmers and governments little time for adjustment. Food prices in 2008 increased from historically low levels. Actually, 'even the highest price levels experienced in 2007 and 2008 [were] substantially below the peaks in the previous world food crisis in 1973–1974. Indeed, real prices in mid-2008 for corn, wheat and rice remain[ed] well below what was considered "normal" until the full impact of the Green Revolution was felt after 1980' (ADB, 2008:73).

These crises affect SSA in a multitude of ways. Not only have they highlighted Africa's heavy dependence on food aid (OECD, 2008) but the global recession also hits SSA through falling exports, enhanced costs of imports, reduced remittances from diaspora communities, lower levels of foreign direct investments and, possibly, reductions in foreign aid (ERD, 2009). It has been estimated that, due to the global financial crisis, SSA lost incomes of over US$50 billion in the year 2009 alone (te Velde, 2009).[4] Not only has this led to a general economic slowdown (African Economic Outlook, 2009; te Velde, 2009), Bakrania and Lucas (2009) suggest that it will also result in lower agricultural investments. It is also widely believed that food prices in SSA are likely to remain high for years to come (Diao *et al.*, 2008; Heady and Fan, 2008). Several observers stress that the recent food crisis has caused riots and troubles in several sub-Saharan African countries (OECD, 2008; Walt, 2008; African Economic Outlook, 2009). Bakrania and Lucas (2009:9) warn: 'should the crisis persist ... the danger of regime-threatening instability will increase dramatically.' With such prospects, it would be easy to despair. The UN Secretary General Ban Ki-moon (2008), however, found this to be 'a perfect storm of new challenges'. So, how are governments in SSA responding to the challenge?

What is New in the 21st Century?

There is a new momentum towards recognizing the importance of agriculture for both food security and development in Africa. This has come about step by step since 2000 and is now amounting to a new situation in which there is reason to assess afresh what the prospects are for African agriculture to move from crisis to development. The following chapters in this book provide empirical evidence from a cross-section of countries to indicate that progress is indeed taking place, even if it is not yet transformational. At this point it is important to acknowledge the following factors as part of the new scene in Africa: (i) a growing commitment to investment in agriculture and agriculture-related activities; (ii) scientific advances in crop varieties; and (iii) new pro-farmer policies.

A growing commitment

The most important thing about what is happening in the beginning of this new century is that the effort to revitalize agriculture in Africa is led by African governments and institutions. During previous decades the interest in these circles was never as explicit and strong as it is now. It was often donor-driven and never embraced very sincerely by African governments at the time. This is not

[4] This figure can be compared with net ODA for sub-Saharan Africa, which amounted to US$34 billion in 2007 (African Economic Outlook, 2009).

to imply that there was no agricultural growth. It was there in countries like Kenya and the Ivory Coast, where policies and other factors helped facilitate an export-oriented agricultural development. Having faced hunger and a failure to get the economies to grow merely as a result of the 1980s SAP, African governments have decided to return to the drawing board, i.e. to see what steps can be taken to make agriculture grow again in ways that serve the triple purposes of ending hunger, reducing poverty and enabling a national development.

The initial step was taken in the very first couple of years of the century and culminated in the Maputo Declaration, which was adopted in September 2004 by African heads of state. Key actors in the process were the New Partnership for Africa's Development (NEPAD)[5] and Ministers of Agriculture of African Union member countries, with external support from the FAO. The most important item in the declaration is the adoption of the Comprehensive Africa Agriculture Development Programme (CAADP) and the commitment made by the African presidents to set aside 10% of the national budget for the agricultural sector.

The work leading up to the Maputo Declaration was the first concrete and substantive contribution to African development by NEPAD. It was the catalyst for action on the African side and could ensure that the process and the outcome were conceived as being under African ownership. The office of CAADP has been charged with monitoring the implementation of the Maputo Declaration. Performance among member countries has so far been mixed. The full effects of these national expenditures are not always possible to trace in the short run and it would be wrong to draw far-reaching conclusions at this stage. Yet, the fact that there is a monitoring mechanism to keep up the pressure on governments and that there is evidence that several countries do commit an increasing percentage to agriculture is an encouraging sign and proof that the momentum is being sustained. Some countries, such as Tanzania,[6] have adopted special programmes to catalyse progress at the national level.

The Maputo Declaration has also helped mobilize interest and resources from other sources. Notable among these are commitments made by the British and US governments (DFID, 2003; USAID, 2004) and an extra push by the Africa Commission, appointed by former Prime Minister Tony Blair (Commission for Africa, 2005). The biggest challenge in this new climate in support of agriculture is to avoid recycling old ideas that never worked. The African ownership is not necessarily a guarantee that mistakes will be avoided, but it is an opportunity to do things differently and – above all – avoid simple 'technical' or 'institutional' fixes that were so often applied in the past.

Governments, however, are not the only actors on the scene. A new feature today is the growth of non-governmental entities at regional as well as national levels. Foremost of these is the Accra-based Alliance for a Green

[5] At the end of 2009, NEPAD was fully integrated into the African Union structure and renamed the NEPAD Planning and Coordinating Agency (NPCA).
[6] *Kilimo Kwanza* (in English: Agriculture First).

Revolution in Africa (AGRA), which receives funding from two US philanthropies – the Rockefeller Foundation and the Bill and Melinda Gates Foundation. According to its own home page, AGRA is 'an African-based and African-led organization charged with sustainably increasing the productivity and profitability of small-scale farms throughout Africa'. It is currently headed by the former Secretary General of the United Nations, Kofi Annan, and serviced by a group of highly competent and committed technical staff.

There are two things with AGRA that are different from past efforts. The first is that the funders are not, as in the past, mainly funding the international research centres in agriculture but try, as much as possible, to support initiatives created and managed by African stakeholders. Thus, funding is channelled directly to those in charge of making a difference at the national or regional level. For instance, in 2009 AGRA joined with NEPAD in a partnership to speed up food production for enhanced food security in Africa.

The second thing is that there is recognition that progress requires involvement by many agencies and organizations. Spreading the resources across several different actors, therefore, is not viewed as a weakness but as strength when it comes to promoting agriculture. It should be added here that a new set of African organizations have emerged in the wake of the collapse of the cooperative movements that were so prominent in the 1960s and 1970s (Gyllström, 1991). They are typically made up of farmers, both small and medium producers, who are engaged in sustained production for sale, whether on the local or export market. These are voluntary efforts and there is a strong sense of ownership, something that had been lost in the old organizations. An example at the regional level is the Pan African Farmers Platform (PAFP), based in Addis Ababa, which argues that 'there are no alternatives to the mobilization of our own human resources and our own financial resources' (PAFP, 2008:3).

It can, of course, be questioned whether Africa's problems can be solved at the regional level. NEPAD as well as AGRA and other regional entities are far removed from the day-to-day realities of small-scale farmers. There needs to be institutional linkages between what happens at the regional, national and local levels. Governments themselves have never been particularly good at taking policies all the way down to the local level. For example, despite having elaborate agricultural extension services connected with research stations, the effects of government interventions at the farm level were indeed very limited. Leonard's classical study of the agricultural extension service in Kenya's Western Province provides the most convincing evidence of the limitations in implementing agricultural policies at the local level (Leonard, 1977). The agricultural extension services throughout Africa are even weaker today than they used to be, but there is a positive difference as well. Unlike in the past, when agricultural extension services were supply-driven in a top-down fashion, these services are today increasingly being demanded. This means that extension officers, even if they are fewer than in the past, are being employed in more rewarding pursuits than in the past. They find satisfaction in their services being requested. This applies first and foremost to those extension officers that work outside the regular ministry organization, but even among those there is a new sense of appreciation that was lacking years ago.

Scientific advances

As part of the commitment to improving African agriculture to promote food security and development, funding for research plays an important role. Even in regard to this topic, things are different in 2010 from how they were before, in at least three respects. The first is that modern information technology is facilitating the growth of research networks that have the potential for making a difference in a way that was difficult, if not impossible, in the past. The second is that the evolution of research on crops at international, as well as national, agricultural centres has become more closely focused on the peculiarities facing the most important food crops on the continent. The third is that, among development researchers also, the focus is increasingly on agricultural productivity and how it can be enhanced.

The networks are especially important in Africa, where the national research systems are small and therefore unable to sustain the effort that is needed on their own. Each region of the continent has its own regional research network. For example, there is the Association for Strengthening Agricultural Research in Eastern and Central Africa (ASARECA). It has its equivalent in southern and western Africa. Together, these three networks form the Forum for Agricultural Research in Africa (FARA). Sandwiched between the international and national research systems, FARA plays advocacy and coordination roles. In addition to these general agricultural research networks, there is a network for pretty much every crop and legume that is grown in Africa. Some of these efforts are funded by private philanthropies from the USA, e.g. McKnight and Rockefeller Foundations, others by bilateral donors like the Canadian International Development Agency (CIDA) and the British Department for International Development (DFID, these days also known as UKAID). All these networks serve an important role in empowering national agricultural research institutes and adding value to their work. Today this is a much stronger support system for agricultural development than the case was some decades ago, when the link between the international and national research systems was much more brittle and the technology to sustain networks was much less developed.

Cassava is an especially compelling case. It was long considered the crop of the downtrodden, but thanks to research spearheaded by the International Institute for Tropical Agriculture (IITA) in Nigeria it is fast becoming a food crop for the elite in Africa also. Two major diseases of cassava – bacterial blight and leaf mosaic – have been controlled through genetic breeding and the incorporation of resistance genes into high-yielding cassava varieties by IITA. Also, thanks to its Africa-wide programme on the biological control of the cassava mealy bug, the institute has waged a successful war on this devastating pest. Having freed Africa's most friendly crop from the vagaries of some of the prevailing diseases and pests, IITA now has many improved cassava varieties available that are high yielding and early maturing. While the older varieties used to yield, at best, 4–6 t/ha, the new ones are late maturing and capable of yielding 30 t or more per hectare in just 12 months.

The 'New Rice for Africa' (NERICA) is another interesting example of scientific and technological progress in African agriculture. It is a cultivar developed by the West African Research Development Association (WARDA)[7] to improve the yield of African rice varieties. Although the vast majority of the population in West Africa relies on rice as the primary source of food energy and protein in their diet, most of this rice has been imported at high cost. The NERICA project, which has received funding from the African Development Bank and the Japanese government, has resulted in new rice varieties suited to the semi-arid lands of the region. As of 2006, these varieties had been distributed and sown on more than 200,000 ha during the previous 5 years in African countries such as Guinea, Ivory Coast and Nigeria, and also in Uganda in eastern Africa. The gains achieved by NERICA is a fourfold increase in grains per head and thus a similar increase in tonnes per hectare, an additional 2% of protein compared with that contained in the original varieties, a taller plant that makes harvesting easier, a greater resistance to pests and a lower uptake of water. Although countries using the new variety are still not self-sufficient in rice production, Africa has already demonstrated a capability to double rice production in a much shorter time than did Asia during its Green Revolution. For example, production in West Africa went from 2.76 million t (milled rice equivalent) in 1985 to 5.75 million t in 2005. Strong government support in Uganda and Nigeria has produced returns that show the continent can yet beat not only the current crisis but also the cumulative 10-year crisis that some experts predict will cripple world cereal supplies (AfricaRice, 2008).

It is also significant that development researchers have begun to return to issues facing agriculture and how it can be made more productive. In the context of the Millennium Development Goals and their emphasis on poverty reduction, these issues have not received the attention that they deserve and need. Social development issues have overshadowed the ones related to how the economic wheels of Africa can begin to spin more effectively. The return to agriculture is perhaps best manifest in the influential *World Development Report*, whose 2008 issue is devoted to this theme (World Bank, 2007). An increasing number of scholars, however, have more recently tried to shift the attention to the role of agriculture in development. Contributions in this direction include a previous volume by many of the authors in this volume (Djurfeldt *et al.*, 2005) and publications on the role of crop science in alleviating poverty (Mosely, 2002; Lipton, 2005), as well as an analysis of the sustainability of agriculture in Africa (Southgate and Graham, 2006). This and other evidence of a reorientation in the outlook on African development are important and encouraging, but much remains to be done, as a recent report commissioned by Oxfam International indicates with its subtitle – 'Turning promise into reality on the ground' (Crola, 2009).

New pro-farmer policies

As already noted in the first section of this chapter, the crisis in African agriculture – and development – was to a very large extent caused by the

[7] Now renamed the Africa Rice Center.

neoliberal economic policies that were forced upon African countries in the 1980s by the IFIs, especially the World Bank. Although they were expected to favour the farmers by offering better price incentives than before, the same policies contributed to taking away the support structures that had been in place since colonial days, notably marketing boards, extension services and farmer subsidies.

With NEPAD and the renewed effort by African governments to seize greater control of the development agenda on their continent, there has been a change in orientation. Leaders have become more assertive and have broken ranks with the mainstream. Two countries that have been in the forefront of this reorientation are Malawi and Rwanda. In the former, it is President Mutharika, who reintroduced seed and fertilizer subsidies to farmers so that they could get better yields and help turn the country from being a food-importing to a food-exporting country. For 4 years running, the country managed a 7% growth in GDP per capita, fuelled in large part by the agricultural sector. In Rwanda, President Kagame has achieved equally impressive results by similar policies. For example, in 2007 food production grew by 15% and by 16% the following year (AGRA, 2009).

Subsidies are not a 'silver bullet' and controversy surrounds the extent to which they should be used. For instance, without targeting the subsidies they may make little difference in the long run. Other issues, whether they relate to promoting better technology, assuring greater market access to small farmers, protecting the environment or strengthening infrastructure, are also important. The most important thing, though, is that African governments begin to shape home-grown agricultural policies that provide comprehensive support to small farmers. An important initiative in this direction has been taken by AGRA. Focusing on five countries – Ethiopia, Ghana, Mali, Mozambique and Tanzania – the initiative is meant to strengthen agricultural policy-making capacity by training agricultural policy analysts, bolstering policy think tanks, establishing data banks to support evidence-based policy development and coordinating national policy hubs. It will focus on policies that support farmers in the areas of seeds, soil health, markets and trade, land rights, women's rights and environmental sustainability. This is a tall order but it is a step in the right direction, and any improvement is likely to make a difference for food security and agricultural development.

What are the Issues that Need to be Resolved?

The reinvigorated political will notwithstanding, there are a set of issues that will face policy makers as they try to put promise into practice. These can be divided into three types: (i) external factors; (ii) domestic capacity; and (iii) agricultural sector issues. Africa's performance with regard to these challenges will, to a large extent, determine how far countries will be able to enhance their food security and promote agricultural development.

External factors

External factors are those over which African governments have little or no control. They may be man-made, like the financial crisis, or natural events, like flooding or drought, that occur on an unpredictable schedule. Some may have positive effects, like foreign investments, but most, including many foreign investments, have negative consequences.

The international donor community is positively inclined towards the idea that Africans need to take greater ownership of their own development. After having tried to dictate it during the past three decades, things have been changing in the light of the 2005 Paris Declaration, which, among other things, calls for a partnership relationship between donors and recipients, in which the latter will take increasing charge. With a greater appreciation on the part of the donors of the need for more sustained attention to agriculture, the global donor stage is positively set for a more assertive leadership by African governments and other local actors in global as well as regional policy circles. NEPAD and the processes that the African Union has initiated in order to improve governance in African countries are structures that will help reinforce this situation.

The clouds on the global horizon are more economic than political. The financial crisis that has adversely affected most countries, rich and poor alike, continues to affect institutions that have been in the forefront of helping to finance African development. Bilateral aid has been declining, albeit not catastrophically, but in the case of some countries, where the ambition is to hold on to the 0.7% of GDP in foreign aid, the decline in economic growth translates into less money also for development purposes in Africa. The credit crunch that has affected many countries in the West is also a factor that cannot be ignored, especially for any partnership that involves private sector investments.

Because the financial crisis has affected Western economies most, it is worth paying attention to those countries, like China, which have been able to ride through the economic crisis with little or no damage. China is already becoming an increasingly important actor in African countries. It is investing in several sectors, including agriculture. So, what can be made of its role in African agriculture? There is no consensus answer.

Some would interpret it as just another example of China's growing appetite for African resources. Others prefer to see it as a genuine effort to help African agricultural development. Feeding the country's 1.3 billion people is a priority of the Chinese government, but with only 7% of the world's arable land and the continued loss of millions of hectares of arable land to pollution and urban growth, it is no surprise if it looks to securing agricultural assets abroad. In such a scenario, analysts are not ruling out the possibility that Chinese investments would be accompanied by Chinese labour and technology at the expense of domestic growth. This interpretation of the Chinese presence in Africa may be too slanted and simplified. Critics say that importing food from Africa on a large scale would never become efficient. According to the country's own Ministry of Commerce, the number of Chinese agricultural experts in Africa is given as 1100. In addition, there are, according to the same source, around 1 million Chinese farm labourers spread around 18 African countries. These experts and labourers help maintain 11 agricultural research stations and over 60 agricultural investments projects, mainly in eastern and southern Africa (Rubinstein, 2009).

It is significant that these investment projects have not led to the acquisition of land leases, an indication that they are meant to address African rather than Chinese food security issues. Examples of what Chinese-funded projects do are an agricultural demonstration project in Mozambique that tests the durability of various staple crops and a fish-farming project in Uganda aimed at reducing the overfishing that goes on in Lake Victoria. These and other projects are important contributions to both food security and agricultural development in Africa, although they tend to be less well known than projects funded by Western donors or international agencies. The challenge facing the Chinese in Africa is to ensure that they become less isolated from other activities and more sensitive to home-grown initiatives on the continent.

Finally, with regard to the external environment, there is the issue of climate change. It will no doubt become a more important and pertinent issue in coming years, but what is already being done in terms of developing high-yield varieties that can withstand drought, flooding and toxins constitutes the most effective answer to these challenges. Continued funding for these efforts, therefore, is likely to have the best pay-off for the purpose of reducing poverty and the risk of hunger.

Domestic capacity

The biggest challenges are likely to be found in the domestic environment, notably in how well the state is capable of managing a process focused on agricultural development. Its record is not particularly encouraging. Its ineffectiveness, including its tendency to excessively tax the small farmers in the 1970s, was a reason for the introduction of the unfortunate structural adjustment policies. The early state policies after independence favoured the consumer over the producer, the urban over the rural resident (Lipton, 1977; Bates, 1981). Compared to India and other countries that went through a transformation of agriculture in the latter part of the 20th century, African attempts at the time to imitate such a state-directed effort fell far short of expectation.

Have state institutions in Africa improved since then? The question is obviously of key importance if agricultural development is going to be resuscitated in a sustainable fashion. Donor-funded initiatives to improve governance have seen some results: free and fair elections are being held on a regular basis in a majority of African countries (Lindberg, 2006); human rights violations continue but on a more modest scale and with greater political penalties or costs attached to such acts; and media and civil society actors are more vocal and keep government on its toes in ways that did not happen before. These are accomplishments that should not be underestimated. Yet, the state in Africa remains 'soft', i.e. it is prone to corrupt practices. Public sector reforms have not had the intended effects of making government departments more efficient and accountable to the public; political patronage continues to be more important than policy development and implementation (Hyden, 2006). To be sure, many of these same features were prevalent in the Asian state during the time of the Green Revolution there, but they were compensated by a level of professionalism and a political will that have been much weaker, if not totally absent, in the African context.

The Green Revolution in Asia was small-farmer-based, market-mediated, and state-driven (Djurfeldt *et al.*, 2005). The exact same scenario is unlikely in Africa, although bits and pieces of it no doubt will be important there also. The Green Revolution in Asia occurred at a time when it was expected that development would be directed by the state in a top-down fashion. The political climate today is different. Development is perceived as a participatory process in which several stakeholders interact. The problem here is that African governments are not always ready to allow other actors to get involved, certainly not on an equal basis. They want to keep civil society and the private sector at arm's length and let them into the process only on their own terms. Politics, although patterns vary, is largely driven by neo-patrimonialism, i.e. with a Big Man controlling the policy scene in a discretionary manner (deGrassi, 2008). Today's challenge in Africa, therefore, may not be to ensure that the state directs the process of transformation alone but instead that it 'opens up' to greater interaction with other stakeholders in a more equal and reciprocal fashion. This does not mean that there is no need for enhancing state capacity. There certainly is, especially with regard to the kind of issues that the AGRA initiative mentioned above involves.

So, can these domestic political hurdles be overcome? The answer, in principle, is yes, but how far it will become reality depends on the willingness of both government leaders and donors to create the necessary policy space. One way of doing that is to encourage a more demand-driven change process through the establishment of agricultural development funds outside of the political realm to which groups and organizations of small-scale farmers have access. Money for these funds could come from internal as well as external sources and they could be managed by boards made up of people with a reputation of professionalism and public integrity. Groups could apply for funds for projects or activities that they have designed and give priority to. Such a 'bottom-up' process would give incentives to groups to seek funding based on their own skills to prepare project proposals. It would help building managerial capacity and would reduce the risk of projects being 'hijacked' by individual politicians for their own interest. These funds, as public institutions catering for competing local groups in a non-political context, would be a more suitable instrument for promoting a Green Revolution in Africa, given the soft nature of the state. The challenge would be to establish a way of managing these funds that takes away the risk of patronage and misappropriation. For that reason, it may be desirable to have representatives of the funders, e.g. the donors, represented on the board so there is accountability to both local actors and those from the outside who help provide support.

This model for funding development was originally launched 15 years ago at an expert consultation in Kampala, Uganda, at which SSA government representatives, members of parliament, civil society leaders and a few academics participated, together with representatives of several major donor agencies (Dag Hammarskjold Foundation, 1995). Many Africans have seen the rationale for this and a few such funds have been established, e.g. in Tanzania and South Africa, with good results. Donors, on their part, however, have remained reluctant because they have failed to accept the two conditions that make these funds different from all other funds the donors have participated in creating. The first is the importance of locating the institution outside of the soft state and insulating it from patronage politics. The second is for them to participate in the management so as to strengthen the prospect that the decisions to allocate funds are made on professionally sound grounds and in the spirit of fairness, qualities that cannot be immediately secured in an environment of competition for scarce resources in countries characterized by horizontal ethnic group relations.

Agricultural sector issues

There are a number of policy issues that affect the performance of the agricultural sector. We have already dealt with the scientific and technological issues above and will return here to three issues that are current in the debate about how to reduce hunger and promote development of African agriculture: (i) should the emphasis lie on food or export crops?; (ii) is agricultural development best pursued through big commercial or small peasant farms?; and (iii) is land tenure reform a prerequisite for development?

Food crops or export crops?

Whether African farmers should opt for producing export crops or food crops[8] is a crucial but also sensitive issue. African governments have often been accused of neglecting the well-being of their subjects when prioritizing production of export crops, since this has been seen to occur at the expense of food security. With few other commodities that could be used to finance necessary imports, there are good arguments for agriculture-based economies to opt for promotion of export crops. Critics have pointed to the fact that, since most smallholders have been primarily subsistence-oriented, they tended to be bypassed by modernization efforts and became locked into low-productivity agriculture (Holmén, 2005).

[8] Whereas it is often impossible to separate cash crops from food crops (peasants selling some maize but retaining part for own consumption), a distinction needs to be made between food and export crops. By export crops here are understood crops that are primarily meant for sale to other countries (spices, fruit, vegetables, flowers). They can also include food staples such as rice in Vietnam or soybeans in Argentina. A distinction, however, needs to be made between crops that are grown in order to be exported and those that are not. Food crops are staples produced in order to be consumed in country, either on farm, locally or in urban areas. Exports of occasional surpluses (as in Malawi in 2007) do not lead to labelling as export crops.

The prioritization of export crops has continued into the 21st century. For example, as late as 2003, the World Bank insisted that high-value, not staple, crops should be a priority (World Bank, 2003:43). This recommendation, based on the premise that a country must always seize on its comparative advantage, is no longer embraced with the same degree of commitment. Experience in the past 30 years has shown that countries in the periphery of the global economy due to poverty and underdevelopment are not always in a position to realize their comparative advantage. In a global economy with highly unequal relations between parties, laissez-faire does not give the weaker party the advantages the theory claims – especially as rich countries do not practise what they preach.

It is increasingly being realized, therefore, that – in the real world – 'free trade is the protectionism of the powerful' (Bové, 2003:xiii). Only a few developing countries have been able to benefit from agricultural high-value exports and 'the many sub-Saharan nations that remain bound to the traditional export/cheap grain import model have fared the worst, as recurring famines and persistently high levels of undernourishment attest' (Weis, 2007:126). It has become obvious that, in order for Africa to prosper, the European Union (EU) and the USA 'must abandon [their] sacrosanct vocation to feed the world' (Herman and Kuper, 2003:96). Even the World Bank has changed tune. It now declares that 'agriculture-dependent countries ... must largely feed themselves' (World Bank, 2007:6). Moreover, to the extent that trade reforms are needed, it finds that 'developing country agricultural trade reforms are estimated to have a much smaller impact on their own terms of trade than developed-country policy changes' would have (World Bank, 2007:108).

The wind in the global arena is definitely turning in a pro-agriculture and pro-growth direction, but equally significant is the recognition that Africa's future relies not only on being able to export high-value crops but also on strengthening its production of food crops. This is an important new orientation, which needs to be followed up in order to allow the countries on the continent to draw on what is being consumed locally. The urban areas may not be the 'engine' of growth that they have been elsewhere and projected to be also for Africa, but it is clear that the linkages between town and village need to be strengthened as part of an approach to development that recognizes the importance of starting from within.

Big or small farms?

Discussions of agricultural development, whether in policy or academic circles, have often been polarized around two seemingly irreconcilable standpoints: 'big is better' and 'small is beautiful'.[9] The 'big is better' paradigm advocates mechanization, modern inputs, specialization and the importance

[9] Big or small does not necessarily refer to the physical size of a farm. A 5-ha farm in intensively cropped, irrigated Asian rice areas would usually be referred to as a big farm, whereas in many parts of the world a 10-ha rain-fed farm would be considered small. Big or small has to do with types of crops grown, intensity of land use, level of mechanization, soil type and climatic conditions.

of scale economy. The 'small is beautiful' paradigm, on the other hand, refers to farm sizes not exceeding the labour requirements that a family can supply. It also emphasizes pluri-culture and sometimes reveals a tendency to romanticize pre-modern village life and the joy of being close to nature. While being a peasant may represent a desirable lifestyle, pre-modern agriculture also means insecurity and hard, often back-breaking work with little material reward.

The first camp claims that large farms are most efficient as they are more rationally operated and have a large output per labour input. They make bulk purchases of inputs and sales of produce and they access credit easier and at lower cost than do small farms. Proponents of small or family farms, on their part, point to the fact that such entities often are more intensively farmed than estates and hence have higher land productivity. Their small amounts of purchased inputs and marketed produce can be compensated through contract-farming arrangements and/or by organizing farmers in cooperatives, etc. Djurfeldt *et al.* (2005:19) underlines that 'in intellectual discussions of agricultural development there has always been a strong bias against smallholders, stressing the importance of scale. Big farms have been seen as necessary for modernized agriculture.' A recent representative of this camp is Paul Collier (2008:5), who calls for 'large, technologically sophisticated agricultural companies' as a solution to the food situation in SSA.

It has been questioned on good grounds whether this would benefit Africa. Prime Minister Meles Zenawi of Ethiopia recently stressed that – considering how numerous smallholders are in Africa – not focusing on small farms 'would be plain stupid' (interviewed in the *Financial Times*, 21 August 2008, quoted in Wolday *et al.*, 2009). Theoretically, a strategy built on big, mechanized farms may solve the food problem but will have few multiplier effects and, hence, is not likely to enhance (or even maintain) former peasants' purchasing power. Jobs may be lost rather than created. Empirically, in contemporary 'developed' countries, big farms were not the trigger of development but one of its outcomes. There is evidence showing a strong, historically positive relationship between productivity increases within the smallholder staple crop sector and broader economic development. As one prominent analyst demonstrates 'there are virtually no examples of mass ... poverty reduction since 1700 that did not start with sharp rises in employment and self-employment income due to higher productivity in small family farms' (Lipton, 2005:viii). In line with this, the World Bank recently concluded 'getting agriculture moving requires ... a smallholder-based productivity revolution centred on food staples' (World Bank, 2007:20).

This is not to imply that smallholding in Africa offers a problem-free scenario, especially since it is often associated with extensive fragmentation, leading to unviable farm sizes. This, in turn, leads to difficulties in obtaining credit, to small volumes of inputs purchased and small volumes of produce sold and therefore comparatively high transaction costs. One often-suggested means to overcome this is that smallholders should organize in cooperatives and similar associations, in order to benefit from economies of scale. However, somewhere there is a threshold below which farm size should not shrink. As Gyllström (1991) showed in his study of cooperatives in Kenya, organizing large numbers of smallholders with miniscule transactions tends to undermine the viability of their effort. Many small transactions and an ensuing myriad of paperwork strain the administrative capacity of these organizations. It also means that transaction costs are transferred from individuals to local organizations. In the long run, many small farms will have to disappear and the average farm size will have to grow.

So, can these limitations be overcome? This is a valid question, given that there has been little success in consolidating farms into larger units. In fact, the opposite has been the trend. Already small farms have been sub-divided further in order to cope with demands for access to land among family and kin relatives. The answer to this question, therefore, takes us to the third policy issue: land tenure.

Customary or individual land tenure?

One important reason why the land tenure issue has become increasingly critical in Africa is the growing interest that other countries show in trying to lease land in order to produce food and other resources for use back home. Several studies have been conducted that show the extent to which African land is being purchased by foreign companies. This may not amount to a 'second scramble for Africa' but it certainly raises issues that are of direct relevance to how Africans can ensure their own food security and promote their agriculture. One recent report titled 'Land Grab or Development Opportunity?' provides interesting evidence on the extent to which land is being transferred to foreign ownership in five African countries: Ethiopia, Madagascar, Mali, Mozambique and Sudan (Cotula *et al*., 2009). It shows that close to 2.5 million ha have been allocated to foreign companies in these five countries in the form of allocations exceeding 1000 ha. Land-based investments have been on the rise in the last 5 years. The five countries included in the study have large tracts of arable land, but most of it is already under use, usually by local people, and pressure is growing on the higher-value land. Investors include companies from EU countries, the Arab Gulf and the Far East.

This expansion of cultivated land in Africa is being questioned not merely on the grounds of posing a threat to domestic food security in individual countries but also on the grounds of constituting a threat to the environment (Economic Commission for Africa, 2009). If these foreign investments are primarily geared towards beneficiaries outside the continent, they are also less likely to stimulate domestic economic activities.

Against this background, the concern about land tenure has grown in various parts of Africa. How can a proper land tenure regime be established that enhances prospects for food security and agricultural development? The question is not new but has yet to be tackled effectively. Land continues to be owned through what is still referred to as 'customary' arrangements. These include a variety of forms. In some cases, it is a local chief that controls ownership. In others, it is an extended family or lineage that exercises control. In whatever shape, their customary nature has been undergoing change over the years and the juxtaposition between customary and statutory (typically private or individual) ownership is by no means as clear cut as it was in early colonial days, when this dichotomy was first introduced.

In high-value areas in particular, there is a definite growth in monetarized land transactions. These are often adjusted 'customary' transactions involving witnesses and the use of contracts. This means that much of what is going on in such transactions is embedded in complex social and often political relations. It is this 'hybrid' type of deal that continues to dominate domestic land markets. They serve short-term needs for seller and buyer alike, but they do not make the transition to a more efficient land market any easier (Cotula, 2007).

Forces driving the change in land regulation at the local level include both demographic and political factors. Migration of people leads to challenges of finding ways to integrate them into local systems of regulating land. This has been a big issue in many West African countries, e.g. Ivory Coast and Nigeria. It also involves state action, whether aimed at directly regulating land sales or at transferring this power to local authorities.

The issue of customary versus statutory land regulation is not going to be resolved in one direction or the other in the near future. There will continue to be incremental changes but these will not necessarily be in a definite direction towards private ownership. Such an evolutionary thinking is too simple. The concern about food security – if not agricultural development – may help accelerate the process of finding ways to resolve the political impasse that is there. Governments may at last be more willing than in the past to take on this tricky issue. No single recommendation for how this should be done is likely to fit the challenges. The solutions will have to come from within each country and involve a wide range of stakeholders in order to become effective. A successful move in this direction is also likely to be important for sorting out the legal framework for large-scale investments in agriculture, whether by domestic or foreign companies, in ways that reduce the tensions that continue to exist in most African countries between small-scale and customary, on the one hand, and large-scale and private, on the other.

An African Green Revolution?

There is no question that Africa needs a Green Revolution, but will it be capable of realizing it? Our answer is at least a cautious yes. It may not translate into a replication of what happened in Asia, but with more auspicious circumstances than before, a definite change is within reach. The positives can be found in the realms of technology, policy and a new stakeholder involvement. The negatives that need to be tackled are, above all, shortage of implementation capacity in state institutions and uncertainties surrounding land ownership and use.

Like the Green Revolutions in Asia, a smallholder orientation in the efforts to enhance food security and promote agricultural development in SSA is a *sine qua non*. The market mediation is also an integral part of these efforts. What is more questionable is the extent to which the state can fulfil its role as leader. It is worth remembering, however, that the Asian states which undertook Green Revolutions in the 1960s and 1970s were also labelled 'soft', i.e. weak, corrupt and severely lacking the capacity to implement plans and policies (Myrdal, 1968). That, however, did not prevent them, in due course, from becoming comparatively effective 'developmental states'.

The triggers for these transformations were to be found in their context, the precarious situations facing many Asian governments at the time. The Asian Green Revolutions were not launched because governments cared particularly about smallholders or the urban poor. But they were under strong pressures – from widespread hunger, food riots and high import prices for staple food, and the threat of popular uprisings – the same challenges that confront many African governments today. In order to survive politically (even literally) these governments felt compelled to opt for reforms that would not only remedy immediate food shortages but also safeguard sustainable domestic food security in the long run. It was imperative to base such reform on the small farmers. Supplementing the spread of new technologies (high-yielding seeds, fertilizer, agrochemicals) with massive support systems (credit, extension services, price policies, infrastructure investments), they largely succeeded.

In recent years, a number of sub-Saharan African countries have seen food riots due to increasing poverty and widespread hunger. This, together with being exposed to the vulnerability of food import dependency, have forced 'African governments [to] turn[] their attention to the potential of domestic agriculture to meet food requirements at home' (OECD, 2008:33). Ejeta (2010:832) asserts that among political leaders in SSA 'we are seeing a new sense of urgency and an increased commitment to making a lasting change in African agricultural development.' For Africa, this is a new situation and, as we have seen above, there are good reasons to believe that this is a correct assessment. In particular, the fact that both NEPAD and the Maputo Declaration preceded the current crisis indicates that African leaders are serious about the reorientation. The political leaderships in African countries have a unique opportunity that they cannot afford to miss.

There is no doubt that the ball is in the African court. The main responsibility lies with African actors and institutions. At the same time, it would be a mistake to overlook the global conditions in which Africa, compared to Asia some 50 years ago, finds itself today. Asian Green Revolutions of the 1960s were strongly backed by Western powers, especially the USA. The 'cold war' was as hot as ever and the fear that the Chinese revolution would spread led to unprecedented support in the form of development aid and free access to technologies (high-yielding seeds), which were treated as a public good and distributed free of charge. The logic was that well-fed peasants don't make revolution. Today, there is little risk of a Communist revolution, and the strategic value of African countries to the West, while not insignificant, is much smaller than it ever was in Asia during the 1960s and 1970s. Furthermore, governments in the West have largely turned crop research over to a small number of transnational corporations, which patent seeds and are unlikely to give products or knowledge away on a grand scale. It has happened, though, but only as an exception. In the 1960s, making technologies available as a public good was a taken-for-granted priority. Today, because of private ownership, access to technologies is a more complicated business. That is why the role that international, regional and national research centres play in providing technological innovations is so important to farmers.

There is no single or easy solution on the horizon, whether one takes an economic, political or technological perspective on how Africa may be able to feed itself and develop its agriculture. There will be many possible answers and solutions, which will come about not as a result of an Africa-wide blueprint but as the ability of a variety of actors to respond to specific contextual challenges. The environmental as well as the political terrain varies from country to country and from crop to crop. It is taking this into consideration and devising solutions accordingly that holds the best prospect for the near to medium future. Even if these steps are adapted to local circumstances they will require significant investments in infrastructure in order to bring inputs to the producer and produce to the consumer at tolerable prices. Subsequent chapters will show in greater detail what is being done or not being done in order to boost food security in a number of sub-Saharan African countries.

References

ADB (2008) *Asian Development Outlook*. Asian Development Bank, Manila, Philippines.

African Economic Outlook (2009) Available at: http://www.africaneconomicoutlook.org (accessed 24 March 2010).

AfricaRice (2008) Can rice crisis be turned into opportunity for Africa? Available at: http://africarice.blogspot.com/2008/05/can-rice-crisis-be-turned-into.html (accessed 24 March 2010).

AGRA (2009) AGRA launches policy initiative to empower Africa to shape home-grown agricultural policies. October 21. Available at: http://www.agra-alliance.org/section/work (accessed 31 October 2010).

Bakrania S. and Lucas, B. (2009) *The Impact of the Financial Crisis on Conflict and State Fragility in Sub-Saharan Africa*. Governance and Social Development Resource Centre, University of Birmingham, Birmingham, UK.

Ban Ki-moon (2008) The New Face of Hunger. *The Washington Post*, Wednesday March 12, A19.

Bates, R.H. (1981) *Markets and States in Tropical Africa: the Political Basis of Agricultural Policies*. University of California Press, Berkeley, California.

Bové, J. (2003) Foreword. In: Herman, P. and Kuper, R. (eds) *Food for Thought: Towards a Future for Farming*. Pluto Press, London, pp. xiii–xiv.

Collier, P. (2008) The politics of hunger. *Foreign Affairs* 87, 67–80.

Commission for Africa (2005) *Our Common Interest: Report of the Commission for Africa*. Available at: http://www.commissionforafrica.org (accessed 24 March 2010).

Cotula, L. (2007) *Changes in 'Customary' Land Tenure Systems in Africa*. International Institute for Environment and Development, London.

Cotula, L., Vermeulen, S., Leonard, R. and Keeley, J. (2009) *Land Grab or Development Opportunity? Agricultural Investment and International Land Deals in Africa*. International Institute for Environment and Development, London.

Crola, J.D. (2009) *Aid for Agriculture: Turning Promises into Reality on the Ground*. Oxfam International Research Report. Available at: http://www.oxfam.org/sites/www.oxfam.org/files/aid-for-agriculture-turning-promises-into-reality-report-12.11.09.pdf

Dag Hammarskjold Foundation (1995) *Autonomous Development Funds in Africa: Report from an Expert Consultation in Kampala, Uganda, 4–6 April 1995*. Dag Hammarskjold Foundation, Uppsala, Sweden.

deGrassi, A. (2008) Neopatrimonialism and agricultural development in Africa: contributions and limitations of a contested concept. *African Studies Review* 51, 107–133.

DFID (2003) *Agriculture and Poverty Reduction: Unlocking the Potential*. Department for International Development, London.

Diao, X., Headey, D. and Johnson, M. (2008) Toward a green revolution in Africa: what would it achieve, and what would it require? *Agricultural Economics* 39, 539–550.

Djurfeldt, G., Holmén, H., Jirström, M. and Larsson, R. (eds) (2005) *The African Food Crisis: Lessons from the Asian Green Revolution*. CAB International, Wallingford, UK.

Economic Commission for Africa (2009) *Economic Report on Africa, 2009*. ECA, Addis Ababa, Ethiopia.

Ejeta, G. (2010) African green revolution needn't be a mirage. *Science (New York)* 327, 831–832.

ERD (2009) *Overcoming Fragility in Africa*. European Report on Development. The European Union.

FAO (2010a) Food price index data. Available at: http://.fao.org/worldfoodsituation/FoodPricesIndex/en/ (accessed 25 March 2010).

FAO (2010b) Update on national food prices situation. Available at http://www.fao.org/isfp/isfp-home/en/ (accessed 25 March 2010).

Godfray, C., Beddington, J., Crute, I., Haddad, L., Lawrence, D., Muir, J., Pretty, J., Robinson, S., Thomas, S. and Toulmin, C. (2010) Food security: the challenge of feeding 9 billion people. *Science (New York)* 327, 812–818.

Gyllström, B. (1991) *State Administered Rural Change: Agricultural Cooperatives in Rural Kenya*. Routledge, London.

Heady, D. and Fan, S. (2008) Anatomy of a crisis: the causes and consequences of surging food prices. *Agricultural Economics* 39, 375–391.

Herbst, J. (2000) *States and Power in Africa: Comparative Lessons in Authority and Control*. Princeton University Press, Princeton, New Jersey.

Herman, P. and Kuper, R. (eds) (2003) *Food for Thought: Towards a Future for Farming*. Pluto Press, London.

Holmén, H. (2005) The state and agricultural intensification in sub-Saharan Africa. In: Djurfeldt, G., Holmén, H., Jirström, M. and Larsson, R. (eds) *The African Food Crisis: Lessons from the Asian Green Revolution*. CAB International, Wallingford, UK, pp. 87–112.

Hyden, G. (2006) *African Politics in Comparative Perspective*. Cambridge University Press, New York.

IFAD (2008) *Soaring Food Prices and the Rural Poor*. Available at: http://www.ifad.org/operations/food (accessed 25 March 2010).

Kydd, J. and Dorward, A. (2001) The Washington consensus on poor country agriculture: analysis, prescription and institutional gaps. *Development Policy Review* 19, 467–478.

Larsson, R. (2005) Crisis and potential in smallholder food production – evidence from micro level. In: Djurfeldt, G., Holmén, H., Jirström, M. and Larsson, R. (eds) *The African Food Crisis: Lessons from the Asian Green Revolution*. CAB International, Wallingford, UK, pp. 113–138.

Larsson, R., Holmén, H. and Hammarskjöld, M. (2002) *Agricultural Development in Sub-Saharan Africa*. Afrint Working Paper No 1. Department of Sociology, Lund University, Lund, Sweden.

Leonard, D.K. (1977) *Reaching the Peasant Farmer: Organization Theory and Practice in Kenya*. University of Chicago Press, Chicago, Illinois.

Lindberg, S. (2006) *Elections and Democracy in Africa*. Johns Hopkins University Press, Baltimore, Maryland.

Lipton, M. (1977) *Why Poor People Stay Poor: Urban Bias in World Development*. Temple Smith, London.
Lipton, M. (2005) *The Family Farm in a Globalizing World: the Role of Crop Science in Alleviating Poverty*. 2020 Discussion Paper No 40. IFPRI, Washington, DC.
Moseley, P. (2002) The African green revolution as a pro-poor policy instrument. *Journal of International Development* 14, 695–724.
Myrdal, G. (1968) *Asian Drama: an Inquiry into the Poverty of Nations*. Penguin Books, Harmondsworth, UK.
OECD (2008) *Turning African Agriculture into a Business: a Reader*. OECD Development Centre, Paris.
Omamo, S.W. and Farrington, J. (2004) *Policy Research and African Agriculture: Time for a Dose of Reality?* Natural Resource Perspectives 90. Overseas Development Institute (ODI), London.
PAFP (2008) *Final Declaration*. Pan African Farmers Platform, Addis Ababa, Ethiopia.
Rubinstein, C. (2009) China's eye on African agriculture. *Asia Times*, 2 October; also available at: http://www.atimes.com.
Southgate, D. and Graham, D. (2006) *Growing Green: the Challenge of Sustainable Agricultural Development in Sub-Saharan Africa*. International Policy Press, London.
te Velde, D.W. (2009) *The Global Financial Crisis and Developing Countries*. Working Paper 306. Overseas Development Institute (ODI), London.
The Economist (2009) Theme: How to feed the world. 'If words were food, nobody would go hungry.' *The Economist*, 21–27 November, pp.61–63.
UN (2003) World population monitoring. Available at: http://www.un.org/esa/population/unpop.htm#new (accessed 25 March 2010).
USAID (2004) *Linking Producers to Markets: a Renewed Commitment to Agriculture. A Strategy for Agricultural Development*. USAID, Washington, DC.
Walt, V. (2008) The world's growing food crisis. *Time Magazine*, Wednesday, 27 February. Also available at: http://www.time.com/time/world/article/0,8599,1717572,00.html.
Weis, T. (2007) *The Global Food Economy: the Battle for the Future of Farming*. Zed Books, London.
Wolday, A., Teketel, A. and Mulat, D. (2009) *Ethiopia. Afrint II Macro Report*. Addis Ababa, Ethiopia.
World Bank (2003) The World Bank: reaching the rural poor – a new strategy for rural development. *Currents* 31/32, 42–45.
World Bank (2007) *Agriculture for Development. World Development Report 2008*. World Bank, Washington, DC.

3 The Millennium Goals, the State and Macro-level Performance – an Overview[1]

Hans Holmén

Department of Geography, Linköping University, Linköping, Sweden

This chapter is based on and summarizes findings from nine *Afrint* macro-level reports focusing on the role of the state in promoting food crop production and smallholder farms in particular. These studies are supplemented with data from relevant studies and Food and Agriculture Organization of the United Nations (FAO) statistics. It seeks to identify actors and factors that may enhance productivity in staple crop production in the small farm sector (accounting for 70–90% of farms in many African countries) and positively impact on food security during the first decade of the new millennium. It draws on a causal and explanatory model of agricultural change, developed through studies of the Asian Green Revolution (Djurfeldt *et al.*, 2005). The Asian Green Revolution was found to encompass much more than technology. The state was commonly the leading agent but did not replace the private sector. On the contrary, it worked with and strengthened the market, and its engagement aimed at enhancing the productive capacity of small farmers. Hence, the Asian Green Revolutions of the 1960s and 1970s derived their impact from a successful combination of *state-drivenness*, *market-mediation* and *small farmer base*.

The primary objective of this chapter is to determine if, or to what extent, the various governments in sub-Saharan Africa (SSA) can and do pursue policies that enhance agricultural productivity and food security – and at the same time promote private sector involvement in inputs and produce in the food crop sector, and progressively address the needs and potential of small farmers.

Background and Preconditions

There is no doubt that SSA will have major difficulties in reaching the Millennium Development Goals of halving, by 2015, the proportion of people who suffer

[1] I am grateful to Christer Gunnarsson for valuable comments on an earlier draft of this chapter.

from hunger and/or extreme poverty. A major reason is the perceived 'underperformance of the agricultural sector' (World Bank, 2007a:xxiii).[2] Even though a slight recovery in per capita food production was noticed in the early 2000s (Badiane, 2008), SSA has, without question, the highest vulnerability to food insecurity in the world (IIASA, 2009). Poverty is widespread, with on average half the population in *Afrint* countries trying to make a living below national poverty lines (World Bank, 2007b). Poverty, moreover, has increased during the past decade and, according to Oxfam, the average number of food emergencies in Africa has almost tripled since the 1980s (ID21, 2007). Hence, the United Nations does not expect the targets of decreasing poverty or hunger to be met by 2015 in SSA (UN, 2007).

There are many reasons that combine to explain this dismal situation. Frequently, however, it is stated that the main reason has been bad governance (excessive centralization, neglect, rent-seeking and corruption). Hence, since around 1980, calls for 'structural adjustments' and a rolling back of the state has dominated development theory and international policy and led to both declining levels of and conditionalities for development aid, meaning that only a few decades after formal independence, African policies were again to be formulated abroad. 'Good governance' became the new war cry but frequently it merely meant less government (Bignante *et al.*, 2007). After massive critique of the Washington Consensus, the meaning of good governance has shifted somewhat and now, to some extent, recognizes the importance of the state but remains suspicious and often focuses on bringing down corruption (e.g. Sida, 2009). Whatever the words used or their precise content, the dominating view among donors – that the African governments are to blame for poverty – remains strong.

In a previous study (Djurfeldt *et al.*, 2005) it was found that tales of rulers' indifference and neglect of agriculture have been exaggerated.[3] Contrary to common messages, food production in SSA has not been without improvements. However, demand for food, due to population growth, has been greater than growth of food production. Instead of neglect, many attempts by African governments to support or improve agriculture were found. Frequently they led to 'spurts of production', which, however, were not sustainable (Holmén, 2005). It is true that many African attempts at state-led agricultural development and the administrative structures (state-led cooperatives, monopolies and marketing boards)

[2] Performance and underperformance can be measured in various ways, not always fair ones. A simple comparison between average yields for major food crops in Africa and the world average reveals an African 'underperformance' of 70% for maize, 40% for sorghum and 16% for cassava (Tsegay *et al.*, 2009). Such comparisons, however, do not take into consideration that agro-ecological preconditions vary significantly from one (sub)-continent to another. In our previous study, based on intra-African and intra-village comparisons, we found a significant untapped agricultural potential in sub-Saharan Africa (Holmén, 2005; Larsson, 2005). This, indeed, confirms that much African agriculture is underperforming.

[3] Also Jayne *et al.* (2002:1977) suggest that 'the commonly accepted "Berg hypothesis" that African governments taxed agriculture, an observation that was drawn mainly from West African experience, was not appropriate for many countries in eastern and southern Africa.'

were both costly and ineffective and could not be maintained in the long run. Sometimes they also displayed (or gradually developed) an urban bias. That such policies were launched were, however, not so much a symptom of bad leadership or 'faulty perceptions' as caused by an absence of alternatives after independence. With meagre financial and managerial resources, the state (readily supported by donors and international financial institutions (IFIs)) took it upon itself to substitute a non-existing market in order to facilitate development. Under structural adjustment this was to be reversed, now with 'a defective market ... substitut[ing] for a defective state' (Hettne, 1992:5).

Food security and the Millennium Development Goals

According to recent FAO data, sub-Saharan Africa is moving (albeit slowly) in the desired direction. Whereas the number of undernourished people increased somewhat between 1990–1992 and 2004–2006, on a general level the proportion has been declining. Although this positive development has been (temporarily?) distorted by the present global economic crisis – originating outside SSA – the trend has been positive.[4] When it comes to meeting the Millennium Development Goals of halving, between 1990 and 2015, the proportion of people who suffer from hunger, the general trend is negative, however (FAO, 2009).

Average yields for four major food staples are, with the exception of rice in Kenya and cassava in half of the investigated countries, low by world standards. The widespread adverse preconditions for agriculture in large parts of SSA (see below) are partly to blame. The low use of improved inputs (seed, fertilizer) – and the common decline in such use after structural adjustment programmes (SAP) – together with non-availability of credit for smallholders explain much of this productivity gap. The latter circumstances are a source of concern, since modern inputs could compensate for the former adversities. These low levels of productivity are all the more disturbing since, in a previous study (Djurfeldt *et al.*, 2005), a considerable unused potential for improved agriculture was detected in most countries studied. In particular, the collapse of fertilizer distribution systems appears to have had devastating effects on poverty and food security in the region.

Trends during the most recent decade (Table 3.1) – which also represents the period after implementation of SAP – are revealing. Even though production of these food staples generally shows upwards trends, much of the increases are due to area expansion. In Ethiopia, maize and sorghum show significant increases in both production and yields, in the case of maize mostly due to improved yields and for sorghum mainly due to area expansion. In the latter case, this growth of a low-yielding crop may indicate an expansion into drier

[4] For all the difficulties caused by the 2007–2008 global food price hike, this situation has 'triggered renewed attention that could boost agriculture and help ... exploit Africa's unused production potential' (Kamara *et al.*, 2009:i).

Table 3.1. Trends (percentage change) in production, yield and area harvested for four food staples (1995–2007). (Adapted from: FAOSTAT data 2009.)

	Cassava			Maize			Rice			Sorghum		
	p	y	a	p	y	a	p	y	a	p	y	a
Ethiopia	–	–	–	74	57	13	–	–	–	83	25	48
Ghana	46	s	57	16	s	16	35	s	28	s	s	s
Kenya	–15	–12	s	15	s	19	23	–15	57	61	38	12
Malawi	643	342	74	57	33	28	46	s	56	16	15	11
Mozambique	49	57	s	83	29	40	s	s	28	s	10	s
Nigeria	30	s	30	s	57	–25	27	–12	45	38	18	24
Tanzania	s	–12	s	42	–35	119	80	s	72	s	–18	25
Uganda	128	98	30	46	s	30	97	s	98	24	s	15
Zambia	32	–11	83	15	s	13	43	s	42	–44	–14	–30

p = production; y = yield; a = area harvested; – = cassava and rice are irrelevant in Ethiopia; s = stagnant (<10% change).

lands that may not be able to support agriculture sustainably. In Zambia, low-yielding sorghum is giving way to considerable area expansion for the other crops. In Uganda, growth in cassava production is due to both higher yields and expanding acreage, while yield stagnation for the three other crops is compensated for by expanding area. In Nigeria, significantly raised maize yields have allowed a reduction in maize area and an increase in areas allocated for the other three crops. This enlargement apparently also reflects the Nigerian government's ambition to 'increase cultivable land by 10 percent annually' (Akande and Ogundele, 2009:25). Almost invariably, contemporary production increases are thus due to extensification rather than productivity enhancements. In general, area expansion appears to compensate for stagnation or decline in yields.

The exception is Malawi, a country that recently has attracted much attention due to its shift from a food-deficit country to a maize exporter in only a few years. It should be noted, though, that maize values in Fig. 3.1 are strongly influenced by the spectacular but short increase between 2005 and 2007. On the other hand, Malawi actually scores positively since 1995 for all the four crops investigated and most strikingly for cassava, where the relative increase in production is sixfold and that for yields is threefold in the same period. This growth in cassava is due to the combined impact of availability of new, cassava mosaic virus-resistant varieties and increasing commercialization and growing urban demand for cassava.

Looking at production and productivity statistics for the most common important food crop – maize – Figs 3.1 and 3.2 reveal that production shows an upward trend in four investigated countries. In a few cases, notably Ethiopia and Malawi, increases in maize production during the early years of the third millennium have been dramatic, mainly due to higher yields. The general picture is diverse, however, and yields remain low overall, with average yields of 1.5 t/ha during the same period (Table 3.2).

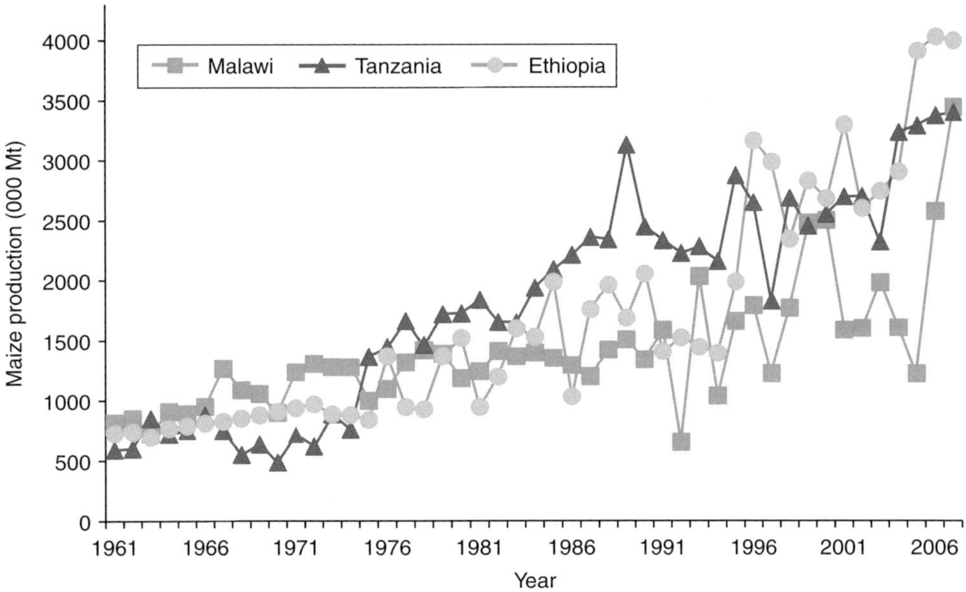

Fig. 3.1. Maize production in Malawi, Tanzania and Ethiopia, 1961–2007. (Adapted from: FAOSTAT data, 2009.)

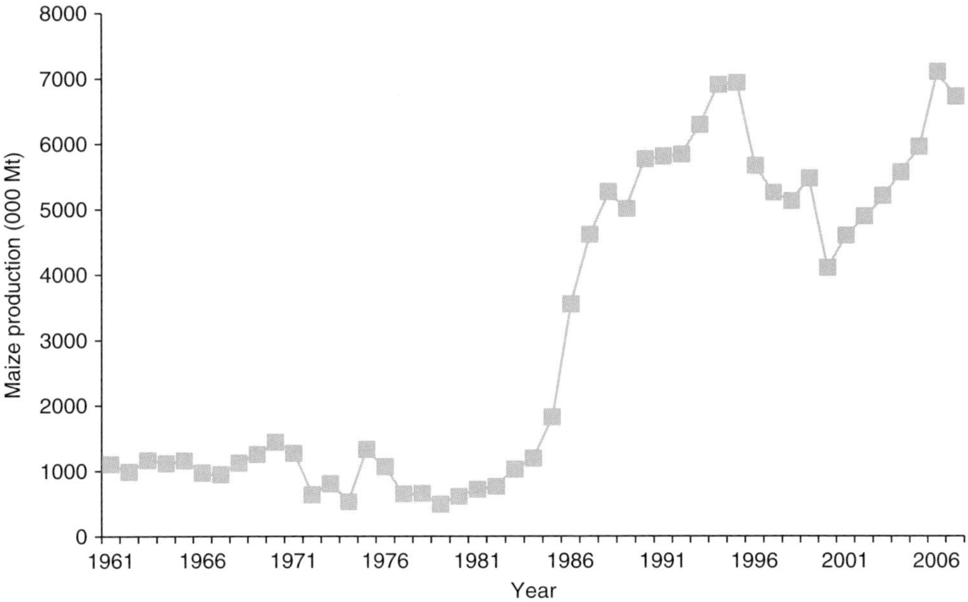

Fig. 3.2. Maize production in Nigeria, 1961–2007. (Adapted from: FAOSTAT data, 2009.)

Of equal importance, Figs 3.1 and 3.2 reveal that fluctuations in maize production post-SAP have become much more dramatic than they were pre-SAP, with significantly wider swings between tops and bottoms in good and bad years. Hence, it has become increasingly hazardous to be a peasant in SSA

Table 3.2. Average yield (t/ha) of four major food crops, 1998–2007. (Adapted from: FAOSTAT data, 2009.)

	Cassava[a]	Maize	Rice[a]	Sorghum
Ethiopia	–	1.9	–	1.3
Ghana	12.3	1.5	2.0	1.0
Kenya	8.5	1.7	3.7	0.8
Malawi	14.0	1.4	1.6	0.7
Mozambique	6.2	1.0	0.9	0.7
Nigeria	10.3	1.5	1.4	1.2
Tanzania	7.6	1.7	1.6	0.9
Uganda	12.8	1.6	1.4	1.5
Zambia	5.7	1.6	1.2	0.7
World	10.9	4.6	3.9	1.4

[a]Rice and cassava are irrelevant in Ethiopia.

during the last decades. Under such circumstances, rural livelihoods are highly volatile and emerging markets develop slowly and haphazardly. No wonder, then, that the number of food emergencies have almost tripled since the introduction of SAP.

The African Dilemma

African governments have many objectives, and economic growth, though important, is only one of them. Immediately after independence it may not even have been the most important objective. Most countries were food secure (at least most of the time) well into the 1970s (Larsson *et al.*, 2002). Nation-building, i.e. consolidating state hegemony over not yet integrated territories inherited from former colonial powers, was initially probably the most important goal. It is still often the case that governments' effective outreach and control is limited only to parts of national territories (Herbst, 2000). Their influence in rural areas is often less than assumed. Hence, the eagerness to roll back the state is perhaps not the most appropriate medicine when nation-building is still an unfinished business (Holmén, 2010).

Moreover, for markets to function – even to develop – suppliers need to be connected to customers, and vice versa. Hence, road and transport infrastructure is of vital importance. Africa is a vast continent, accounting for one-fifth of the earth's landmass but being the home of only 12% of its population. In large parts of SSA, population densities are low and the subcontinent has sometimes been characterized as 'under-populated' (Amin, 1972). The peripheries in SSA are huge. This not only renders governmental outreach problematic, it also makes infrastructure development and service provision extremely costly. The preconditions for integration and (market) development are therefore comparatively unfavourable in SSA. It has, for example, been found that in the 1990s, the density of sub-Saharan Africa's road network was only one-sixth of what

India's had been in the 1950s, prior to its Green Revolution (Hazell and Fan, 2002). Any comparison of Africa's dismal agricultural performance with those in other parts of the world needs to be done with this fact in mind. It is still the case that, in terms of road and railroad density and quality, SSA lags behind all other major regions of the world (Goldstein and Kaufmann, 2006). According to one source, on average, only 27% of SSA's rural population has access to all-season roads (Tsegay *et al.*, 2009).

Africa is not only huge, it is also an ancient continent, with some of the oldest, most depleted soils in the world (ECAPAPA, 2003). Although some highly fertile lands can be found, notably along the Rift Valley and in some mountainous areas in East Africa, these are limited in extent and soils generally are leached and of low fertility (de Blij and Mueller, 1997; Morris *et al.*, 2007). These unfavourable preconditions are exacerbated by shortage and unreliability of rainfall. Dry climates are more extensive in Africa than in any other continent (Hodder, 1992). High temperatures and strong solar radiation cause high levels of evaporation and evapotranspiration. This 'atmospheric thirst' further reduces water availability and soil moisture. Lele and Stone (1989:19) found that 'solar radiation in east Africa is the highest in the world'. The FAO (1987:4) notes that 'Africa has ... higher evaporation per unit area than any other region' in the world. At the same time, the potential for irrigation development is limited (FAO, 1987; Rosegrant and Perez, 1997). Farming in such an environment and under such circumstances is not only difficult, it is often also a high-risk undertaking. It is mostly practised by small, highly subsistence-oriented farmers with low use of modern technology. In Ghana, smallholders represent 80% of the farmers (Danku *et al.*, 2009), in Malawi 85% (Mulwafu, 2009) and in Tanzania 99% (Isinika and Ashimogo, 2009) and so on.

Who then are the small farmers? Which capacities do they have to invest for growth and productivity enhancement? Coughlin (2009a:12), writing from Mozambique, amply illustrates a situation that (apart from land availability) appears to have wide relevance:

> Despite the general availability of additional land for farmers who want it, most farmers ... only cultivate 0.5 to 1.5 hectares, few ... have access to irrigated land and few use fertilizer on maize, often their principal food crop. Many farmers ... think they could not cultivate any more land at all even if they wanted to, while [the majority] think they might manage to farm another quarter hectare, ... another half hectare [or even] another hectare. But they do not expand their cultivated land. Why? The farmers' technological, capital and labour constraints are usually too severe to permit them to cultivate more land. For distant, isolated villages, marketing or transport costs are further problems. Stuck with using traditional storage techniques with high losses, the farmers have little incentive to store crops for later sale during the high-price hungry season. Fearing losses, they sell soon after the harvest despite the low prices received, low prices that make high-input high-yielding farming impossibly risky and comparatively less profitable than traditional techniques.
>
> Fear, risks and poverty comprise a vicious cycle of low prices and incomes, low-yield, low-risk techniques, while urgent, untimely sales depress prices and incomes. Meanwhile, traditional extension services focusing on simple, low cost

technologies and improved marketing – but meagre capital investment – have proved insufficient to break the poverty cycle for most small farmers.

It is against this background that efforts to boost development in general and food production in particular must be seen and evaluated. It is in this environment that donors and African governments experiment with new policies in order to reach out, to integrate smallholders and improve agricultural productivity and eventually reach the Millennium Development Goals. Which, then, are their policies and priorities – and which resources are they able to mobilize in order to enhance smallholder productivity and food security?

Priorities in Agricultural Policy

Policy goals

All the countries included in the *Afrint* study face widespread poverty and more or less severe food security problems, either transitory or structurally, or both. Hence, poverty reduction is generally mentioned as the overarching goal (e.g. Bashaasha, 2008; Akande and Ogundele, 2009; Mulwafu, 2009; Wamulume, 2009; Wolday *et al.*, 2009). Although not always explicitly declared, all investigated governments have also made food security and the Millennium Development Goals their official policy objectives. At the time of writing, in a short-time perspective, these goals may seem illusory and, for example, Malawi 'plans to prioritise food sufficiency' (Mulwafu, 2009:24) while, in Tanzania, food security is implied, 'not mentioned explicitly' (Isinika and Ashimogo, 2009:ii), and Ghana – among others – has 'no single, well-coordinated official policy towards food security' (Danku *et al.*, 2009:9).

This is also reflected in the attention given (or not) to the role of women in agriculture, food production in particular. Although their roles may differ from one region or society to another, in sub-Saharan Africa women make up a substantial part of the agricultural labour force and are commonly responsible for 70–80% of food production. Women also face a number of constraints that limit their potential as farmers, e.g. less access to land, extension services and credit. Hence, an indication of the importance of food security and enhanced food crop productivity can be found in the way emphasis is given to women's roles and needs as producers. Information on this topic is scant, and although the issue seems to be high on some official agendas, progress apparently is slow. Mulwafu (2009:20) asserts that, in Malawi, 'the government has adopted a strategy of mainstreaming gender issues in national development policies and programmes', but is rather silent on if and how this is implemented.

Likewise, in Ghana, there is a lack of information on women in agriculture and, although attention to women's needs is given in the Poverty Reduction Strategy, there appears to be no special policy formulated (Danku *et al.*, 2009). In Zambia, women's situation 'is mentioned' in policy documents, but so far the prioritized task is on 'creating gender awareness among policy makers and farmers' (Wamulume, 2009:8). In Kenya, the state has reportedly 'played a minimal role when it comes to gender and agriculture' (Olouch-Kosura,

2009:28). And in Ethiopia, while the government is officially committed to the promotion of gender equality and to the improvement of the status of women, this seems to be more rhetoric than practical policy. The Women's Affairs Department within the Ministry of Agriculture has very limited resources in terms of budget and manpower and its activities are barely integrated into key decision-making and planning processes (Wolday *et al.*, 2009).

It may seem paradoxical that poor societies can abstain from promoting the productive capacity of this large proportion of the farming population. As mentioned, central authorities' ability to manipulate local communities are limited, and age-old traditions and well-established institutions are not easily changed – at least not until new opportunities unfold. Poverty reduction and food security face some heavy competitors when it comes to the practical realization of political and administrative agendas.

Of all the different development options mentioned in Chapter 2, this volume, all investigated governments prioritize growth in agriculture. How this is to be accomplished – whether through own food production, by policies targeting small or big farmers, men or women, through promotion of export crops or by other means – is, however, less clear. While all governments agree on the need to commercialize agriculture, the sought role for agriculture varies. Whereas Ethiopia seeks to realize an 'agricultural and rural development-led strategy of development' (Wolday *et al.*, 2009:15), Ghana's overriding ambition is to 'become an agro-industrial economy by the year 2010' (Danku *et al.*, 2009:9).

Budget allocations for agriculture

Whatever the end goals, development of agriculture will be imperative. This was also realized by the African Union (AU) and, at the African Heads of State and Government meeting in Maputo in 2003, it was decided that African governments, within 5 years, should allocate 'at least ten percent of national budgetary resources to agriculture and rural development' (AU, 2008:1). In 2008, of 34 countries in Africa, only six had reached this target (Tsegay *et al.*, 2009). Of the nine countries included in the *Afrint* study, by 2007–2008 this target had only been met in one country, notably Malawi.[5] In some *Afrint* countries, budget allocations remain very low (Ghana 2.5%, Uganda 3.1%, Nigeria 4%), whereas in Ethiopia, Kenya, Tanzania and Zambia, allocations are about 6% (Afrint background papers).[6] Only in Malawi have allocations been above 10% since 2005–2006, and they reached 14% in 2008–2009 (Mulwafu, 2009).

[5] Figures are somewhat inconclusive. Whereas *Afrint* country reports reveal that only Malawi had reached 10% budget allocations by 2007–2008, Tsegay *et al.* (2009) claim that, by 2008, Ethiopia had also reached the target and that Mozambique almost reached it. This discrepancy does not change the overall picture, however.

[6] Akande and Ogundele, 2008; Bashaasha, 2008; Coughlin, 2009a,b; Danku *et al.*, 2009; Isinika and Ashimogo, 2009; Mulwafu, 2009; Olouch-Kosura, 2009; Wamulume, 2009; Wolday *et al.*, 2009.

It would be easy to conclude that the commitments to developing agriculture are less than policy declarations would lead us to believe. However, due to previous commitments, such budget reorientations may take time to realize. The fact that allocations do increase points in the desired direction, even if targets cannot be immediately met. Also, even if allocations so far are lower than recommended, they seem generally to be slowly increasing. However, considering the generally low levels of budget allocations for agriculture, due to the fact that in, for example, Ghana (2.5% in 2007) and Nigeria (4% in 2007) a large proportion is absorbed by personnel costs, little is left for development investments (Akande and Ogundele, 2009; Danku *et al.*, 2009). The requirements of the Maputo Declaration are still valid.

Arguably, they are even more valid when the total budget is shrinking. The contemporary global recession is a case in point. Moreover, with the implementation of SAP came reduced development aid. It is no mere coincidence that between 1980 and 2004 public spending on agriculture fell from, on average, 7% to 4% of total public spending (World Bank, 2007b). On a general level, aid to SSA decreased by one-third during the 1990s (Earthscan, 2000) but aid to agriculture declined 50% (Haggblade, 2005). On the one hand, 10% of a declining budget may not be much and the effect may be meagre.[7] On the other hand, the AU's call for allocating at least 10% of budget to agriculture reflects the urgency of the matter – 'do not sacrifice agriculture when resources are declining!' It is increasingly realized that, as long as markets and the private sector are undeveloped, the public sector has a 'crucial role to play in promoting the overall investment in agriculture required to achieve the full socio-economic potential of agricultural growth' (FAO, 2009:40). It was recently noted that 'official development assistance (ODA) falls short of the minimum amounts required even for achieving the Millennium Development Goals' (WCSDG, 2004:xi). Even if overseas development aid to Africa has gradually increased since then, it is still the case that G8 countries are off their 2010 target of doubling aid to Africa by 27%, on average (Tsegay *et al.*, 2009). Hence, it seems as if Africa will have to do it alone. Also, whereas it remains disputable that governments in SSA neglect agriculture, it is quite clear that Western donors fail to honour their commitments to support African development.

Besides, available figures further reveal that tales of government neglect of agriculture pre-SAP often have been ill-founded or at least exaggerated. In Ghana, for example, agricultural expenditures were above 10% until implementation of SAP in 1984 (Danku *et al.*, 2009). In Zambia, they averaged 20% until 1994, when they dropped significantly (Wamulume, 2009). Also, in Kenya, allocations were above 10% until the late 1980s, when they rapidly dropped to reach a lowest level of 2.4% in 1997–1998, a level from which they have only partially recovered (Olouch-Kosura, 2009).

[7] Actually, even 10% of government budget for agriculture may not be enough. Fan *et al.* (2008, quoted in Tsegay *et al.*, 2009:38) estimate that, to meet the first Millennium Development Goal, 'African countries will need to spend 20% of their total government spending in the agriculture sector.'

Policies towards the private sector

Policies towards agriculture can be analysed at several levels, reforms aimed at farmers directly and those addressing their overall institutional and organizational environment at large. Invariably, the ambitions are to transform subsistence agriculture to commercial agriculture. This is also often reflected in language, which frequently talks of the private sector as if smallholders were not part of it. Governments and donors and much development literature tend to treat the private sector as a third sector, separated from both the public and the subsistence sectors. This is problematic because subsistence-oriented smallholders can be said to be more private than all others. We will return to issues directly aimed at small farmers below, but first a look at general policies is warranted.

Officially, the role and importance of the private sector in agricultural development is accentuated everywhere, although with highly diverging emphasis, reflecting the various points of departure in different countries. In Ghana the private sector is seen as 'the engine of growth' (Danku *et al.*, 2009). In Tanzania, the aim is to create 'a favourable climate for commercial activities' (Isinika and Ashimogo, 2009:51; see also Bashaasha, 2008 on Uganda). Nigeria officially opts for a public–private partnership (Akande and Ogundele, 2009), as does Malawi, although the relation between state and private sector 'has sometimes been fraught with contradictions' (Mulwafu, 2009:30), most clearly evidenced through the ambiguous position of the parastatal ADMARC[8] and by the fertilizer support programme (see below). Also, in Zambia, relations between state and the private sector have, at times, been ambiguous, reflecting the government's view that the private sector failed to fulfil its role in the 1990s (Wamulume, 2009).

Whether this is due to the Zambian government's 'failure to adjust' – and thus represents a deliberate effort to crowd-out private traders (Minde *et al.*, 2008) – or a 'failure of adjustment', i.e. the slow emergence of markets (Jayne *et al.*, 2002; Holmén, 2005) and, hence, a need for governments to step in, is a matter of controversy. To some extent, the issue is about prioritizing between immediate needs and long-term objectives, where the call for quick solutions tends to override strategic objectives. Crowding-out – intentional or unintentional – is reported from a number of SSA countries, e.g. Ghana (Dorward, 2009), Malawi (Mulwafu, 2009), Ethiopia (Wolday *et al.*, 2009) and Zambia (Minde *et al.*, 2008; Wamulume, 2009).

In Ethiopia, emerging from a highly centralized system, the ambition is likewise 'to develop a free market economic system' (Wolday *et al.*, 2009:15). While this may be a slower and more ambiguous process than usually acknowledged, agricultural commercialization is to be accomplished by way of a 'big push' (Wolday *et al.*, 2009:70). A 'big push', arguably, requires a top-down approach and is probably difficult to realize when state involvement is to be reduced. Significantly, a large number of traders withdrew from the fertilizer market, complaining about lack of a level playing field compared to the

8 State-owned agricultural input and marketing organization.

government agency and some party- and government-affiliated operators (Wolday *et al.*, 2009). This leads to doubts concerning the sincerity of the Ethiopian government's declaration that it 'sees its role mainly in terms of facilitation' (Wolday *et al.*, 2009:88).

Apart from this ambiguity in relations between state and market and their possible effects on smallholders, invariably there seems to be a cash-crop bias, as policies tend to emphasize export competitiveness (Bashaasha, 2008; Mulwafu, 2009; Wamulume, 2009). It is yet unclear what this will mean for national food security and food sovereignty, let alone for smallholders (see further below). It is, however, worth reminding of Diao and Dorosh's conclusion that 'the best prospects for agriculture-led growth remain in the food sector, where domestic demand represents a large and growing market' (Diao and Dorosh, 2007:275).

Presence of market agents in agriculture

In principle, governments have withdrawn from direct engagement in the productive sectors. Monopolies, parastatals and marketing boards have been abolished or divested. Trade in agricultural inputs and marketing of produce has been privatized since implementation of SAP. These withdrawals are almost nowhere complete. In, for example, Kenya, maize markets were fully liberalized by 1999, but rice has remained under the mandate of the National Irrigation Board (Olouch-Koshura, 2009). From the 1990s, both Malawi and Zambia liberalized all input and output prices except for maize (Mulwafu, 2009; Wamulume, 2009). Subsidies, likewise, were, in principle, abolished during SAP, and the state is no longer the only provider of extension services. Hence, it would seem as if the space has been opened up for private merchants and service providers. However, abundant evidence now shows that withdrawal of the state is not enough for development to 'take off'. On one hand, there has hardly been a private sector ready to fill the gap of a retreating state (Holmén, 2005). On the other hand, the down-scaling of public services has created difficulties for emerging private entrepreneurs who depend on such services, especially in rural areas.

Private sector development is constrained by a number of internal and external circumstances, e.g. limited financial resources, lack of information, inexperience and insufficient capacities among operators, in combination with insufficient transport and storage infrastructure, inadequately developed financial markets, frequent policy shifts and administrative inconsistencies. Hence, agricultural traders and service providers are generally small-scale and with narrow capacities and limited presence. Moreover, agriculture, especially food crop agriculture, may not present the most attractive business opportunities. Wolday *et al.* (2009:115) thus found that, while there is 'a positive trend in private sector development in Ethiopia', there is also a private sector bias against agriculture and 'most of the investment projects so far tended to concentrate in urban areas and non-agricultural activities' (Wolday *et al.*, 2009:116). Overall, development of the private sector in agriculture is geographically selective and tends to be concentrated in high-potential areas (Olouch-Kosura, 2009) and/or to areas with good roads (Mulwafu, 2009; Wamulume, 2009).

There are also significant differences between private sector engagement in input and output trade. In Ghana, agricultural input markets have a significant number of formal institutional participations, whereas the opposite is the case in output markets, where the informal sector dominates (Danku *et al.*, 2009). A similar situation characterizes Zambia, where the government continues to play a significant role in input supply (Wamulume, 2009). Whereas the formal banking system offers little in terms of credit to smallholders and small traders alike (Bashaasha, 2008; Wamulume, 2009), and for trade in food crops in particular, private traders sometimes provide credit to their clients (Danku *et al.*, 2009), although few seem to have that capacity. It is even the case that small cash-constrained traders resort to barter trade and offer, for example, blankets, shoes or other commodities in exchange for farmers' crops (Wamulume, 2009).

Constraints to private sector participation

On a general level, the biggest obstacles to private sector development in SSA are reportedly inadequate infrastructure, burdensome regulations and frequent domination of emerging markets by ethnic minorities, which inhibits competition and improvement of the business environment (Ramachandran, 2009). The *Afrint* country studies further reveal that private traders often lack both capital and storage facilities (e.g. Bashaasha, 2008; Danku *et al.*, 2009; Olouch-Kosura, 2009). The paucity of non-farm sources of income and employment, together with tiny industrial sectors, leads to small volumes traded and high transaction costs, in turn often leading to monopolistic distribution systems (Bashaasha, 2008; Akande and Ogundele, 2009; Danku *et al.*, 2009; Wolday *et al.*, 2009). These circumstances, in combination with inadequate transport infrastructure, further limit traders' presence to more easily accessible and high-potential areas, often near big cities and/or along main roads (Mulwafu, 2009; Olouch-Kosura, 2009; Wamulume, 2009). Hence, private sector participation in agricultural development remains low everywhere. In Ethiopia, government policies have been 'exclusively concentrated on increasing agricultural production ... and ignored the marketing issues, particularly the very real needs of the private marketing firms and entrepreneurs' (Wolday *et al.*, 2009:137). Markets do not develop overnight, and whether governments have been negligent or not, the private sector cannot yet shoulder the burden of rural development alone. Writing from Malawi but with apparent wide applicability, Mulwafu (2009:31) concludes 'private traders are nowhere near to offering peasants opportunities for regular markets, stable prices, inputs, credit'. In fact, 'farmers have been calling for the return of ADMARC', the former parastatal (Mulwafu, 2009:31).

Administrative decentralization

Apart from rolling back the state in favour of the market, SAP also meant political and administrative decentralization and devolution of authority to lower levels of the public sector: regional and local governments. The idea was invariably that such reforms would bring development closer to people, facilitate implementation and improve services, since local and regional actors have better knowledge of local constraints and opportunities. They are also believed to be more sensitive and able to adapt more easily to local needs, opinions and unforeseen circumstances. This turnaround has been executed with varying degrees of enthusiasm and with varying degrees of success. The limited effects of the reform package reflect circumstances, which are often overlooked when policy recommendations are aired.

Throughout SSA, reports abound about lack of administrative and managerial capacity at local and regional levels to contribute effectively to, let alone take the lead in, development (Crook, 2003; BTI, 2006c; Akande and Ogundele, 2009; Danku *et al.*, 2009; Isinika and Ashimogo, 2009; Wamulume, 2009; Wolday *et al.*, 2009). Despite decentralization, administrative structures are still commonly characterized as top-down (BTI 2006a,b,c; Danku *et al.*, 2009; Wolday *et al.*, 2009). Local governments often suffer from a lack of legal backing (BTI, 2006c; Wamulume, 2009) and also everywhere depend heavily on financing from the central government, a circumstance that might reduce their flexibility in project prioritization (BTI, 2006c; Bashaasha, 2008; Akande and Ogundele, 2009; Danku *et al.*, 2009; Isinika and Ashimogo, 2009; Wolday *et al.*, 2009).

Whether this is a sign of central governments' unwillingness to let go of control or rather is an effect of local governments' lack of competence and limited capacity for self-financing is a matter of dispute. There is probably more than one answer to that question. In, for example, Tanzania 'there are districts which have returned funds to the treasury because they could not use them, thus negating previous complaints that funding was the most limiting factor' (Isinika and Ashimogo, 2009:iii). Low capacities among local and regional governments often limit their efforts to participate in government-led initiatives (Bashaasha, 2008) or presidential initiatives for rural development.[9] Decentralization is a process and not a one-shot intervention. Hence, it is not uncommon to find overlap of responsibilities, or even competition and rivalry, between different tiers of government (African Development Fund, 2005; Kabumba, 2007; Wadala, 2007). It is also sometimes the case that 'lower levels of the administration are able to obstruct to some degree the open market policy of the [central] government' (BTI, 2006e:14; Isinika and Ashimogo, 2009).

To a not insignificant degree, these 'clashes of administration' occur because donors and IFIs have pushed for decentralization at breakneck speed. But they also occur because central governments are not yet in full control over 'their' territories. In, for example, Kenya, 'large parts of the country are not under government control' (BTI, 2006c:6). In Nigeria, 'the state's monopoly on the use of force is limited in several parts of the country' (BTI, 2006d:4). The same applies to several other countries in SSA

[9] For example, the presidential initiative on cut flowers in Ethiopia (Wolday *et al.*, 2009) or the many presidential initiatives (on cassava, vegetable oil, rice, food security and others) in Nigeria (Akande and Ogundele, 2008).

(Herbst, 2000). In Ghana, 'particularly in the rural areas, the visibility of state administration is limited' (BTI, 2006b:4). In Uganda, where 'the state's administrative structures are generally present throughout the country ... they are operative to very different degrees' (BTI, 2006f:6). Hence, some 50 years after formal independence indirect rule is still a common feature (Holmén, 2005), reflecting both the weakness of central governments and the 'persistent influence and power of traditional elites' (Kaarhus and Nyirenda, 2006). In order to reach out, central governments have co-opted and/or made alliances with traditional leaders of various kinds. The irony of the matter is that this has often led to more corruption, not less (Ferguson and Mulwafu, 2007). Traditional elites have often been found to be 'resistant or indifferent to pro-poor policies' (Crook, 2003:1). Instead they often use their positions to preserve neo-patrimonial structures and privatize public resources (Tambulasi and Kayuni, 2007). Not surprisingly, based on a large number of village-level interviews, Ellis and Freeman (2004:17) found that, while fulfilling other functions well, traditional leaders are 'rarely mentioned in a developmental capacity'.

In many ways, therefore, downsizing 'had adverse consequences on service delivery – especially in rural areas' (Foerster, 2009:384). This is particularly serious as local governments generally have been assigned greater responsibility for development, infrastructures and extension. Obviously, and for many concurrent reasons, the rolling back of the state was not a magic bullet able to cure most or all of Africa's ills. 'Good governance', national integration and private sector bonanza are not automatic outcomes of a reduced role for central government. It rather appears that the state has been used as a convenient scapegoat, diverting attention from more pressing issues such as nation-building, limited capacities, donors' self-interests and superpower agendas. What Africa needs is not so much policy prescriptions as capacity building, but that requires much longer time horizons than those that have so far guided donors and IFIs. *Afrint* background papers reveal that, post SAP, many central governments in SSA do try to reassert themselves in various ways and have also taken a more active role in poverty reduction and as facilitators of development (see also Holmén, 2005). Three themes (market-supporting investments, extension services and infrastructure) that are of fundamental importance for Africa's future ability to feed itself and have far-reaching implications for inclusion or exclusion of smallholders in development are looked upon below.

Policies towards market promotion, rural infrastructure and farmer mobilization

In the aftermath of SAP, the role of the state is 'changing but not diminished' (Mulwafu, 2009). African governments are in various ways reasserting themselves and are assuming a more active role as promoters of development. However, rather than being directly involved in agriculture, they now increasingly assume a 'regulatory role' (Danku *et al.*, 2009:24) and seek to 'provide an enabling environment' for other actors (Olouch-Kosura, 2009:28). Among infrastructures, the focus here is on storage, transport and communication systems, including extension and organization development. Also fertilizer subsidies are touched upon, since it is essential that this basic input is made available at affordable prices.

Fertilizer subsidies

Even though subsidies may appear to contradict the above claim of hands-off policies, they are now applied in a number of countries investigated. In virtually all countries covered in the *Afrint* study, agricultural subsidies were abolished under strong donor pressures during SAP because they were considered costly and inefficient, as subsidies are thought to distort markets and send the wrong messages to traders and farmers alike. Immediately after this reform, fertilizer prices skyrocketed, leaving them unattainable for most farmers. Since then, 'farm-level fertilizer prices in SSA are among the highest in the world, due to high transport costs and limited market development' (Thangata and Blackie, 2006:1). Therefore, subsidies for agriculture, notably fertilizers, have been reintroduced in a majority of countries studied. The rationale behind this move is threefold. First, soils in many places are getting exhausted. Secondly, donor countries compete unfairly with African farmers when exporting subsidized agricultural products, thus undermining market development in SSA. This renders productivity-enhancing investments futile unless countervailing measures are taken. Thirdly, with fertilizer markets collapsing after SAP, subsidies are a substitute for missing markets, at least temporarily. Fertilizer subsidy programmes fulfill two important roles: they enhance food security and contribute to poverty alleviation, and they act as stepping stones towards a more commercialized agriculture. As such, fertilizer subsidy programmes may well be regarded as basic infrastructure.

Zambia has maintained a fertilizer support programme (representing a 50–60% income transfer subsidy) for small farmers since the mid-1990s (Wamulume, 2009). In Nigeria, a 25% fertilizer subsidy was introduced in 2000 (Akande and Ogundela, 2009), and a limited fertilizer subsidy was launched in 2002 in Tanzania (Isinika and Ashimogo, 2009). Likewise in both Ghana and Uganda, fertilizer subsidy programmes were introduced in 2008 (Danku *et al.*, 2009; Kretzman, 2009). By far the most well-known subsidy programme is in Malawi. Having promoted fertilizer use in various limited programmes (Starter Pack, Targeted Input Program) since the 1990s, in 2005/6 the government introduced the nationwide Agricultural Input Support Program. Initially, it did so under great opposition from donors and IFIs, some of which also pulled out of the country in response to the government's disobedience. However, with the apparent success (disputed) of the programme, 'funding partners are [now] lining up to support [the] programme' (Riungu, 2008). Apparently, they are eager to share the credit for something that seems to work – despite them.

It is undisputable that the impact of the AISP (Agricultural Input Support Program) has been 'significant' (Wamulume, 2009). *The Economist* (2009:62) contends that in just a few years 'the programme has turned Malawi into a breadbasket'. Others, however, suggest that the government's production figures have been exaggerated (Dorward *et al.*, 2008). Moreover, the success of the programme was at least partly due to favourable weather and timely rains (Minde *et al.*, 2008). Thus Chinsinga (2007:15) found the programme to be 'fairly successful'. Even so, since the 2005/6 season Malawi, frequently plagued by food deficits and crises, has achieved 100% food self-sufficiency and has even exported maize. For this, President Mutharika has earned accolades from various quarters for outstanding policies. It has been observed that private merchants have, to varying degrees, been crowded-out and have been assigned a smaller role in fertilizer supply than they have capacity for (Chinsinga, 2007; Dorward *et al.*, 2008; IRIN, 2008). This may be a temporary shortcoming though, since donors, who now increasingly support the programme, also demand a greater role for the private sector. Of possibly greater significance are the ever-increasing costs for the programme. Initially, the value of the extra production achieved significantly outweighed the cost of the programme (Dorward *et al.*, 2008). However, the global food, fuel and financial crises all occurred after the AISP was introduced. Not only have fertilizer prices increased on the world market, transport costs have escalated as well. Hence, the costs for the programme have repeatedly exceeded predictions, forcing the government to adjust budget allocations upwards at a magnitude that might jeopardize the sustainability of the programme (Wamulume, 2009).

Subsidy programmes should not be seen in isolation, though. They must be placed within a framework of other policies that eventually may render them unnecessary. And governments in SSA have, indeed, taken a range of measures to improve agriculture and food production.

Promotion of producer organizations

One such measure is the promotion of independent cooperatives and rural producer organizations as players on liberalized markets (Bashaasha, 2008; Akande and Ogundele, 2009; Danku *et al.*, 2009; Isinika and Ashimogo, 2009; Olouch-Kosura, 2009; Wolday *et al.*, 2009). Farmers' associations of various kinds are thought to enable smallholders to access credit and information more easily, to enhance their bargaining power and benefit from economies of scale through bulk purchases of inputs and sales of produce, and hence to have a greater influence on the economic environment in which they operate.

To facilitate the creation of voluntary farmers' organizations is all the more important as farmers in SSA generally remain poorly organized (Isinika and Ashimogo, 2009; Holmén, 2010). Previously, cooperatives in SSA used to be controlled by the state. SAP led to a 'collapse of government support to the cooperative sector in most countries' (ICA, 2003:4). Many associations still suffer from a 'dependency syndrome', which hampers the realization of their full potentials as voluntary entities (COPAC, 2000). Whereas organizations can be extremely valuable during a period of transition (Holmén, 1990), they are hardly the driver of transition. Hence, 'it is not simply a matter of local communities organizing themselves effectively' (Jayne *et al.*, 2005:2). Organizations mirror their environments much more than they shape them (Holmén, 2010), and where markets are weak or absent, organizations tend to be weak or absent. In SSA, peripheral and community-oriented organizations often take a preservationist (bonding) role, whereas successful cooperatives and 'progressive' (bridging) rural producer organizations are almost invariably found near big cities and in areas with good transport infrastructure. They also tend to be dominated by big, commercial farmers (Holmén, 2010) and are, until now, of less value for smallholders. Consequently, the temptation is sometimes great to try to speed up formation among smallholders, which, however, often leads to excessive external involvement and/or an urge to use cooperatives as a means to further governments' immediate interests.

In Uganda, the governments' ambition is to establish cooperative societies in every parish (Flygare, 2006), partly for the above-mentioned reasons but also because they can be used 'to deliver Government programs' (Government of Uganda quoted in Flygare, 2006:73). In Ethiopia, likewise, the government's ambition is to establish 'at least one cooperative in each Kebele [district] by the end of 2010' (Wolday *et al.*, 2009:113). By 2005, cooperatives in Ethiopia existed in only one-third of the districts and served only 9% of the farmers (Wolday *et al.*, 2009:113), a circumstance that may tempt the government to assume more tight control of the cooperatives. In Zambia, a major weakness of cooperatives is that they are still 'government-led rather than member driven' (ILO/ICA, 2003:13). Organization is a slow process and, as far as viable farmers' organizations go, the present situation is ambiguous.

Market information systems

Other measures implemented, either by governments themselves or in partnership with the private sector, are the establishment of market information systems, e.g. agricultural commodity exchanges and warehouse receipt systems. For decades, weak marketing and information systems have been major constraints to the realization of African agricultural potentials. Commodity exchanges are intended to facilitate access to market information so as to reduce transaction costs and constitute a meeting place for farmers and traders. The aim is to further boost producers' returns and to improve their access to credit, and therefore commodity exchanges are intended to work in close relation to producers' associations. With the use of radio and mobile phones, which are presently spreading rapidly in SSA, farmers and traders can have nationwide access to up-to-date price information. Once a buyer and a seller agree on a deal 'the exchange acts as a clearing house and (for a commission) arranges the financial and logistical aspects of the sale' (Okalla, 2002). Agricultural commodity exchanges have recently been set up *inter alia* in Ethiopia, Kenya, Malawi, Uganda and Zambia (ICA, 2003; IDEAA MIS, 2009; Wamulume, 2009; Wolday *et al.*, 2009).

In a similar vein, warehouse receipt systems have been established or are under construction in, for example, Ethiopia, Ghana, Malawi, Tanzania, Uganda and Zambia (Onumah, 2002; NRI, 2006; Bashaasha, 2008; Danku *et al.*, 2009; Isinika and Ashimogo, 2009). Even though they are part of the private sector, governments have facilitated their emergence by passing Warehouse Receipts Systems Bills, for example in Tanzania in 2005 and in Uganda in 2006 (NRI, 2006). Warehouse receipt systems are expected to reduce seasonal price variability, to reduce the risk of cheating, to improve access to finance for all operators and to improve the enforcement of contracts, and thus stabilize and expand markets. They operate at various geographical levels, from community warehouses in Ghana (Danku *et al.*, 2009) to a regional presence in Zambia (Onumah, 2002) and nationwide coverage in Uganda (NRI, 2006). They handle both food crops and traditional cash and export crops. Farmers deliver crops to a warehouse, run by a certified operator, who stores the produce until the farmer is ready to sell it. If warehouses are well run, they reduce the risk of postharvest losses. In return for crops delivered, farmers can use the receipt as collateral to obtain credit from financial institutions. The loan is then paid back when the crop is sold. It is believed that commodity exchanges and warehouse receipt systems not only will boost productivity but also, despite their initial attraction for larger and commercial farmers, will also indirectly benefit smallholders through their 'aggregate impact on price stability and the transparency of price formation' (Coulter and Onumah, 2002:335).

Policies towards roads and transport development

Producers' organizations and market information systems will be of little avail if markets cannot be reached. In virtually all countries studied, transport systems are inadequate and roads and railroads are of poor standard. This is particularly the case in rural areas, where large tracts often are 'not passable during the rainy season when farmers need inputs' (Isinika and Ashimogo, 2009:58). This has been a persistent problem in SSA, but in many cases roads deteriorated due to governments' budget cuts when implementing SAP (Mulwafu, 2009; Olouch-Kosura, 2009). In recent years, improvements are commonplace. However, SSA has attracted only a small share of private sector investments in infrastructure, and that share 'has been heavily tilted toward telecommunications' (Sheppard *et al.*, 2006:1; UN, 2008).

While roads and transport infrastructure thus remains a public affair, governments have enacted quite different policies towards this sector. In both Tanzania and Uganda, budget allocations for national road construction and maintenance appear to have stagnated lately (Bashaasha, 2008; Isinika and Ashimogo, 2009). On the other hand, Ethiopia, Ghana, Kenya, Malawi and Zambia have seen consistent increases in budget allocations, with expansion, rehabilitation and upgrading of road networks (Danku *et al.*, 2009; Mulwafu, 2009; Olouch-Kosura, 2009; Wamulume, 2009; Wolday *et al.*, 2009). Sometimes, e.g. in Ghana, the contribution from donors has been significant (Danku *et al.*, 2009). On other occasions, as in Malawi, road investments have been through government funding, 'most of which have been made available through the debt cancellation program' (Mulwafu, 2009:29). More generally, China has 'emerged as an important source of finance of infrastructure development in Africa' (UN, 2008:3). In whichever case, road and transport systems have generally been put at the forefront of development agendas after many years of neglect. In Zambia, 'the road sector has been identified as the single most important vehicle for the realization of the PRSP [Poverty Reduction Strategy Paper] program' (Wamulume, 2009:40). In Ethiopia, the national road authority was 'given a 2008 Good Practice Award by the World Bank's Independent Evaluation Group' (Wolday *et al.*, 2009:179).

Policies towards extension

Extension services – and organizations for their dissemination – are part of the general infrastructure needed for rural transformation and agricultural productivity enhancement. It is essential that farmers, smallholders in particular, are informed about new technologies and practices and on how markets and externally oriented associations function. Recent policies on extension differ greatly, however, from those on transport systems. Whereas many governments in SSA are prioritizing roads and transport systems, much less generally seems to be invested in agricultural advisory services. In part, this is an effect of SAP, where decentralization and privatization of governmental agencies was believed to be particularly important in the agricultural sector, well known for innumerable tales about top-down, inefficient and lethargic extension services. Instead, it is expected that private companies, local governments and/or non-governmental organizations (NGOs) and farmers' associations will be more efficient and more responsive to the needs of their clients.

Invariably, extension services have become more diverse during the early 2000s, with more providers on the scene. Officially, all governments are reorienting themselves towards more pluralistic, participatory and demand-driven systems for extension provision. However, governments are still often seen as important agents, even as direct providers of services. Nevertheless, governments have implemented quite diverse policies, ranging from almost full liberalization in Uganda (Bashaasha, 2008) to enhanced engagement in Ethiopia (Wolday *et al.*, 2009). In some countries, liberalization was rather abrupt and actually meant a collapse of the then prevailing extension system (Mulwafu, 2009), whereas in other cases it has been less dramatic (Akande and Ogundele, 2009; Danku *et al.*, 2009; Olouch-Kosura, 2009). But it is not only a matter of degree, government engagement is also becoming more pronounced in some sectors while being lessened in others. In Zambia, the extension system – especially that which is targeted to small-scale farmers – remains publicly funded (Wamulumu, 2009). In Kenya, the government, while striving to commercialize and privatize extension services, still offers 'free extension services for food crop production' (Olouch-Kosura, 2009:44). Likewise, in Ghana, Ethiopia and Nigeria, while the aim is to commercialize agricultural extension, it is so far provided by government free of charge (Akande and Ogundele, 2009; Danku *et al.*, 2009; Wolday *et al.*, 2009).

Downsizing and commercialization of agricultural extension services are often interpreted as an abandonment of the small farmers. For one thing, remaining governmental extension services are still sometimes criticized for being of low quality (Coughlin, 2009b; Danku *et al.*, 2009; Wamulume, 2009), also 'top-down, supply-driven and non-participatory' (Wolday *et al.*, 2009:170).

Moreover, while smallholders are numerous (and thus seem to provide a large enough market) they are widely spread and costly to reach. They also have limited capacity to pay for services. And local governments have limited capacity to provide such services – especially if services are to be demand-driven. The 'capacity of the private sector to provide services is still weak' (Isinika and Ashimogo, 2009; see also Bashaasha, 2008; Coughlin, 2009b; Olouch-Kosura, 2009). This is also so among NGOs and farmers' associations (Holmén, 2010). Hence, downscaling of public services has not been compensated for by a sufficient increase of alternative providers. In fact, the numbers of extension agents have decreased dramatically since SAP.[10] In Ghana, the number of extension agents is below requirement (Danku *et al.*, 2009); in Kenya the gap is 29% (Olouch-Kosura, 2009), in Malawi about 45% (Mulwafu 2009), and in Tanzania, 38% – but fully 80% at village level (Isinika and Ashimogo, 2009). Hence, there is an acute risk that food crop-oriented smallholders will be left without new technologies and advisory services, with potentially disastrous effects for food security.

Contract farming

In order to fill this void, IFIs, donors and governments have placed great hope on contract farming and out-grower schemes as means to reach the small farmers, not only with advisory services but also with credit and market links. Basically, contract farming and out-grower schemes are the same thing under different names. Both involve a contract. Under contract farming, the farmer signs a contract with a company (trader, processor) to grow a certain crop according to specified conditions. The company usually provides credit, inputs and technical services and the farmer is obliged to sell the crop (or a specified part of it) to the company, usually at a predetermined price. An out-grower scheme usually means that small farmers link up with an adjacent big farmer (who already has a contract) to access credit and extension and in order to reach economies of scale for input supply and marketing of produce. Both the big and the small farmers are expected to benefit from such arrangements.

Most investigated governments are encouraging contract farming and out-grower schemes, often with the expressed hope that they will reach small producers more easily and enhance commercialization of smallholder agriculture (Bashaasha, 2008; Wamulume, 2009; Wolday *et al.*, 2009). However, none has yet an official policy on the subject (*Afrint* background papers). Contract farming is still very limited and is currently almost exclusively used for cash crops and export crops such as plantation crops, horticulture, tobacco and cotton.[11] While there have been occasions where farmers cheated on the contract and

[10] Ethiopia is the exception. Here, the number of extension workers 'has increased tremendously since the mid-1990s' (Wolday *et al.*, 2009:173). However, this fast expansion of numbers has been at the expense of quality, fostering 'a dilettante attitude ... in extension management' (Wolday *et al.*, 2009:174).

[11] Bashaasha (2008) mentions out-grower schemes also for sorghum and barley in Uganda. In these cases, however, the crops are contracted with the beer industry and therefore remain outside conventional food crop classification.

'sold to the highest bidder no matter who supplied them with seeds, fertilizer and chemicals' (Bashaasha, 2008:29), the system appears to have worked rather well in Uganda (Bashaasha, 2008:29).

In Tanzania, experiences are mixed, and while there apparently are many positive examples, there is also instances where 'farmers were left high and dry when buyers changed their minds and renegade on the contract' (Isinika and Ashimogo, 2009:50). Suppliers in contract schemes sometimes charge exorbitant prices for services and inputs, a possibility emanating from their monopoly position in such schemes (Coughlin, 2009b; see also Liebhardt, 2005). Also, in Malawi, performances are mixed, with both good and bad experiences. 'In the worst scenario, contract farming has tended to assume a negative connotation as it is reminiscent of the colonial forced labour system' (Mulwafu, 2009:17).[12]

With thin and volatile markets characterizing much of rural SSA, uncertainty in transactions is commonplace and constitutes a strong barrier against smallholder commercialization and development. Contract farming appears to be an attractive alternative for linking smallholders to emerging markets, especially as the state has been rolled back and no longer substitutes for missing markets. Since no government has formulated a policy for contract farming, this is, however, a strategy by default rather than the result of active choice. It has been concluded that 'under appropriate enabling environments, the potential advantages of contracting ... tend to outweigh the potential disadvantages' (da Silva, 2005:1). When such an environment is missing, the effect can be very different from what was expected.

Contract farming can function in different ways. Eaton and Shepherd (2001:10) find that 'when efficiently organized and managed, contract farming reduces risk and uncertainty for both parties'. It can constitute a means to link smallholders to a market they would not be able to reach on their own and thus provide security in external relations. But it can also be initiated as a means of outsourcing and then also involves the transfer of risk from the company to the farmer, hence leading to enhanced insecurity for the latter.[13] Also, with the scaling down of government extension services, if governments rely too much on contract farming, which is almost exclusively oriented towards cash and export crops, there is a risk that food staples will be left without extension services and technical support. This would be detrimental for the large majority of farmers in SSA: the smallholders. It could also have very negative repercussions for future food security.

So far, little has been done to investigate long-term social and economic effects of contract farming and out-grower schemes, or their possible contributions to agricultural growth and development in SSA. The system apparently

[12] While there is no reason to doubt that exploitation and abuse take place, it has been pointed out that 'exploitative arrangements by managers are likely to have only limited duration [because] it can jeopardize agribusiness investments. Similarly, ... honoring contractual arrangements is likely to be to [the farmers'] long-term benefit' (Eaton and Shepherd, 2001:3).

[13] It has, for example, been reported that multinationals in Central America have 'moved away from plantation production of bananas to contracts with individual producers' (Agriculture21, 2001:1).

is here to stay but is, at the time of writing, unregulated and without guidance. Normally, contract schemes follow the logic of the market and are confined to locations near big cities or major roads (Havnevik *et al.*, 2007). They often constitute a top-down, take-it-or-leave-it approach, with limited educational qualities. Hence, they cannot substitute for normal extension services. And there are entry barriers for small-scale farmers, limiting their participation (Sautier *et al.*, 2006). Contractual arrangements are often vague or merely oral agreements. A system that theoretically would be beneficial for all parties often suffers from contractual arrangements that are skewed in favour of the bigger party (Isinika and Ashimogo, 2009). Contract farming, so far, provides little protection for smallholders and 'it is in fact quite common to find that patron–client relationships . . . extend to relations between projects and participating farmers' (Kudadjie-Freeman *et al.*, 2008:11).

Hence, governments will need to formulate policies on the issue and to define their role in relation to it. If contract farming is to have a broad positive impact, it will be necessary for governments to strengthen the institutional and legal framework to protect smallholders and ensure that contracts are fair. At the same time, it may be wise to lessen expectations on the virtues of contract farming. Successful contract farming requires a pre-existing market (Eaton and Shepherd, 2001) and may not be suited to establish markets where none exist. It can still be beneficial in some sense and in some locations but it is not an omnibus solution for improving smallholder agriculture and neither will it automatically eradicate rural poverty. Contract farming can be an important complementary device but will hardly substitute for normal public advisory services.

Implications for smallholders and food security

This résumé of trends at macro level shows diverging trends, some of which may be called progressive and others not. On a general level, average yields for four important food crops remain low and productivity increases are moderate at best. In fact, most gains in production are due to expansion of cropped area and only in rare cases to higher yields. The use of improved inputs is still very low by world standards. This is not to say that nothing positive is happening (see Chapter 5, this volume) but on this level of aggregation progress appears to be modest.

Actually, a lot is happening. In the aftermath of SAP, the state is frequently becoming more engaged and supportive and is increasingly taking on a role that in some senses resembles that of the Asian developmental states in the 1960s and 1970s. However, this is so only in some senses. On a rhetorical level all governments claim to prioritize agriculture but their spending patterns often do not reflect that ambition. Budget allocations for agriculture development in most cases are (far) below those recommended by the African Union. On the other hand, they seem generally to be increasing. Compared to the Asian model referred to – *state-driven, market-mediated and small farmer-based development* – we can detect both similarities and divergences. Most policy rhetoric also indicates that governments claim to strive to create a favourable climate for commercial activities and in most cases confine themselves to a supportive and regulatory role. However, this is not done overnight. With central governments starved of resources and local governments lacking capacity to spearhead development, the private sector remains weak and markets are slow to develop. Transaction costs are generally high. This is especially so for food crops, where attractive business opportunities are comparatively scarce.

In this situation, most governments have made renewed efforts to subsidize food crop agriculture, i.e. to support smallholders, who easily become bypassed by a 'free' market. This has sometimes (if not always) been in direct opposition to donors' and IFIs' policy prescriptions.[14] Efforts to promote independent cooperatives and farmers' associations have been half-hearted at best. This, however, does not necessarily have to be because governments have an ambition to control them, even if that sometimes may be the case. Small farmers in SSA remain largely subsistence-oriented and poorly organized and are unfamiliar with emerging new rules of the game. Rural organizations with a commercial orientation are few and mainly promote the interests of big farmers in limited geographical areas.

In order to facilitate integration, commercialization and market development (something that will also provide incentives for farmers to organize), governments have supported the creation of market information systems (commodity exchanges and warehouse receipt systems) and they have, in recent years, with few exceptions, invested heavily in expansion and upgrading of roads and transport infrastructure. Extension systems have become liberalized and pluralistic, except in some countries for food crops and/or smallholders, where they remain a public service free of charge. These are all policy measures fully in line with the Asian model.

On the other hand, extension services have deteriorated and, although there are today more providers, the shortage of extension workers is high – sometimes extremely high. NGOs have limited capacities and limited spatial coverage. Private traders are commonly more oriented towards export crops than food crops, which largely leaves the food-producing smallholders bereft of advisory services. The fact that many governments explicitly promote export crops points in the same direction. Instead, governments have pinned their hopes on contract farming and out-grower schemes, with the hope that they will also be beneficial for small farmers – a hope that largely seems to be frustrated. In this sense, current developments do not seem to have a strong smallholder orientation – to the contrary. There are reasons to believe that, while contemporary policies may well lead to some aggregate growth, smallholders risk being bypassed by 'development', while food security and the Millennium Development Goals are in jeopardy.

[14] It may be worth recalling that when Asian governments choose to opt for Green Revolution interventions in the 1960s, this was also often done against the advice of major donors and IFIs (Djurfeldt *et al.*, 2005). Maybe there is something to be learnt here?

References

African Development Fund (2005) *Federal Republic of Nigeria – Agriculture and Rural Institutions Support Project. Appraisal Report.* Agriculture and Rural Development Department, Central and West Region, OCAR, Nigeria.

Agriculture21 (2001) Agribusiness and small farmers. *Agriculture21*, FAO, Rome.

Akande, T. and Ogundele, F. (2009) *The Millennium Development Goals and the African Food Crisis: Nigeria Macro Study*. Nigerian Institute of Social and Economic Research (NISER), Abidjan, Ivory Coast.

Amin, S. (1972) Underpopulated Africa. *Maji Maji* 6.

AU (2008) *10 Percent National Budget Allocation to Agriculture Development*. African Union (AU), Addis Ababa. Available at http://www.africa-union.org (accessed 4 April 2010).

Badiane, O. (2008) *Sustaining and Accelerating Africa's Agricultural Growth Recovery in the Context of Changing Global Food Prices*. Policy Brief No. 9, IFPRI, Washington, DC.

Bashaasha, B. (2008) *Afrint II: Macro-Level Study for Uganda*. Dept of Agricultural Economics & Agribusiness, Makerere University, Kampala, Uganda.

Bignante, E., Dansero, E. and Scarpocchi, C. (2007) The global/local nexus in local development strategies in developing countries. Paper presented at the Sixth Pan-European Conference on International Relations, University of Turin, Turin, Italy.

BTI (2006a) Bertelsmann Transformation Index 2006, Ethiopia. Available at: http://www.bertelsmann-transformationindex.de (accessed 4 April 2010).

BTI (2006b) Bertelsmann Transformation Index 2006, Ghana. Available at: http://www.bertelsmann-transformationindex.de (accessed 4 April 2010).

BTI (2006c) Bertelsmann Transformation Index 2006, Kenya. Available at: http://www.bertelsmann-transformation-index.de (accessed 4 April 2010).

BTI (2006d) Bertelsmann Transformation Index 2006, Nigeria. Available at: http://www.bertelsmann-transformation-index.de (accessed 4 April 2010).

BTI (2006e) Bertelsmann Transformation Index 2006, Tanzania. Available at: http://www.bertelsmann-transformation-index.de (accessed 4 April 2010).

BTI (2006f) Bertelsmann Transformation Index 2006, Uganda. Available at: http://www.bertelsmann-transformation-index.de (accessed 4 April 2010).

Chinsinga, B. (2007) *Resurrecting the Vestiges of a Developmental State in Malawi? Reflections and Lessons from the 2005/2006 Fertilizer Subsidy Programme*. Future Agricultures, Chancellors College, Zomba, Malawi.

COPAC (2000) *Strategic Alliances – Cooperatives, Farmers' and Rural Workers' Organizations*. Committee for the Promotion and Advancement of Cooperatives (COPAC), Geneva.

Coughlin, P. (2009a) Agricultural Intensification in Mozambique – Lessons from Ten Villages. Afrint II micro study. EconPolicy Research Group, Maputo, Mozambique. Draft.

Coughlin, P. (2009b) Is Crisis an Opportunity? Rethinking Agricultural Development Strategy in Mozambique. Afrint macro study. EconPolicy Research Group. Maputo, Mozambique. Draft.

Coulter, J. and Onumah, G. (2002) The role of warehouse receipt systems in enhanced commodity marketing and rural livelihoods in Africa. *Food Policy* 27, 319–337.

Crook, R.C. (2003) Decentralization and poverty reduction in Africa: the politics of local–central relations. *Public Administration and Development* 23, 77–88.

Danku, F., Aryeetey, E. and Sarpong, D.B. (2009) *Afrint II Macro Study: Ghana Report*. ISSER, Legon, Ghana.

da Silva, C. (2005) *The Growing Role of Contract Farming in Agri-food Systems Development: Drivers, Theory and Practice*. FAO, Rome.

de Blij, H.J. and Mueller, P.O. (1997) *Geography: Realms, Regions and Concepts*. John Wiley & Sons, New York.

Diao, X. and Dorosh, P. (2007) Demand constraints on agricultural growth in East and Southern Africa: a general equilibrium analysis. *Development Policy Review* 25, 275–292.

Djurfeldt, G., Holmén, H., Jirström, M. and Larsson, R. (eds) (2005) *The African Food Crisis – Lessons from the Asian Green Revolution*. CAB International, Wallingford, UK.

Dorward, A. (2009) *Rethinking Agricultural Input Subsidy Programmes in a Changing World*. Centre for Development, Environment and Policy, SOAS, University of London, London.

Dorward, A., Chirwa, E., Kelly, V., Jayne, T., Slater, R. and Boughton, D. (2008) *Evaluation of the 2006/7 Agricultural Input Subsidy Programme*. Ministry of Agriculture and Food Security, Lilongwe, Malawi.

Earthscan (2000) *The Reality of Aid 2000*. Earthscan Publications Ltd, London.

Eaton, C. and Shepherd A.W. (2001) *Contract Farming: Partnership for Growth*. FAO, Rome.

ECAPAPA (2003) Electronic Newsletter 28 March 2003, Vol. 6, No. 6. Eastern and Central Africa Programme for Agricultural Policy Analysis. Available at: http://www.asareca.org/ecapapa (accessed 4 April 2010).

Ellis, F. and Freeman, A. (2004) Rural livelihoods and poverty reduction strategies in four African countries. *The Journal of Development Studies* 40, 1–30.

FAO (1987) *Irrigation and Water Resources Potential for Africa*. FAO, Rome.

FAO (2009) *The State of Food Insecurity in the World*. FAO, Rome.

FAOSTAT data (2009) Food and Agriculture Organization of United Nations, http://faostat.fao.org/ (accessed 4 April 2010).

Ferguson A.E. and Mulwafu, W.O. (2007) *Decentralization, Participation and Access to Water Resources in Malawi*. BASIS CRSP, University of Wisconsin, Madison, Wisconsin. Available at: http://www.basis.wisc.edu (accessed 4 April 2010).

Flygare, S. (2006) *The Cooperative Challenge – Farmer Cooperation and the Politics of Agricultural Modernisation in 21st Century Uganda*. Uppsala University, Uppsala, Sweden.

Foerster, A. (2009) Donors must be patient. *Development and Cooperation* 36, 384–385.

Goldstein, A. and Kaufmann, C. (2006) *Is More Money Enough to Fix Africa's Transport Infrastructure?* Policy Insights 21. OECD Development Centre. Available at http://www.oecd.org/dev/insights (accessed 4 April 2010).

Haggblade, S. (2005) From roller coaster to rocket ships: the role of technology in African agricultural successes. In: Djurfeldt, G., Holmen, H., Jirström, M. and Larsson, R. (eds) *The African Food Crisis: Lessons from the Asian Green Revolution*. CAB International, Wallingford, UK, pp. 139–159.

Havnevik, K., Bryceson, D., Birgegård, L.-E., Matondi, P. and Beyene, A. (2007) *African Agriculture and the World Bank*. The Nordic Africa Institute, Uppsala, Sweden.

Hazell, P. and Fan, S. (2002) The importance of public investments and institutions for the functioning of food markets in developing countries. In: Lutz, C. (ed.) *Food Markets and Food Security in West Africa*. CDS Research Report 15, pp. 99–116.

Herbst, J. (2000) *States and Power in Africa: Comparative Lessons in Authority and Control*. Princeton University Press, Princeton, New Jersey.

Hettne, B. (1992) *The Future of Development Studies*. Paper presented at the 40th anniversary of ISS, Institute of Social Studies, The Hague, the Netherlands.

Hodder, B.W. (1992) Future prospects. In: Gleave, M.B. (ed.) *Tropical African Development – Geographical Perspectives*. Longman, Harlow, UK.

Holmén, H. (1990) *State, Cooperatives and Development in Africa*. Research Report No. 86. The Scandinavian Institute of African Studies, Uppsala, Sweden.

Holmén, H. (2005) The state and agricultural intensification in sub-Saharan Africa. In: Djurfeldt, G., Holmén, H., Jirström, M. and Larsson, R. (eds) *The African Food Crisis: Lessons from the Asian Green Revolution*. CAB International, Wallingford, UK, pp. 87– 112.

Holmén, H. (2010) *Snakes in Paradise: NGOs and the Aid Industry in Africa*. Kumarian Press, Sterling, Virginia.

ICA (2003) *Proceedings of the 3rd ICA Regional Workshop on Agricultural Marketing*. Kenya School of Monetary Studies, International Cooperative Alliance (ICA), Nairobi.

ID21 (2007) Ugandan NGOs act as sub-contractors for international development agencies. Available at: http://www.id21.org (accessed 4 April 2010).

IDEAA MIS (2009) *Malawi Agricultural Commodity Exchange*. Available at: http://www.ideaamis.com (accessed 4 April 2010).

IIASA (2009) Poverty and food insecurity: a threat to billions. *Options*, Summer. International Institute for Applied Systems Analysis, Laxenburg, Austria, pp.14–17.

ILO/ICA (2003) *The Role of Cooperatives in Designing and Implementing Poverty Reduction Strategies*. Report on a regional workshop held in Dar es Salaam, Tanzania.

IRIN (2008) *Malawi: Subsidising Agriculture is Not Enough*. Available at: http://www.irinnews.org (accessed 4 April 2010).

Isinika, A. and Ashimogo, G. (2009) *Tanzania Macro Report: Addressing National Food Self Sufficiency*. Sokoine Agricultural University, Morogoro, Tanzania.

Jayne, T.S., Govereh, J., Mwanaumo, A., Nyoro, J.K. and Chapoto, A. (2002) False promise or false premise? The experience of food and input market reform in eastern and southern Africa. *World Development* 39, 1967–1985.

Jayne, T.S., Mather, D. and Mghenyi, E. (2005) *Smallholder Farming In Difficult Circumstances: Policy Issues for Africa*. Michigan State University, East Lansing, Michigan. Draft.

Kaarhus, R. and Nyirenda, R. (2006) *Decentralisation in the Agricultural Sector in Malawi – Policies, Processes and Community Links*. Noragric Report 32. Norwegian University of Life Sciences, Aas, Norway.

Kabumba, I. (2007) Conflicts between elected and appointed officials in districts. In: Asiimwe, D. and Nakanyike, B.M. (eds) *Decentralisation and Transformation of Governance in Uganda*. Fountain Publishers, Kampala, Uganda.

Kamara, A.B., Anyanwu, J.C., Osei, B., Rahji, T. and Vencatachellum, D.J.M. (2009) *Soaring Food Prices and Africa's Vulnerability and Responses: an Update*. Working Paper Series, 97. African Development Bank Group.

Kretzman, S. (2009) Questioning Old Traditions. *Inter Press News Agency*. Available at: http://www.ipsnews.net (accessed 4 April 2010).

Kudadjie-Freeman C., Richards, P. and Struik P.C. (2008) *Unlocking the Potential of Contract Farming: Lessons from Ghana*. Gatekeeper 139, IIED, London.

Larsson, R. (2005) Crisis and potential in smallholder food production – evidence from micro level. In: Djurfeldt, G., Holmén, H., Jirström, M. and Larsson, R. (eds) *The African Food Crisis: Lessons from the Asian Green Revolution*. CAB International, Wallingford, UK, pp. 113–138.

Larsson, R., Holmén, H. and Hammarskjöld, M. (2002) *Agricultural Development in Sub-Saharan Africa*. Afrint Working Paper No. 1. Department of Sociology, Lund University, Lund, Sweden.

Lele, U. and Stone, S.B. (1989) *Population Pressure. The Environment and Agricultural Intensification Variations on the Boserup Hypothesis*. The World Bank, Washington, DC.

Liebhardt, J. (2005) White gold or fool's gold: what will a rollback of U.S. cotton subsidies mean for farmers in Burkina Faso? *Multinational Monitor* 26, No. 5–6. Available at http://www.multinationalmonitor.org.

Minde, I., Jayne, T.S., Crawford, E., Ariga, J. and Goverath, J. (2008) *Promoting Fertilizer Use in Africa: Current Issues and Empirical Evidence from Malawi, Zambia, and Kenya.* Regional Strategic Agriculture Knowledge Support System (Re-SAKSS) for Southern Africa.

Morris, M., Kelly, V.A., Kopicki, R.J. and Byerlee, D. (2007) *Fertilizer Use in African Agriculture: Lessons Learned and Good Practice Guidelines.* The World Bank, Washington, DC.

Mulwafu, W. (2009) *Afrint II: Malawi Macro Study Country Report on Agricultural Intensification.* Centre for Social Research, University of Malawi, Zomba, Malawi.

NRI (2006) *Warehouse Receipts Systems. What's New?* National Research Institute, University of Greenwich, Chatham, UK. Available at: http://www.nri.org (accessed 4 April 2010).

Okalla, A.W. (2002) *The Kenya Agricultural Commodity Exchange.* ICT Update, December. Available at http://ictupdate.cta (accessed 4 April 2010).

Olouch-Kosura, W. (2009) *The Millennium Development Goals and African Food Crisis: Implications of Macroeconomic Policies on Agricultural Intensification.* AERC, Nairobi.

Onumah, G.E. (2002) *Improving Access to Rural Finance through Regulated Warehouse Receipt Systems in Africa.* National Resources Institute, University of Greenwich, Chatham, UK.

Ramachandran, V. (2009) *Africa's Private Sector: What's Wrong with the Business Environment and What to Do About It?* Centre for Global Development, Washington, DC.

Riungu, C. (2008) *How the Government Defied Donors on Subsidies.* Available at: http://www.allafrica.com (accessed 4 April 2010).

Rosegrant, M.W. and Perez, N.D. (1997) Water Resources Development in Africa: a Review and Synthesis of Issues, Potentials and Strategies for the Future. EPTD Discussion paper 28. IFPRI, Washington, DC.

Sautier, D., Vermeulen, H., Fok, M. and Biénabe, E. (2006) *Case Studies of Agri-Processing and Contract Agriculture in Africa.* RIMISP, Santiago de Chile, Chile.

Sheppard, R., von Klaudy, S. and Kumar, G. (2006) Financing infrastructure in Africa. *Gridlines* 13, Sept.

Sida (2009) http://www.sida.se

Tambulasi, R.I. and Kayuni, H.M. (2007) Decentralization opening a new window for corruption. An accountability assessment of Malawi's four years of democratic local governance. *Journal of Asian and African Studies* 42, 163–183.

Thangata, P. and Blackie, M. (2006) Can fertilizer subsidies help farmers out of poverty? Available at: http://www.ID21.org (accessed 4 April 2010).

The Economist (2009) If words were food, nobody would go hungry. 17– 21 November, pp. 61–63.

Tsegay, Y., Rusare, M. and Ndlovu, B. (2009) Development Support Monitor 2009: Africa in our hands. African Monitor.

UN (2007) Ban Ki-moon convenes 'unprecedented' meeting to boost African development. UN News Service. Available at http://www.un.org/apps/news/printnewsAr.asp?nid=23780 (accessed 6 April 2010).

UN (2008) Infrastructure: strengthening the foundations for Africa's development. High-level meeting on Africa's development needs, UN, New York. Available at: http://www.un.org/AR (accessed 4 April 2010).

Wadala, A. (2007) The politics of decentralization in Uganda. In: Asiimwe, D. and Nakanyike, B.M. (eds) *Decentralization and Transformation of Governance in Uganda.* Fountain Publishers, Kampala, Uganda.

Wamulume, M. (2009) *Afrint II – the Millenniumm Development Goals and the African Food Crisis: Zambia Macro Study.* Institute of Economic and Social Research, University of Zambia, Lusaka, Zambia.

WCSDG (2004) *World Commission on the Social Dimension of Globalization*. ILO, Geneva.
Wolday, A., Teketel, A. and Mulat, D. (2009) *Ethiopia. Afrint II Macro Report*. Addis Ababa, Ethiopia.
World Bank (2007a) *World Bank Assistance to Agriculture in Sub-Saharan Africa – an IEG Review. Independent Evaluation Group*. The World Bank, Washington, DC.
World Bank (2007b) *World Development Report*. The World Bank, Washington, DC.

4 Smallholders Caught in Poverty – Flickering Signs of Agricultural Dynamism

Magnus Jirström,[1] Agnes Andersson[1] and Göran Djurfeldt[2]

[1]*Department of Human Geography, Lund University, Lund, Sweden;*
[2]*Department of Sociology, Lund University, Lund, Sweden*

This chapter presents findings from two farm household surveys, carried out in 2002 and 2008, offering an up-to date picture of the current trends of production, area productivity, levels of commercialization and sources of cash income among some 3000 farming households in eight sub-Saharan African (SSA) countries. The focus is set on four major African staples – maize, cassava, sorghum, and rice – the production of which dominates among the different activities and income sources for the great majority of households in the sample.

In Chapter 3, production and productivity trends based on aggregated national statistics were analysed for the 1995–2007 period (see Holmén, Chapter 3, this volume). For the four crops under study it was shown that increases in production have primarily been driven by area expansion and not yield growth, although important examples of the latter were also identified. The period, in other words, does not seem to show any significant change in the long-term historic pattern of the region's output, growth being the result of area expansion. An important change, however, is the past decade's agricultural growth rate of above 3.5%, outpacing the continent's population growth, at approximately 2% (Binswanger-Mkhize, 2009). These developments at the aggregate level are, however, not reflected in the findings from this study of local-level production trends in the approximately 100 villages where the two surveys were carried out.

Instead, the general picture emerging from the findings is one of a continued crisis in the smallholder sector, with decreasing farm sizes, low levels of output per farm, low productivity and a high degree of subsistence farming. On an aggregated level, including all countries, the difficult situation facing the majority of farm households in the survey villages does not seem to have improved significantly since the results of the first survey were presented (Larsson, 2005). Not surprisingly, however, there also continues to be marked differences between countries. Furthermore, the 2008 survey confirms the results of the 2002 survey by identifying wide gaps in productivity and commercialization among farmers within the same region and village.

Background

Small-scale family farming dominates rural employment and still forms the backbone of the rural economy in most SSA countries. Accounting for 70–90% of all farms in many SSA countries, most of which can be defined as agriculture-based, the development of the smallholder sector is, in the view of many, at least in the short and medium term, a key factor for overall development (World Bank, 2007; FAO, 2009; Wiggins, 2009). According to IFAD, more than 70% of the subcontinent's poor are living in rural areas and are, to a large extent, dependent on incomes from agriculture (IFAD, 2001). These are all conditions contributing to the recently growing attention, as emphasized by the launching of the *World Development Report 2008 – Agriculture for Development*, to agricultural development in developing economies in general and the smallholder sector in SSA in particular.

The new emphasis is, however, being questioned on different grounds. Some doubt the overall role of agriculture for the region's development and ask whether Africa should be doing any agriculture at all, by pointing at lacking comparative advantages for the sector in a globalized world (Rosenzweig, 2004, quoted in Timmer, 2005). Others point at what they conceive as a process of 'de-agrarianization', according to which rural economic diversification in SSA has resulted in agriculture being a rapidly shrinking sector (Bryceson, 2002). Rural households, according to Ashley and Maxwell (2001) and Ellis (2005), already have diverse and geographically dispersed sources of income, and the potential for agriculture to drive rural growth is low. Instead of investments in agriculture, the recommendation is to support a more holistic livelihoods approach or even, as commented by Hazell (2006), to support exit strategies – public investments should ease migration to urban centres.

Among those who do foresee a future important role for agriculture in Africa's development, there are also widely differing views on what strategies to pursue. One debate has revolved around the question of whether African small-scale family farms can be expected to be in the centre of a process of agricultural transformation based on rapid productivity growth. Collier and Dercon express scepticism against any 'exclusive focus on smallholders' and call for 'a much more open-minded approach to different modes of production' (Collier and Dercon, 2009:1). Collier suggests that the most realistic way to increase the global food supply is to invest in 'large-scale commercial agriculture' using, for example, 'large swathes of Africa... that have good land that could be used far more productively if it were properly managed by large companies'. Betting on smallholders is, according to Collier, a mistake, as 'peasant farming is not well suited to innovation and investment' (Collier, 2008).

Collier's arguments could perhaps be sorted in among the set of arguments forwarded by the camp of the 'smallholder pessimists', which, according to Timmer (2005), are engaged in a sharp debate with the opposing camp, made up of 'smallholder optimists'. The smallholder pessimists, who also point at the new global reality, add a number of conditions that they claim make a broad-based and smallholder-driven agricultural development process unlikely. In a world of open borders for trade and continued agricultural protection in the rich countries, small-scale African farms cannot, allegedly, compete. Additional arguments refer to the prohibitively high cost of necessary investments in infrastructure due to low population densities; lack of technology suitable for Africa's cropping systems and lack of water control. Given the globalization of food trade and the revolution in supply chains, with a rapid growth of supermarkets transforming food retail markets, a strategy aiming at increased productivity in small-scale agriculture is, according to the pessimists, doomed to failure. Instead it is the large-scale farms with state-of-the-art technology that will be able to respond to the domestic and international markets for high-value crops and products. The world has changed so dramatically since the 1960s and 1970s that the positive experiences in many Asian countries of rapid agricultural growth are not very useful when designing policies for Africa today (Ellis, 2005).

The small-scale optimists, however, believe that the historically positive relationship between small-scale family farming and economic growth still holds (Lipton, 2005). The situation of low agricultural productivity and farm profitability characterizing most smallholder agriculture in SSA can and must, according to this stand, be changed if a broad-based process of poverty reduction is to be achieved (Jayne *et al.*, 2006). Two major initiatives, the Comprehensive Africa Agriculture Development Programme (CAADP) under The New Partnership for Africa's Development (NEPAD) and the Alliance for an African Green Revolution (AGRA), emphasize the importance of productivity growth in the smallholder sector.[1] According to this line of reasoning, only with a substantial increase in the productivity of staple food agriculture will the great majority of Africa's population, the millions of semi-subsistence smallholders, be able to invest in more education, in a more diversified output mix, including high-value crops, and in new economic activities outside the farm. Only if labour demand increases due to a more productive and labour-intensive agricultural sector will the rural economy employ a larger share of a growing rural population, for which the traditional alternative of opening up new land is fast closing (Lipton, 2005; Hazell, 2006).

The widely diverging views accounted for in the previous sections may, to some extent, depend on the relative weakness in agricultural statistics in Africa, a problem that makes evidence-based analysis difficult. It is the aim of this chapter to throw light on some of the realities on the ground, as reported by some 4000 smallholders in the eight SSA countries under study. By smallholders we refer to family farms, mainly worked by family members, producing partly for subsistence and generally cultivating small areas, on average 2 ha in 2008. As will be shown in more detail in subsequent sections, however, the notion of the smallholder sector being 'unimodal' in terms of land distribution

[1] In November 2009, NEPAD and AGRA entered into a partnership with a focus particularly on plans to develop high-potential breadbasket areas of African countries.

is not supported by this study, which instead points at a relatively high degree of inequality of land access.

Following the next section on data and methodological considerations, the remaining parts of the chapter start with a presentation and analysis of production data for the four main staple crops, including the findings of significant, village-level, yield gaps. This leads on to two sections focusing on technology adoption and market integration in the smallholder sector. The chapter focus is subsequently broadened to include an analysis of non-farm income sources also and, finally, in a section focusing on socio-economic differentiation in terms of gender and access to land, production and income data are analysed from these perspectives.

Data Collection, Sampling Strategies and Cross-section Analysis

The data presented in this chapter is based on the two *Afrint* household surveys, conducted in 2002 and 2008, respectively (see Andersson *et al.*, Chapter 5, this volume). Households interviewed in 2002 were revisited in 2008. The cross-sectional database obtained consists of roughly 3500 households for the 2002 round and 4800 for the one in 2008. Approximately 100 villages in eight countries are covered.[2]

Original sampling aimed to shed light on the stalled intensification process within the African staple food crop sector and was based on a multi-stage sampling strategy. The potential for intensification was perceived to be higher in areas that meet basic conditions in terms of agro-ecological potential and access to markets (infrastructure) than in places that are peripheral in this respect. Twenty agro-ecological/market regions were purposively sampled to account for variation in the agricultural potential of the regions in which the households resided. The (purposively) sampled villages are typical of the type of environment in which a majority of the smallholder population in SSA reside and diverse enough to yield information about crucial conditions responsible for farmer performance.

The 2002 survey round prioritized reliability and focused on production volumes rather than prices and incomes. Data related to technology use and the institutional environment for technology adoption was a crucial component of the 2002 survey. During the second wave (2008), detailed questions relating the cash income sources of the households (both farm and non-farm related) and a number of more detailed questions in relation to marketing of the major staple crops have been added to the survey.

Each wave of data collection has been conducted to assure cross-sectionally representative data for the farming population of the sampled villages for the respective years. Attrition due to non-traceability has been tolerable (18%)

[2] Surveys were also carried out Uganda in 2002 and 2008. Due to data-quality problems in the 2002 survey comparisons over time are not possible and hence Uganda is not included in this chapter.

during the second round of data collection, while descendant households have been sampled and traced in the case of partition and one of the descendants sampled to replace the original household. Where village in-migration has been sizeable, in-migrant households have been sampled to complement the re-interviewed households.

A note on the challenges associated with the collection of smallholder production data is warranted. In the two surveys, enumerators interviewed the farm manager, in nearly all cases this person was also the head of the household. Production and related data was not collected for different farm plots but instead recorded crop by crop. In cases of intercropping the respondent assessed the pure stand equivalent areas for the different crops. This means that the data collected do not reflect exact areas under cultivation nor quantities produced, consumed, sold or used for other purposes. In the analysis of production data we have, apart from excluding extreme values, also excluded production taking place on areas of less than 0.1 ha, as these generally indicated unrealistic area productivity figures.

The presentation and analysis of the results of the two survey rounds offers a statistically representative picture of the production and area productivity at the village level at the two points in time. However, in order to understand patterns of dynamism or stagnation as well the drivers of change, a panel approach is necessary. Chapter 5 offers such an analysis for the case of maize (Andersson *et al.*, Chapter 5 in this volume; see also Djurfeldt *et al.*, 2008).

Decreasing Farm Sizes – Relatively Stable Production Portfolios

The mean farm size, here defined as area under cultivation, has declined from 2.42 ha in 2002 to 2.16 ha in 2008, confirming earlier studies anticipating and reporting on rapidly declining farm sizes in several SSA countries (Table 4.1) (Lipton, 1989; Ellis, 2005; Jayne *et al.*, 2006). In Mozambique, Ghana and Tanzania the mean and the median farm size have declined during the period, while in Nigeria, farms have been substantially expanded.

Breaking down the farm size data further shows that the per capita access to land is very small in absolute numbers – 0.12 ha per capita or less for the bottom 25% of the 2008 sample in all countries but Ethiopia and Nigeria (Table 4.2). In Kenya, the level is only a third of that figure – 0.04 ha per capita. Jayne *et al.* (2006) present data for five countries – Ethiopia, Kenya, Rwanda, Mozambique and Zambia – showing similar conditions and conclude that 'the bottom 25% of agricultural households are virtually landless, having access to 0.10 hectares per capita or less in each of the five countries examined' (Jayne *et al.*, 2006:iv).[3] Considering that the production systems under study are overwhelmingly rainfed rather than irrigated, the land constraints facing an increasing share of smallholders are indeed serious.

[3] The data presented in Jayne *et al.* (2006) are, unlike our data, based on nationwide surveys carried out between 1995 and 2002 (Jayne *et al.*, 2006:28).

Table 4.1. Land under cultivation, ha (mean and median), 2002 and 2008.

	Ethiopia	Ghana	Kenya	Malawi	Mozambique[a]	Nigeria	Tanzania	Zambia	Group total
Mean farm size 2002	2.20	2.40	0.97	1.24	2.02	4.47	1.96	3.01	2.42
Mean farm size 2008	2.40	2.06	0.98	1.23	1.33	6.13	1.72	2.72	2.16
Change (%)	9	−14*	2	0	−34***	37***	−13*	−9	−11***
Median farm size 2002	2.00	2.00	0.80	1.01	1.50	2.80	1.60	2.40	1.70
Median farm size 2008	2.00	1.60	0.75	1.01	1.00	4.20	1.40	2.00	1.42
Change (%)	0	−20	−6	0	−33	50	−13	−17	−17
No. of cases 2002	322	413	298	387	272	490	399	465	3046
Missing cases 2002 (%)	0	0.7	0	3.40	4.00	1.00	1.00	4.50	1.40
No. of cases 2008	316	414	298	394	400	237	392	418	2869
Missing cases 2008 (%)	0	0	0	0	0	0	0	0	0

T-test for paired samples: ***Significant at the 0.1% level; *at the 5% level

[a] Mozambique figures are from 2003–2005 and 2006–2008.

Table 4.2. Household landholding size per capita.

Country	Mean household farm size (ha) per capita, total sample			Means for household quartiles ranked by per capita farm size (ha) by village, 2002				Means for household quartiles ranked by per capita farm size (ha) by village, 2008			
	2002	2008	% change	Q1	Q2	Q3	Q4	Q1	Q2	Q3	Q4
Ethiopia	0.33	0.39	17**	0.16	0.25	0.34	0.54	0.17	0.27	0.37	0.61
Ghana	0.43	0.44	2	0.11	0.22	0.36	1.00	0.12	0.23	0.41	1.14
Kenya	0.22	0.18	–20	0.06	0.11	0.19	0.48	0.04	0.09	0.16	0.38
Malawi	0.29	0.29	–3	0.10	0.20	0.30	0.54	0.11	0.18	0.27	0.57
Mozambique[a]	0.37	0.28	–24***	0.12	0.23	0.35	0.72	0.09	0.16	0.25	0.79
Nigeria	0.45	0.73	61***	0.16	0.28	0.42	0.73	0.38	0.48	0.76	1.57
Tanzania	0.40	0.34	–15	0.15	0.26	0.39	0.75	0.11	0.18	0.30	0.83
Zambia	0.41	0.34	–15*	0.16	0.28	0.40	0.76	0.11	0.21	0.33	0.66
Group Total	0.37	0.36	–1	0.13	0.23	0.34	0.70	0.12	0.21	0.34	0.74
No. of cases[b]	2547	2604		575	629	635	708	556	687	675	686
Missing cases %	22	12									

T-test for paired samples: ***Significant at the 0.1% level; **at the 1% level; *at the 5% level

[a]Mozambique figures are from 2003–2005 and 2006–2008

[b]The number of cases in each quartile group varies as all households with the same per capita farm size are automatically placed in the same group in this SPSS routine.

Differences between the four per capita farm size quartiles are quite large, with the highest quartile of the 2008 sample cultivating, on average, between four to nine times as much land as the bottom one (Table 4.2). This reminds us of the need to view the smallholder farm sector as quite heterogeneous in terms of access to land. Looking at change over time for the period 2002–2008, Zambia and Kenya show a clear decrease in the mean per capita farm size, while in Nigeria the pattern of area expansion can be discerned for all quartiles. Inequality in land distribution within the villages has grown in three countries – Mozambique, Tanzania and, to a lesser extent, Zambia. This confirms Jayne *et al.*'s (2006) contention that land inequality within the smallholder sector in a number of countries in eastern and southern Africa are by now 'comparable to or higher than those estimated for much of Asia during the 1960s and 1970s' (Jayne *et al.*, 2006:4). In parts of the subcontinent these signs of land scarcity and polarization are not new.[4]

Turning to the allocation of land to different crops, the general pattern seems to suggest relative stability in land use (Table 4.3). The four staples in focus are, on average, grown on approximately a half to one hectare (0.55–1.05 ha). Land allocated to cassava has dropped by 17% in the period. Maize remains the single most commonly cultivated crop, with approximately 85% of all farms in the sample growing it on around half their land. The share of households growing sorghum, 'other food crops' (e.g. vegetables, beans, potatoes, etc.) and 'non-food crops' (e.g. coffee, tobacco, tea, sugarcane, etc.) has dropped, and in the cases of sorghum and 'non-food crops' the fall is substantial.

As shown in Tables 4.4–4.10, the aggregates presented in Table 4.3, however, mask important variation both between countries and within the study sites, to be further explored in the subsequent sections, focusing in detail on the main staple crops.

Production and Area Productivity

Maize

The average household production and yield of maize – SSA's most important food staple – remain very low in an international perspective. Mean production has declined from 1.61 to 1.25 t per farm, while the median values of 0.67 t in 2002 and 0.64 t in 2008 indicate that the decrease in production has primarily taken place among the larger producers (Table 4.4). For the majority of households, production stops at 600–700 kg, not allowing much, if anything, to be sold. The overall mean yield (three-season average) of 1.43 t/ha in 2000–2002 had fallen to 1.20 t/ha for the 2006–2008 period (Table 4.5). In other words, the aggregate picture is one of continued stagnation in maize production and even decline in area productivity.

[4] As remarked two decades ago by Lipton (1989:354) 'already in the mid-1970s, 40% of SSA populations lived in countries where land was judged "scarce" or "moderately scarce" (given its quality) in respect of capacity to feed their populations'.

Table 4.3. Land under cultivation (total and per crop) and percentages of households cultivating by type of crop.

	Maize	Cassava	Sorghum	Rice	Other food crops	Non-food crops	Total
Mean farm size/crop area (ha) 2002	1.05	0.66	0.98	0.77	0.63	0.85	2.42
Mean farm size/crop area (ha) 2008	0.97	0.55	0.94	0.83	0.63	0.87	2.16
Change in mean farm size (%)	–7	–17***	–4	8	0	2	–11***
Median farm size (ha) 2002	0.6	0.50	0.75	0.60	0.40	0.50	1.70
Median farm size (ha) 2008	0.61	0.25	0.70	0.60	0.40	0.50	1.42
Change in median farm size (%)	1	–50	–7	0	0	0	–17
Households cultivating (%) 2002	84	38	28	30	82	39	98
Households cultivating (%) 2008	86	36	23	32	75	28	100
Change in proportion cultivating (%)	2	–2	–5***	2	–7***	–11***	2
No. of cases 2002	2580	1066	835	588	2418	1117	3046
Missing cases 2002 (%)	1	4	4	3	5	7	1
No. of cases 2008	2462	878	644	572	2083	736	2869
Missing cases 2008	39	47	15	20	80	73	0
Missing cases 2008 (%)	2	5	2	3	2	9	0

T-test for paired samples: ***Significant at the 0.1% level.

Table 4.4. Maize production and cultivated maize area seasons 2000–2002 and 2006–2008.[a]

		Ethiopia	Ghana	Kenya	Malawi	Mozambique	Nigeria	Tanzania	Zambia	Group total
Proportion of sampled farmers growing maize (%)	2002	57	49	100	99	90	98	89	91	84
	2008	60	67	96	99	86	99	88	96	86
	Change (%)	3	18	–4***	0	–4	1	–1	5***	2
Farm area under maize (ha), three-season average (2000–2002 and 2006–2008)	2000–2002	1.06	1.07	0.54	0.30	0.91	1.68	1.06	1.41	1.06
	2006–2008	0.69	0.67	0.60	0.75	0.75	3.04	0.92	1.28	1.08
	Change (%)	–34***	–38***	12	148***	17**	81***	–14*	–9	2
Three-seasons' mean maize production (t/farm)	2000–2002	1.48	0.88	1.26	0.70	0.48	3.73	0.98	1.63	1.61
	2006–2008	0.94	0.62	0.99	0.82	0.48	2.88	1.15	2.04	1.25
	Change (%)	–37**	–30**	–21	18**	–2	–23*	17	26*	–22***
Three-seasons' median maize production (t/farm)	2000–2002	0.88	0.63	0.42	0.53	0.35	1.10	0.63	0.85	0.67
	2006–2008	0.40	0.37	0.43	0.62	0.33	1.83	0.68	1.1	0.64
	Change (%)	–55	–42	2	17	–5	67	8	29	0.0
No. of cases	2000–2002	136	167	208	377	211	467	332	396	2294
Missing cases (%)	2000–2002	0	2	0	0	0	1	0	1	0
No. of cases	2006–2008	175	242	198	391	296	238	331	373	2244
Missing cases (%)	2006–2008	0	0	0	0	0	1	2	0	0

T-test for paired samples of mean proportion maize growers, mean area for maize, and mean production. ***Significant at the 0.1% level; **at the 1% level; *at the 5% level.

[a] Based on a subsample including maize growers who cultivated an average area of at least 0.1 ha during the 2000–2002 and 2006–2008 periods. The sample also excludes cases with a sixfold or higher yield increase between the two periods. Mozambique figures are from 2003–2005 and 2006–2008.

Table 4.5. Maize yields.

		Ethiopia	Ghana	Kenya	Malawi	Mozambique	Nigeria	Tanzania	Zambia	Group total
Three-seasons' mean maize yield (t/ha)[a]	2000–2002	1.20	1.08	1.80	2.22	0.54	1.80	0.97	1.13	1.43
	2006–2008	1.07	0.96	1.41	1.15	0.75	1.49	1.29	1.45	1.20
	Change (%)	–10	–11	–22**	–48***	39***	–17**	34***	29***	–16***
Three-seasons' median maize yield (t/ha)	2000–2002	0.96	0.71	1.17	1.95	0.47	1.35	0.79	0.95	1.06
	2006–2008	0.87	0.87	1.05	0.99	0.57	1.25	1.08	1.27	0.99
	Change (%)	–10	22	–10	–49	21	–7	37	34	–7
No. of cases	2000–2002	136	167	206	375	211	466	332	396	2289
Missing cases (%)	2000–2002	0	2.3	1.0	1	0	1.1	0	0.8	0.7
No. of cases	2006–2008	175	241	198	391	295	238	332	373	2243
Missing cases (%)	2006–2008	0	0.4	0	0	0.3	1.2	2	0	0.5
5% best-performing farmers' yield (t/ha) maize[b]	2000–2002	3.24	4.24	4.99	5.17	1.41	4.23	2.68	2.74	3.71
	2006–2008	2.45	2.74	3.32	2.85	2.25	2.83	3.23	3.36	2.91
	Change (%)	–24	–36	–34*	–45***	60**	–33**	21	23***	–22***
95% lowest-performing farmers' yield (t/ha) maize[b]	2000–2002	1.05	0.92	1.52	2.00	0.47	1.59	0.87	1.03	1.26
	2006–2008	0.98	0.84	1.25	1.04	0.64	1.39	1.17	1.32	1.08
	Change (%)	–7	–9	–17*	–48***	37***	–12*	34***	27***	–14***

T-test for paired samples of mean yields all maize growers, mean yields 5% best performing, and 95% lowest performing: ***Significant at the 0.1% level; **at the 1% level; *at the 5% level.

[a] Based on a subsample including maize growers who cultivated an average area of at least 0.1 ha during the 2000–2002 and 2006–2008 periods. The sample also excludes cases with a sixfold or higher yield increase between the two periods. Yields above 10 t/ha at farm level have been excluded. Mozambique figures are from 2003–2005 and 2006–2008.

[b] Based on village aggregates.

There is, however, important variation among countries, as well as within countries and among farmers within the same area. Nigeria stands out, with a mean production of 2.9 t per farm per year for the 2006–2008 period, a figure reflecting the relatively large average area under maize (3 ha). The sharp fall in mean farm production stands in clear contrast to the growing median production and reflects a shift out of maize among the larger, commercially oriented Nigerian farms.

In our Malawi survey sites, higher maize production per farm seems to be related to a significant increase in the proportion of farmland allocated to maize and not to improvements in average yields. This is in contrast to the national-level data, pointing at a combination of area expansion and yield growth as an explanation of the dramatic increase in production associated with the much-debated subsidized seed fertilizer packages run by the Malawian government since 2005 (see Table 3.1 in Holmén, Chapter 3, this volume).[5] Only in three countries – Tanzania, Zambia and Mozambique – have mean maize yields in the surveyed villages increased. Looking at the median values, Ghana can be added to the group, this in spite of the serious floods affecting especially the Upper East Region, in which four of the eight sampled Ghana villages are situated (Table 4.5). In the case of Kenya, political instability in 2008, especially in the west, could explain some of the fall in production.

Differences in maize yields among maize growers at the village level – between the majority and the 5% best-performing farmers within the same village – gives further evidence of significant variation in the production conditions facing farmers. It is also an important reminder of the fact that at the farm level there is presently a clear potential for increasing area productivity, given existing technology.

The yield gaps, which are to be discussed in more detail later in this chapter (see Table 4.11), may, to some extent, depend on differences in agro-ecological conditions such as soil quality. There is, however, as shown by Jayne *et al.* (2006) for Kenya, Mozambique and Zambia, significant yield heterogeneity that is not explained by agro-ecological potential but rather by differences among households in crop input and management factors.[6]

Andersson *et al.* (Chapter 5, this volume) seek explanations for such variation in maize production, trying to identify the drivers behind dynamism in the maize sector. Here we stop by concluding that, important variation notwithstanding, average production and yields remain very low, both in an international comparison and in relation to what this study defines as the local-level potential.

[5] As commented upon in Chapter 5 (Andersson *et al.*, this volume), there may be a bias in our sample villages as these include a seemingly not representative share of households not having had access to the subsidized seed fertilizer packages.

[6] By stratifying maize-growing households in Kenya, Mozambique and Zambia by quintiles of maize yields and then comparing the median household yields with the median district maize yield, Jayne *et al.* (2006) identify a high degree of variability that cannot be explained by differences in agro-ecological potential.

Cassava

Cassava is Africa's second most important food staple crop. It was long referred to as a 'famine crop', suggesting that it was a foodstuff shunned under normal conditions and resorted to only during periods of food shortages. This was a much too narrow view of the importance of cassava – it has long been a popular part of West African diets. Moreover, the past decades' rapid growth of cassava as a major commercial crop for food and industry purposes in West Africa has gradually changed this picture. Today efforts to promote a development in line with what is now more broadly recognized as the cassava revolution in West Africa can be noticed in other parts of the continent. In Zambia and Malawi, for example, cassava production has surged dramatically in response to the introduction of modern cassava varieties and growing markets (Haggblade and Nyembe, 2007).

In the 2002 *Afrint* survey we encountered difficulties in the collection of cassava production data. Production data for cassava are subject to large measurement error because farm households typically harvest cassava year round and over a period of several years. Annual production aggregates based on small daily harvests produce wide variation in estimated cassava output, both nationally and locally. For this reason we refrained from collecting production data in the 2008 survey. Consequently, Table 4.6 only depicts data on the area cultivated and the proportion of households growing cassava. It is noteworthy that farmland under cassava has grown markedly in the Nigerian villages, while in Ghana and Mozambique, where it is also a principal crop, land allocated to cassava seems to have diminished.

Sorghum

Sorghum is SSA's second largest cereal crop and withstands harsh climatic conditions where most other crops do not thrive. In light of climate change impacts, which are expected to cause drier conditions in parts of the subcontinent, the recent breakthrough in the breeding of high-yielding hybrid sorghum varieties by Malian scientists is much welcomed (AGRA, 2009).

Despite such promising developments, sorghum seems to have experienced the most pronounced decline in terms of production and yields when compared with the three other main staple crops.[7] On average, the mean three-season production in 2006–2008 is only about 40% of that in 2000–2002, falling from 1.3t per farm to 0.5t per farm (Table 4.7). The fall in production can be observed in all six countries for which we have data for both periods. Yields have fallen by 31%, with Mozambique being the exception, showing major yield growth (Table 4.8).

[7] Our survey data do not, however, tally well with the FAO estimates. According to these, sorghum yields have remained stable at 0.8t/ha, while there has been an overall expansion of the area under sorghum, leading to an increase in production (FAOSTAT data, 2009).

Table 4.6. Cassava area and proportion of cultivating households.[a]

		Ethiopia	Ghana	Kenya	Malawi	Mozambique	Nigeria	Tanzania	Zambia	Group total
Proportion of sampled	2002	na	49	11	24	75	65	20	24	38
farmers growing cassava (%)	2008	na	49	31	23	63	61	9	22	36
	Change (%)	na	–	−20***	−1	−12***	−4	−11***	−2	−2
Farm area under cassava (ha),	2000–2002	na	0.92	0.12	0.23	0.65	0.86	0.27	0.41	0.66
three-season average	2006–2008	na	0.53	0.11	0.26	0.37	1.50	0.23	0.40	0.55
(2000–2002 and 2006–2008)	Change (%)	na	−42***	−10	12	−43***	74***	−13	−2	−17***
No. of cases	2000–2002	na	201	6	94	210	317	79	159	1066
Missing cases (%)	2000–2002	na	1	81	1	1	1	1	5	4
No. of cases	2006–2008	na	203	89	90	252	138	26	80	878
Missing cases (%)	2006–2008	na	0	5	1	1	10	28	13	5

T-test for paired samples of mean farm area under cassava: ***Significant at the 0.1% level.

[a] Based on a subsample including sorghum growers who cultivated an average area of at least 0.1 ha during the 2000–2002 and 2006–2008 periods. Mozambique figures are from 2005 and 2008.

Table 4.7. Sorghum production and cultivated area.[a]

		Ethiopia	Ghana	Kenya	Malawi	Mozambique	Nigeria	Tanzania	Zambia	Group total
Proportion of sampled farmers growing sorghum (%)	2002	55	50	11	3	41	42	1	25	28
	2008	48	49	8	3	0	39	na	6	23
	Change (%)	–7	–1	–3	–	–41	–3	na	–19***	–5***
Farm area under sorghum (ha), three-season average (2000–2002 and 2006–2008)	2000–2002	1.08	0.99	0.15	0.22	0.51	1.41	0.25	1.02	1.03
	2006–2008	1.29	0.99	0.10	na	0.34	3.06	na	0.54	0.52
	Change (%)	20	0	–31	na	–33***	118***	na	–47***	–50***
Three-seasons' mean sorghum production (t/farm)	2000–2002	1.11	0.37	0.12	0.34	0.31	3.42	0.31	0.61	1.28
	2006–2008	0.95	0.14	0.09	na	0.21	1.68	na	0.26	0.52
	Change (%)	–15	–60***	–27	na	–35**	–51***	na	–57***	–59***
Three-seasons' median sorghum production (t/farm)	2000–2002	0.90	0.30	0.11	0.33	0.23	1.59	0.31	0.41	0.52
	2006–2008	0.72	0.11	0.09	na	0.15	1.17	na	0.24	0.18
	Change (%)	–20	–64	–16	na	–36	–26	na	–42	–65
No. of cases	2000–2002	175	203	9	8	115	205	2	188	835
Missing cases	2000–2002	1	5	23	2	0	1	2	4	38
Missing cases (%)	2000–2002	1	2	72	20	0	0.5	50	3	4
No. of cases	2006–2008	150	206	23	na	147	85	na	23	634
Missing cases	2006–2008	0	0	0	na	0	12	na	2	24
Missing cases (%)	2006–2008	0	0	0	na	0	12	na	8	3.6

T-test for paired samples of mean proportion of sorghum growers, mean area for sorghum, and mean production. ***Significant at the 0.1% level; **at the 1% level.

[a] Based on a subsample including sorghum growers who cultivated an average area of at least 0.1 ha during the 2000–2002 and 2006–2008 periods. The sample also excludes cases with a sixfold or higher yield increase between the two periods. Mozambique figures are from 2003–2005 and 2006–2008.

Table 4.8. Sorghum yields.[a]

		Ethiopia	Ghana	Kenya	Malawi	Mozambique	Nigeria	Tanzania	Zambia	Group total
Three-seasons' mean	2000–2002	1.14	0.46	0.626	0.54	0.37	1.24	0.47	0.53	0.78
sorghum yield (t/ha)	2006–2008	0.80	0.167	0.339	0.48	0.74	0.85	na	0.58	0.53
	Change (%)	-30^{***}	-64^{***}	–46	–11	100^{***}	-32^{***}	na	9	-31^{***}
Three-seasons'	2000–2002	1.05	0.395	0.506	0.15	0.3	1.17	0.47	0.45	0.6
median sorghum	2006–2008	0.64	0.139	0.3	0.15	0.566	0.70	na	0.42	0.35
yield (t/ha)	Change (%)	–39	–65	–41	0	89	–40	na	–6	–42
No. of cases	2000–2002	151	194	8	7	89	156	2	103	710
Missing cases (%)	2000–2002	3	1	0	0	1	10	0	0	4
No. of cases	2006–2008	122	199	17	10	105	69	na	21	543
Missing cases (%)	2006–2008	0	0	0	0	6	0	na	0	0.8
5% best-performing	2000–2002	2.34	1.192	1.238	1.25	0.90	2.12	0.47	0.83	1.54
farmers' yield	2006–2008	2.04	0.513	0.45	1.37	1.37	1.54	na	0.95	1.17
(t/ha) sorghum[b]	Change (%)	–13	-57^{***}	–63	10	53^{**}	-27^{*}	na	14	-24^{***}
95% lowest-performing	2000–2002	1.05	0.418	0.562	0.14	0.33	1.13	na	0.51	0.71
farmers' yield (t/ha)	2006–2008	0.72	0.146	0.291	0.25	0.68	0.71	na	0.43	0.46
sorghum[b]	Change (%)	-32^{***}	-65^{***}	–48	76	104^{***}	-37^{***}	na	–15	-34^{***}

T-test for paired samples of mean yields all sorghum growers, mean yields 5% best-performing, and 95% lowest-performing: ***Significant at the 0.1% level; **at the 1% level; *at the 5% level.

[a] Based on a subsample including sorghum growers who cultivated an average area of at least 0.1 ha during the 2000–2002 and 2006–2008 periods. Yields above 3 t/ha at farm level have been excluded. Mozambique figures are from 2003–2005 and 2006–2008.

[b] Based on village aggregates.

Rice

During the last 10–15 years rice has become the most rapidly growing food source in SSA, with demand outstripping production to the extent that imports have risen from 2.5 million tonnes in 1980 to 7.2 million t in 2005 (CGIAR, 2006). During the 1985–2003 period, rice production grew by approximately 4% per annum, and in recent years (2001–2005) the annual growth rate in the region was 6%, an increase that has been driven mainly by land expansion (Somado *et al.*, 2005; CGIAR, 2006).

In the *Afrint* samples, the average area used for rice production varies greatly among the six countries in which rice is grown in the sampled villages. In Nigeria, the mean rice area has doubled from 1 to 2 ha over the 6-year period (Table 4.9). Also, in Tanzania, the area (approximately 1 ha) allocated to rice among rice growers is relatively large. Production per farm is also highest in these two countries.[8] Over the period, Ghana shows a dramatic fall in production, something which, as shown in Table 4.10, can be explained by the similarly large fall in yield (approximately 50%). In Nigeria, area expansion drives production increases, while yield improvements can be noted in Mozambique and Tanzania.

The share of rice growers benefitting from irrigation on all or part of their rice land was 10% and 15% in 2002 and 2008, respectively. As shown in Table 4.10, in the 2006–2008 period, this group reached yield levels that were 36% higher than those of upland/rain-fed rice growers.

Yield gaps

Farmers growing staple crops in SSA countries generally attain yields far below those of farmers in other developing countries. One out of numerous sets of explanations used when analysing these differences points to the more difficult agro-ecological conditions (irregular rainfall, poorer soils, etc.) characterizing many sub-Saharan farming regions. Another common explanation points to the lack of improved technology, especially improved seeds and plant materials. Such explanations are undoubtedly valid and form part of a more complex analysis of productivity development in SSA. However, placing too much emphasis on these differences leads to the risk of diverting attention from the existing potential given prevailing agro-ecological conditions and available agricultural technology. As shown in previous sections, there is wide variation in country mean yields but also among farmers within the same regions and villages. A small minority of farmers produce yields substantially above those of the majority and much closer to the yield levels typical of developing countries in Asia and Latin America.

We have defined the yield potential as the mean yield of the 5% best-performing farmers per crop and village (outliers excluded). Unlike the more

[8] The FAO country estimates for our nine sample countries show a clear increase in the rice area harvested during the period 2000–2007. On a continental level (Africa+ in the classification), area expansion during the period has been approximately 20% (FAOSTAT data, 2009).

Table 4.9. Rice production and cultivated area.[a]

		Ghana	Malawi	Mozambique	Nigeria	Tanzania	Group total
Proportion of sampled farmers	2002	42	24	13	21	46	30
growing rice (%)	2008	37	21	23	27	48	32
	Change (%)	–5	–3	10***	6*	2	2
Farm area under rice (ha),	2000–2002	0.68	0.47	0.41	1.11	1.04	0.82
three-season average	2006–2008	0.65	0.52	0.29	2.16	0.96	0.88
(2000–2002 and 2006–2008)	Change (%)	–4	10	–30*	94***	–8	8
Three-seasons' mean	2000–2002	0.52	0.81	0.16	2.25	1.55	1.15
rice production (t/farm)	2006–2008	0.27	1.05	0.24	2.69	1.79	1.21
	Change (%)	–47***	29*	49	20	16	4
Three-seasons' median	2000–2002	0.34	0.72	0.11	1.52	1.23	0.72
rice production (t/farm)	2006–2008	0.16	0.94	0.15	1.73	1.32	0.64
	Change (%)	–54	31	42	14	8	–11
No. of cases	2000–2002	159	77	30	87	160	514
Missing cases	2000–2002	1	0	1	0	0	2
Missing cases (%)	2000–2002	1	0	3	0	0	0
No. of cases	2006–2008	152	80	61	68	191	552
Missing cases	2006–2008	0	0	0	0	1.00	1
Missing cases (%)	2006–2008	0	0	0	0	1	0.2

T-test for paired samples of mean yields: ***Significant at the 0.1% level; *at the 5% level.

[a]Based on a subsample including rice growers who cultivated an average area of at least 0.1 ha during the 2000–2002 and 2006–2008 periods. The sample also excludes cases with a sixfold or higher yield increase between the two periods. Mozambique figures are from 2003–2005 and 2006–2008.

Table 4.10. Rice yields.[a]

		Ghana	Malawi	Mozambique	Nigeria	Tanzania	Group total	Partly or fully irrigated rice
Three-seasons' mean rice yield (t/ha)	2000–2002	0.99	1.91	0.41	2.32	1.53	1.49	1.94
	2006–2008	0.43	2.18	0.98	1.31	1.87	1.35	1.92
	Change (%)	–56***	14	142***	–43***	22**	–9	–1
Three-seasons' median rice yield (t/ha)	2000–2002	0.76	1.69	0.29	1.83	1.47	1.23	1.86
	2006–2008	0.30	2.18	0.63	1.30	1.55	1.00	1.31
	Change (%)	–61	29	118	–29	6	–18	–30
No. of cases	2000–2002	159	77	30	86	160	513	54
Missing cases	2000–2002	1	0	1	1	0	3	125
Missing cases (%)	2000–2002	1	0	3	1	0	1	0
No. of cases	2006–2008	152	80	61	66	190	549	89
Missing cases	2006–2008	0	0	0	2	2	4	49
Missing cases (%)	2006–2008	0	0	0	3	1	0.7	0
5% best-performing farmers' yield (t/ha) rice[b]	2000–2002	2.95	3.97	0.81	3.54	3.48	3.01	4.27
	2006–2008	1.16	4.24	3.09	1.53	3.87	2.82	3.88
	Change (%)	–61*	7	282**	–57**	11	–6	–9
95% lowest-performing farmers' yield (t/ha) rice[b]	2000–2002	0.95	1.39	0.39	1.95	1.35	1.21	1.75
	2006–2008	0.38	2.01	0.66	1.29	1.73	1.23	1.92
	Change (%)	–60***	44***	68**	–34*	28***	1	10

T-test for paired samples of mean yields all rice growers, mean yields 5% best performing and 95% lowest performing: ***Significant at the 0.1% level; **at the 1% level; *at the 5% level.

[a] Based on a subsample including rice growers who cultivated an average area of at least 0.1ha during the 2000–2002 and 2006–2008 periods. Yields above 8t/ha at farm level have been excluded. Mozambique figures are from 2003–2005 and 2006–2008.

[b] Based on village aggregates.

common use of the term, we use the concept yield gap to capture local-level conditions and realities and do not refer to the agronomic yield potential of the crop as measured under controlled conditions at experiment stations. The aggregates based on such village means are presented in Tables 4.5, 4.8 and 4.10. In Table 4.11, the summary yield gaps for maize, sorghum and rice are given for the two periods. As shown, the gaps are substantial, varying between 54% and 66%, where the percentages indicate the discrepancy between the yield potential and the average of the farmers that are not among the 5% best performing.

Intra-village agro-ecological differences (soil quality, sloping land, etc.) may form part of the explanation for the yield gaps. But as mentioned and as shown by others, crop input and management factors can be assumed to form an important part of the explanation (Jayne *et al.*, 2006). As shown for maize in Andersson *et al.* (Chapter 5, this volume), the reasons for the observed yield gaps can be related to differences in a number of economic and political conditions. These, in turn, affect farmers' access to yield-improving technologies as well as their ability to invest in surplus production under highly uncertain market conditions.

Technology adoption

The introduction and adoption of productivity-increasing technologies has been one of the key components of the intensification processes that have characterized the past decades' generally successful agricultural development in Asia and Latin America. Africa lags behind in technology adoption but is catching up and the oft-repeated description of a continent largely bypassed by new technologies seems to be increasingly outdated. Focusing in this section on the adoption of seed and planting material technology and the use of chemical fertilizer, we can show that adoption rates of improved/hybrid seeds were quite high in 2008, reaching 53% for maize and 35% for rice but only 8% for sorghum.[9] The recent breakthroughs in sorghum yields mentioned earlier may, of course, come to change the situation.

Table 4.11. Summary of yield gaps 2000–2002 and 2006–2008.

	2000–2002 Mean yield (t/ha)	2000–2002 Potential yield (t/ha)	2000–2002 Yield gap (%)	2006–2008 Mean yield (t/ha)	2006–2008 Potential yield (t/ha)	2006–2008 Yield gap (%)
Maize	1.26	3.71	66	1.08	2.91	63
Sorghum	0.71	1.54	54	0.39	1.04	63
Rice	1.21	3.01	60	1.23	2.82	57
Partly or fully irrigated rice	1.75	4.27	59	1.92	3.88	51

[9] The adoption rates of modern varieties of maize and rice in this sample point to adoption levels on par with or even higher than the adoption rates among smallholders in Asia in the early 1980s (Evenson and Gollin, 2003).

Some caution is warranted in the interpretation of the figures presented in Table 4.12, as farmers reporting on the use of improved/hybrid seed and planting material sometimes refer to re-circulated seeds, and in many cases the use of old and/or a mixture of varieties results in lower production than the use of fresh and clean seeds. On the other hand, farmers sometimes refer to improved seed and/or planting material as traditional because they have used it for several years and consequently label it traditional technology.

A related and important question is that of fertilizer use – without the application of fertilizers many of the disseminated seed technologies do not deliver their potential benefits. As depicted in Table 4.12, the use of fertilizers is common among maize growers, with an adoption rate above 40%. While the share of maize farmers applying fertilizers has remained stable, it has fallen markedly among sorghum growers, possibly contributing to the overall stagnation of sorghum production and yields discussed previously.

Commercialization and market integration

Access to productivity-raising technologies is a key factor for agricultural development, and currently the lack or insufficient use of such technologies constrains agricultural growth in SSA. Technology adoption is, however, not only limited by lack of availability and knowledge about its use. For farmers to apply such technologies there must be a commercial incentive and the risks associated with the investments must be reasonable. Farmers must be able to market their crops making a profit and there needs to be a certain level of predictability about demand and prices. Underdeveloped and ill-functioning markets constitute a major constraint for farm households seeking to improve their situation. As shown in Table 4.13, much, if not most, of what is produced on the farms never enters the markets.

Only about half of the growers of maize, cassava and rice sell some of their crop output, and for sorghum the share is approximately a quarter (28%). Looking at all types of crops, we can see that the share of completely non-commercialized crop farming has increased from 17% to 21% over the period. With the exception of maize, which will be explored in more detail in a following

Table 4.12. Percentage of farmers using seed and fertilizer, 2002 and 2008.

Seed/plant material	Maize 2002	Maize 2008	Change (%)	Sorghum 2002	Sorghum 2008	Change (%)	Rice 2002	Rice 2008	Change (%)
Traditional	47	47	–	88	92	4*	74	65	–9**
Improved /OPV	23	16	–7***	11	6	–5**	26	35	9**
Hybrid	31	36	5***	1	1	–	na	na	na
Total	100	100		100	100		100	100	
Fertilizer use	43	44	1	33	11	–22***	29	24	–5

T-test for paired samples of adoption rates: ***Significant at the 0.1% level; **at the 1% level; *at the 5% level.

Table 4.13. Percentage selling and amount marketed by type of crop.

	Maize	Cassava	Sorghum	Rice	Other food crops	Non-food crops	Any type of crop
Proportion of growers who sold the crop in 2002 (%)	43	57	41	64	60	100	83
Proportion of growers who sold the crop in 2008 (%)	48	51	28	54	76	100	79
Change (%)	5**	–6**	–13***	–10***	16***	–	–4***
Average amount sold, all growers 2002 (t)	0.97	3.58	0.69	0.83	na	na	na
Average amount sold, all growers 2008 (t)	0.81	2.52	0.49	0.95	na	na	na
Change (%)	–16	–29*	–30*	15	na	na	na
Median amount sold, all growers 2002 (t)	0.30	1.50	0.30	0.39	na	na	na
Median amount sold, all growers 2008 (t)	0.23	0.91	0.25	0.50	na	na	na
Change (%)	–25	–39	–17	28	na	na	na
Average proportion of total production sold 2002, all farmers (%)	22	na	18	32	na	na	na
Average proportion of total production sold 2008, all farmers (%)	25	na	14	28	na	na	na
Change (%)	14***	na	–19**	–11	na	na	na

T-test for paired samples of proportion selling, amount sold and mean amount sold. ***Significant at the 0.1% level; **at the 1% level; **at the 5% level.

chapter (see Andersson *et al.*, Chapter 5, this volume), the level of commercialization has fallen for all the staple crops.

The amounts sold are low, averaging 0.8t for maize, 0.5t for sorghum, 1.0t for rice and 2.5t for cassava (fresh weight) in 2008, all growers considered. Comparing means and medians, the skewed distribution of marketed output implies that, for the great majority, the amounts marketed are even more modest and in most cases limited to a few bags. For many, the sale of staples is followed by later purchases, eventually turning many sellers into net buyers.

The weakness of the staple crop markets is part of the explanation behind the growing importance of and increasing attention directed towards markets for high-value crops ('other food crops' and 'non-food crops').[10] In this respect, it is interesting to note that the proportion of growers selling such crops has increased (Table 4.13). As depicted in Table 4.14, cash income from other food crops (vegetables, beans, potatoes, etc.) and non-food cash crops (coffee,

[10] Important drivers behind the development are, of course, also an increasing demand for higher-value products, not least by the urban population.

Table 4.14. Composition of cash income, average share of different income sources in total cash income 2008, all households (%).

	Ethiopia	Ghana	Kenya	Malawi	Mozambique	Nigeria	Tanzania	Zambia	Total
1. Sale of food staple	50	28	11	18	25	23	36	23	27
2. Sale of other food crops	10	18	23	26	12	24	13	23	19
3. Sale of non-food crops	13	2	22	8	7	19	6	11	10
4. Sale of animals/animal produce	15	22	16	6	4	10	6	7	11
Farm income (1–4)	89	70	71	59	47	76	61	65	66
5. Leasing out machinery/ equipment[a]	0	0	0	2	1	0	1	1	1
6. Work on others' farms/agricultural labour	1	1	11	9	4	2	6	7	5
7. Non-farm salaried employment	3	9	9	12	6	14	9	7	8
8. Micro-business	5	8	3	13	27	6	19	11	12
9. Large-scale business	0	1	0	0	0	1	0	0	0
10. Rent interests	0	0	0	0	0	0	1	0	0
11. Pensions	0	1	1	0	3	0	0	1	1
12. Remittances[b]	1	10	5	5	11	1	3	7	6
Non-farm income (5–12)	11	30	29	41	53	24	39	35	34

[a] Including oxen, push carters, etc.
[b] From absent household members, children, etc.

tobacco, tea, sugarcane, etc.) indeed make up important sources of cash income (on average 19% and 10% respectively), often generating immediate contributions throughout the year for cash-strapped households. It is, however, also important to note that in five of the eight countries the sale of staple crops forms the single most important source of farm cash income. In this respect, the two extremes in the sample are Kenya and Ethiopia. In Kenya the sale of staples only constitutes 11% of cash income, while in Ethiopia 50% of household cash income is generated through the sale of staple crops.

Off-farm linkages

During the past decade there has been an important debate around the importance of non-farm incomes in SSA. Coining the term 'de-agrarianization' to describe a process of rapid livelihood diversification, Bryceson claims that African smallholders have become much less dependent on farm income and that non-agricultural activities may account for as much as 60–80% of household income (Bryceson, 2002:730). Our data on cash income sources for 2008 do not support such claims, however (Table 4.14). Rather, at the level of 34% of total cash income, our figure is more in accord with other estimates pointing to a non-farm share of around 35% (Reardon *et al.*, 2007:117). Jayne *et al.* report non-farm income shares for four of our sampled countries (Jayne *et al.*, 2006:4). While in the case of Kenya (30.5%), Ethiopia (8.1%) and Zambia (28.5%) their findings tally reasonably well with those from our sites, the Mozambique figure (27.3%) deviates markedly, which is why a comment on the relatively high share of non-farm income in the Mozambique sample is required. The selection of villages may have introduced a bias in the sample in terms of non-farm incomes. Among the sampled farm households a high proportion of those residing in the southern part of the country have male members who migrate to South Africa or Maputo, generating important non-farm household incomes. The sampled villages in the centre of the country are situated close to the Sofala–Manica road, which is why many household members are both farming and engaged in commerce. In the north, in the Nampula and Zambézia provinces, farm income dominates (Coughlin and Givá, 2009).

Cash income, of course, only forms part of total household income for households retaining part of their agricultural output for their own consumption, payment for hired labour, seeds, etc. If this part of the production was to be valued at the price of marketed output, the overall share of farm income would increase markedly. In our samples, non-cash incomes can be expected to be of major importance, particularly for the poorer segments retaining a relatively higher share of the output. On the whole, therefore, we can conclude that our findings do not provide evidence of a situation of 'de-agrarianization' in our study areas.

The proportion of households engaged in non-farm activities does not seem to have changed much in the period. Approximately half of the surveyed households in both 2002 and 2008 did not report any income from non-farm sources (Table 4.15). This finding adds to the impression that, although

Table 4.15. Share of households having different sources of non-farm income, all households (%), 2002 and 2008.[a]

Non-farm income source	2002	2008	Change (%)
Leasing out machinery and or equipment[b]	na	3	na
Work on others' farms/agr. labour	na	15	na
Non-farm salaried employment	19	17	–2
Micro-business	34	29	–5***
Large-scale business	1	1	–
Rent interests	3	1	–2***
Pensions	2	2	–
Remittances[c]	16	17	1
Any non-farm income source	49	53	4***

T-test for paired samples of change in proportion having a certain non-farm income source. ***Significant at the 0.1% level.

[a] A household can have more than one source of income.

[b] Including oxen, push carters, etc.

[c] From absent household members, children, etc.

non-farm activities continue to be of great importance to many African smallholders, the mainstay of the household economy continues to depend on income generated on the farms.

Production and income according to land endowments and gender

Earlier sections have pointed at variation in the sample in terms of access to land (Table 4.2) and also in terms of area productivity/yield gaps (Table 4.11). This section will explore these differences further by focusing on different groups of the sample, being defined by their access to land and the sex of the farm manager of the household.

Inter-household heterogeneity in terms of asset endowments is commonly referred to as an important explanation of differences in production, productivity and income diversification (Barret *et al.*, 2005; World Bank, 2007). Poverty prevents farmers from investing in new technologies and entering markets characterized by high price fluctuations and high transaction costs. In the SSA context, in particular, women farmers face a number of gender-specific challenges and constraints relative to male farmers. In the following sections data on production, productivity and income sources are presented for households, divided into five groups based on the ranking, at the village level, of the household per capita land size. A division is also made according to the sex of the farm managers in the household. The distribution of the sample according gender is given in Table 4.16. The sex of the farm manager as reported by the respondents is, in the vast majority of cases, equivalent to that of the household head.

Using farm size as an indicator of wealth or economic resources is problematic, not least in the case of many SSA countries, where access to land may

Table 4.16. Gender (farm manager) distribution of sampled households.

Sex of farm manager	2002 (%)	2008 (%)
Male	78.6	77.3
Female	21.4	22.7
Total	100	100

be a relatively less constraining factor than, for example, access to capital or credit and thereby to different production inputs. In both the 2002 and 2008 surveys an effort was made to rank households according to wealth based on a qualitative and subjective assessment made by the interviewers. The reliability of this wealth indicator turned out to be weak. Lacking complete data on total household income as well data on household expenditure, we are confined to using land endowments as a proxy, although we realize that, at best, it can only provide a rather crude picture of intra-village disparities.

The results, as depicted in Table 4.17, illustrate the harsh realities facing the great majority of households. Comparing the bottom and top groups, we can see that the production per consumption unit (PCU)[11] of all staple crops for which we have collected data is consistently higher for the better endowed. The level of PCU was between 32% and 40% higher for sorghum and rice in the top quartile compared to the most land-constrained group (first quartile), and in the case of maize the difference, 200% in 2008, was even more pronounced.

Differences in yields are small between the quartiles, and the fall in yields affecting all three crops seems to have affected the more, as well as the less, land-endowed households, to approximately the same extent in the case of maize. For sorghum the top 10% biggest landholders experienced a dramatic fall in yields, whereas in the case of rice, the bottom quartile seems to have fared relatively worse.

Turning to the differences between male- and female-managed farm households in terms of staple crop production, Table 4.17 shows that total farm output on female-managed farms, which generally have lower access to land and family labour, is markedly lower. The PCU level for all three cereal crops was higher in male-managed households than in female-managed ones in 2008. Yield differences are not as pronounced as production differences but are consistently somewhat higher (4–12%) in male-managed households.

Differences in production between household categories divided according to land endowment and gender are also reflected in the patterns of income diversification, as presented in Table 4.18. There is an overlap between the

[11] Following Sukhatme (1970) we have assumed that 220 kg of grain equivalents per person (consumption unit) and year is the approximate minimum food and calorie intake required to keep a person alive, corresponding to 2200 kCal or 600 g of grain per day. In calculating grain equivalents, the weight used for paddy was 0.8. Consumption units: adults (15–60 years) 1.0; children (< 15 years) 0.5; old (>60 years) 0.75.

Table 4.17. Production and productivity by wealth and gender.

	Means for households quartiles[a] ranked by per capita farm size (ha) by village					Gender		
	Q1 0–25	Q2 25–50	Q3 50–75	75–90	Top 10 90–100	Male-managed farms	Female-managed farms	Total sample
Mean farm size 2002 (ha), (n = 3037)	0.92	1.62	2.32	3.60	5.78	2.59	1.54	2.41
Mean farm size 2008 (ha), (n = 2869)	0.81	1.46	2.20	3.25	5.43	2.44	1.26	2.16
Mean farm size per capita 2002 (ha), (n = 2547)	0.13	0.23	0.34	0.51	0.95	0.38	0.31	0.37
Mean farm size per capita 2008 (ha), (n = 2604)	0.12	0.21	0.34	0.53	1.06	0.38	0.29	0.36
Production (t), PCU (kg) and yield (t/ha)[b]								
Average maize production/farm 2000–2002 (t), (n = 1901)	0.74	1.22	1.22	1.28	1.92	1.30	0.87	1.22
Average maize production/farm 2006–2008 (t), (n = 2241)	0.81	0.99	1.24	1.64	2.06	1.43	0.73	1.24
Change (%)	9	–19	1	28*	7	10	–16	1
Average maize PCU 2000–2002 (kg), (n = 1901)	117	191	230	275	542	239	270	245
Average maize PCU 2006–2008 (kg), (n = 2005)	132	161	238	355	537	269	196	251
Change(%)	13	–15	3	29**	–1	13*	–27	2
Average maize yield 2000–2002 (t/ha), (n = 1888)	1.41	1.46	1.43	1.31	1.37	1.42	1.35	1.41
Average maize yield 2006–2008 (t/ha), (n = 2004)	1.23	1.21	1.15	1.22	1.12	1.22	1.10	1.19
Change (%)	–13*	–17***	–20***	–7	–19**	–14***	–18***	–15***

Average sorghum production/farm 2000–2002 (t), (n = 553)	0.56	0.84	1.18	1.06	1.17	1.00	0.59	0.96
Average sorghum production/farm 2006–2008 (t), (n = 538)	0.33	0.42	0.42	0.58	0.73	0.49	0.27	0.46
Change (%)	–42*	–50***	–64*	–45**	–37	–51***	–55	–52***
Average sorghum PCU 2000–2002 (kg), (n = 553)	71	114	191	159	251	157	116	153
Average sorghum PCU 2006–2008 (kg), (n = 538)	43	63	75	107	182	83	70	81
Change (%)	–40***	–45***	–60	–32*	–27	–48***	–39	–47***
Average sorghum yield 2000–2002 (t/ha), (n = 539)	0.71	0.82	0.68	0.64	0.84	0.74	0.69	0.73
Average sorghum yield 2006–2008 (t/ha), (n = 537)	0.56	0.60	0.50	0.53	0.38	0.55	0.43	0.53
Change (%)	–20	–27**	–26**	–17	–55***	–25***	–38**	–27***
Average rice production per farm 2000–2002 (t), (n = 446)	0.79	0.85	1.24	1.48	1.07	1.11	0.94	1.08
Average rice production per farm 2006–2008 (t), (n = 420)	0.64	1.04	0.90	1.15	1.39	1.08	0.59	1.00
Change (%)	–18	23	–27*	–22	30	–2	–37	–8
Average rice PCU 2000–2002 (kg), (n = 446)	172	157	248	315	303	229	240	230
Average rice PCU 2006–2008 (kg), (n = 420)	96	162	168	234	352	197	155	189
Change (%)	–44*	3	–32*	–26	16	–14	–36*	–18*
Average rice yield 2000–2002 (t/ha), (n = 445)	1.39	1.40	1.59	1.50	1.10	1.45	1.31	1.43
Average rice yield 2006–2008 (t/ha), (n = 419)	1.02	1.32	1.20	1.23	1.09	1.20	1.13	1.19
Change (%)	–27*	–6	–24*	–18	–1	–17**	–13	–17**

T-test for paired samples of proportion selling, amount sold and mean amount sold. ***Significant at the 0.1% level; **at the 1% level; *at the 5% level.
[a]The fourth quartile has been divided into two groups (75–90% and 90–100%) in order to highlight the characteristics of the 'elite' group.
[b]Production and productivity figures for the total sample differ in comparison with the figures presented in Tables 4.4–4.10. The discrepancy is due to the difference in populations, which, in turn, is due to the number of missing cases related to variable household size used to calculate the per capita farm size categories used in this table.

two categories, as female-managed farms, on average, are only approximately half the size of male-managed farms (Table 4.17). Table 4.18 shows that smaller farms and female-managed farms are relatively less dependent on farm activities for their cash income and, correspondingly, more dependent on non-farm income sources to obtain cash. Not surprisingly, in comparison with the top land size group, the bottom quartile derives approximately 20% more of their total cash income from non-farm activities. Especially important in relative terms for this category are agricultural wage labour income, micro-business and non-farm salaried employment. It is noteworthy that, in this group of small farm households, income from the sale of non-staple food crops is relatively less important than for the larger farms. Growing high-value crops such as vegetables and other cash crops seems, in other words, to be relatively more important for bigger farms.

As could be anticipated, female-managed farms receive more remittances than male-managed ones and are also generating relatively more cash income

Table 4.18. Composition of cash income by wealth and gender – average share of different income sources in total cash income 2008 (%).

	Means for households quartiles[a] ranked by per capita farm size (ha) by village					Gender		Total sample
	Q1 0–25	Q2 25–50	Q3 50–75	75–90	Top 10 90–100	Male	Female	
Number of cases	496	638	618	383	239	2013	597	2610
1. Sale of food staple crops	24	29	29	29	32	32	23	28
2. Sale of other food crops	19	20	21	23	22	22	20	21
3. Sale of non-food crops	7	10	12	13	17	13	8	11
4. Sale of animals/animal produce	12	11	13	12	15	14	9	12
Farm income (1–4)	57	66	70	72	79	71	54	67
5. Leasing out machinery/ equipment[b]	1	1	1	1	0	1	1	1
6. Work on others' farms/ agricultural labour	10	6	5	3	3	6	9	6
7. Non-farm salaried employment	12	10	7	10	6	10	11	9
8. Micro-business	16	13	13	11	9	14	15	13
9. Large-scale business	0	0	0	0	1	0	0	0
10. Rent interests	0	0	0	0	1	0	0	0
11. Pensions	1	1	0	2	1	1	1	1
12. Remittances[c]	7	7	8	5	6	5	14	7
Non-farm income (5–12)	43	34	30	28	21	29	46	33

[a]The fourth quartile has been divided into two groups (75–90% and 90–100%) in order to highlight the characteristics of the 'elite' group.
[b]Including oxen, push carters, etc.
[c]From absent household members, children, etc.

from agricultural wage labour, although the overall importance of this source of income is quite low for the entire sample.

Having identified quite large differences in the composition of household cash income among different categories of household, the previous conclusion on the relatively low importance of non-farm income sources seems, nevertheless, to hold even for the smallest farm households. Although non-farm income sources constitute some 43% of total cash income for the most land-restricted, total income includes retained production, suggesting that this is still the most important source of household income, even for this group.

With a non-farm cash income share of 21–43% in the five land size groups – implying a much lower share of total household income, as the latter includes retained agricultural production – even the very land-scarce quartile continues to depend on agriculture as the clearly most important source of household income.

Conclusion

The general picture emerging from the findings presented is one of a continued crisis in the smallholder sector, characterized by low levels of output per farm, low area productivity and a high degree of subsistence farming. Changes during the period 2002–2008 have not been substantial in the areas under study, and although variation between countries and within regions have been shown to be great, they do not support an overall impression of an agricultural growth process on a par with the past decade's positive GDP growth rate, which, on average, has surpassed 5%. While increasing productivity in the agricultural sector is becoming a goal for national and international organizations trying to promote growth in the region's agricultural sectors, efforts taken to achieve such change have not yet had any clear impact on area productivity in the village areas covered in this study. The historical pattern according to which output growth, by and large, has been driven, explained by extensification strategies, does not seem to have been reversed during the past decade in the approximately 100 villages studied.

For the staple crops we identify a mixed picture. Maize – the most important crop in SSA and in our sample – has done well in terms of increasing farm production in Malawi and Zambia. Cassava farm areas have expanded significantly in Nigeria, while for sorghum the general trend seems to be one of declining production. In the case of rice, Ghana is the only country showing a negative trend in production, a fall explained by lower yield levels in the 2006–2008 period. Although Nigeria also experienced falling area productivity, an almost doubling of the farm area under rice among rice growers compensated for this. In Mozambique, yield levels increased dramatically, as did the share of the sampled households growing rice. Also, for Tanzania, a statistically significant positive growth (22%) in yields was registered.

There continue to be wide gaps in productivity, as measured by village-level yield gaps between a small minority of high-performing farms and the majority of farm households. Ranging between 54% and 66% for the different staple crops, this suggests that there continues to be an untapped potential in the SSA staple crop sector. Furthermore, the 2008 data confirm the findings of 2002, pointing to a relatively broad use of modern varieties of seeds and planting materials for all crops but sorghum. In the case of maize and rice, adoption rates have increased, whereas they have fallen sharply for sorghum. The use of fertilizer remains common for maize but has decreased significantly for sorghum.

The only one of the four staple crops having experienced a positive change in terms of increased commercialization is maize. On the whole, both the absolute and the relative level of commercialization is low, with only about half of maize, rice and cassava farmers selling any amount of their crops and only about a fourth in the case of sorghum. The median volume sold ranges between 200 and 500 kg for the grain crops and 900 kg (wet weight) for cassava. The proportion of farm households who do not market any of their crop output has increased since 2002 and by now amounts to 21%.

The low level of commercialization in the farm economy does not seem to be compensated through any dynamism in the non-farm economy share of the household economy. The share of farm households lacking any such sources of income remains at approximately 50% – quite low. Calculated as a share of total cash income, non-farm income accounts, on average, for 34%. Total household income, however, includes the value of agricultural output retained. Thus, the actual share of non-farm income in total income is clearly lower than 34%. On the whole, our findings do not concur with the notion of an ongoing process of 'de-agrarianization'.

This chapter has also pointed at the heterogeneity in the smallholder sector in terms of access to land and income composition for different groups. The per capita access to land is very small in absolute number, 0.12 ha per capita or less for the 25% smallest farms in all countries but Nigeria and Ethiopia. In Kenya, the per capita farm size was 0.04 ha in 2008. Female-managed households and smaller farms (often overlapping categories) are relatively more dependent on non-farm sources of income, but although differences in these respects can be clearly distinguished, the general conclusion is that even the categories being the most dependent on non-farm cash income source remain, by far, more dependent on farm income sources for their total household income when taking non-marketed production into consideration.

While several glimpses of dynamism were detected in the analysis of the survey data, by and large, the situation facing the great majority of smallholders calls for major changes, including investments in the sector and also more smallholder-friendly policies, creating better incentives for technology adoption and market integration. To the extent that any change in this direction has been set in motion by the new comprehensive initiatives under NEPAD and AGRA, it had not, by 2008, resulted in any marked changes recorded by this study. The smallholders of sub-Saharan Africa are desperately waiting for change.

References

AGRA (2009) *Breakthrough in Sorghum's Yield Barrier (Mali). Alliance from a Green Revolution in Africa*. Available at http://www.agra-alliance.org/content/story/detail/1042 (accessed 4 April 2010).

Ashley, C. and Maxwell, S. (eds) (2001) Overview rethinking rural development. *Development Policy Review* 19, 395–425.

Barret, C.B., Bezuneh, M., Clay D.C. and Reardon, T. (2005) Heterogeneous constraints, incentives and income diversification strategies in rural Africa. *Quarterly Journal of International Agriculture* 44, 37–60.

Binswanger-Mkhize, H.P. (2009) Challenges and opportunities for African agriculture and food security: high food prices, climate change, population growth and HIV and AIDS. *Proceedings of the Expert Meeting on How to Feed the World in 2050*. FAO Headquarters, Rome.

Bryceson, D. (2002) The scramble in Africa: reorienting rural livelihoods. *World Development* 30, 725–739.

CGIAR (2006) *Transforming Sub-Saharan Africa's Rice Production through Rice Research*. Consultative Group on International Agricultural Research, Story of the Month: September 2006. Available at: http://www.cgiar.org/monthlystory/september2006.html (accessed 4 April 2010).

Collier, P. (2008) Food shortages: think big. Available at http://www.timesonline.co.uk/tol/comment/columnists/guest_contributors/article3746593.ece (accessed 15 April 2008).

Collier, P. and Dercon, S. (2009) African agriculture in 50 years: smallholders in a rapidly changing world. *Proceedings of the Expert Meeting on How to Feed the World in 2050*. FAO Headquarters, Rome.

Coughlin, P.E. and Givá, N. (2009) *Agricultural Intensification in Mozambique: Lessons from Ten Villages, AFRINT 2*. EconPolicy Research Group, Ltd, Maputo, Mozambique.

Djurfeldt, G., Larsson, R., Holmquist, B., Jirström, M. and Andersson, A. (2008) African farm dynamics and the sub-continental food crisis – the case of maize. *Food Economics, Acta Agriculturae Scandanavica C* 5, 75–91.

Ellis, F. (2005) *Small Farms, Livelihood Diversification, and Rural-Urban Transitions: Strategic Issues in Sub-Saharan Africa, the Future of Small Farms*. Proceedings of a Research Workshop, Wye, UK. Available at http://www.ifpri.org/events/seminars/2005/smallfarms (accessed 4 April 2010).

Evenson, R. and Gollin, D. (eds) (2003) *Crop Variety Improvements and its Effects on Productivity: the Impact of International Agricultural Research*. CAB International, Wallingford, UK.

FAO (2009) The special challenges for sub-Saharan Africa. Issues briefs for the Expert Meeting on How to Feed the World in 2050. FAO Headquarters, Rome.

FAOSTAT data (2009) Food and Agriculture Organization of United Nations. Available at: http://faostat.fao.org/site/567/default.aspx#ancor (accessed 4 April 2010).

Haggblade, S. and Nyembe, M. (2007). *Commercial Dynamics in Zambia's Cassava Value Chain, Cassava Transformation in Southern Africa (CATISA), Startup Task 3*. Report on Zambia's Cassava Value Chain, Department of Agricultural Economics, Michigan State University in its series International Development Collaborative Working Papers with number ZM-FSRP-WP-32, East Lansing, Michigan.

Hazell, P. (2006) The role of agriculture in pro-poor growth in sub-Saharan Africa. In: Hårsmar, M. (ed.) *Workshop Proceedings from the Workshop Agricultural Development in Sub-Saharan Africa*. Expert Group on Development Issues (EGDI), Ministry for Foreign Affairs, Sweden, pp.17–35.

IFAD (2001) *Rural Poverty Report 2001 – The Challenge of Ending Rural Poverty*. International Fund for Agricultural Development, Rome. Available at http://www.ifad.org/poverty/ (accessed 2 April 2010).

Jayne, T.S., Mather, D. and Mghenyi, E. (2006) *Smallholder Farming under Increasingly Difficult Circumstances: Policy and Public Investment Priorities for Africa*. MSU International Development Working Paper 86. Department of Agricultural Economics, Michigan State University, East Lansing, Michigan.

Larsson, R. (2005) Crisis and potential in smallholder food production – evidence from micro level. In: Djurfeldt, G., Holmén, H., Jirström, M. and Larsson, R. (eds) *The African Food Crisis: Lessons from the Asian Green Revolution*. CAB International, Wallingford, UK, pp. 113–137.

Lipton, M. (1989) *New Seeds and Poor People*. Unwin Hyman, London.

Lipton, M. (2005) *The Family Farm in a Globalizing World – The Role of Crop Science in Alleviating Poverty*. 2020 Discussion Paper 40. International Food Policy Institute, Washington, DC. Available at: http://www.ifpri.org/2020/dp/vp40.pdf (accessed 4 April 2010).

Reardon, T., Berdegué, J., Barret, C.B. and Stamoulis, K. (2007) Household income diversification into rural nonfarm activities. In: Haggblade, S., Hazell, P.B.R. and Reardon, T. (eds) *Transforming the Rural Nonfarm Economy: Opportunities and Threats in the Developing World*. International Food Policy Research Institute, Washington, DC and The Johns Hopkins University Press, Baltimore, Maryland.

Somado, E.A., Guei, R.G. and Nguyen, N. (2005) Overview: rice in Africa. In: Somado, E.A., Guei, R.G. and Keya, S.O. (eds) *NERICA: the New Rice for Africa – a Compendium*. Africa Rice Center (WARDA), Cotonou, Benin.

Sukhatme, P.V. (1970) Incidence of protein deficiency in relation to different diets in India. *British Journal of Nutrition* 24, 447–487.

Timmer, P. (2005). *Agriculture and Pro-Poor Growth: an Asian Perspective*. Working Paper Number 63, July 2005, Center for Global Development, Washington, DC.

Wiggins, S. (2009) Can the smallholder model deliver poverty reduction and food security for a rapidly growing population in Africa? *Proceedings of the Expert Meeting on How to Feed the World in 2050*. FAO Headquarters, Rome.

World Bank (2007) *World Development Report 2008: Agriculture for Development*. The World Bank, Washington, DC.

5 A New Era for Sub-Saharan African Agriculture? Changing Drivers of Maize Production[1]

Agnes Andersson,[1] Göran Djurfeldt,[2] Björn Holmquist,[3] Magnus Jirström[1] and Sultana Nasrin[3]

[1]*Department of Human Geography, Lund University, Lund, Sweden;* [2]*Department of Sociology, Lund University, Lund, Sweden;* [3]*Department of Statistics, Lund University, Lund, Sweden*

Since the turn of the millennium the African policy environment has shifted with respect to agriculture in general and more specifically in relation to the smallholders that constitute the majority of farmers on the subcontinent. The emergence of the New Economic Partnership for African Development (NEPAD) and its Comprehensive Africa Agricultural Development Programme (CAADP) are tangible and promising outcomes of such policy changes. The global interest in smallholder futures, moreover, has received growing attention through the *World Development Report 2008* (World Bank, 2007). Growing anxiety over global warming, coupled with rising food security concerns in the more populous countries of the world, has directed attention towards agricultural land reserves in Africa as sources of both biofuel production and food. Recent rises in food prices reinserted the food security issue at the top of the global agenda with a great deal of urgency in early 2008. The world food price crisis, as it was labelled at the time, has since abated but has to some extent reconfigured the global markets for staple crops, with national food self-sufficiency re-emerging as a political objective in many countries both within and outside Africa. The long-term effect of the world food price crisis appears to be a higher level of global food prices.

Meanwhile, the post-millennial period has, until recently, been characterized by rapid economic growth in a number of African countries. Since late 2008, what is believed to be the worst global financial crisis since the Depression of the 1930s has altered the growth prospects of the continent radically, however.

[1] Thank you to Robina Ang for sharp-eyed observations of technical details in relation to modelling.

The strong growth witnessed in many African countries since the early years of the millennium has, in some cases, given way to recession. The domestic ability to feed not only the smallholder population but also urban consumers becomes even more vital in a situation where expensive imports can hardly compensate for shortfalls in national production.

The political motives for ensuring food sufficiency are thus increasingly shaped by global processes outside the control of national policy makers, while the necessity of encouraging an African Green Revolution is increasing by the day, partly as a result of such changes. In this context, the historical lessons from the Asian experience of agricultural transformation constitute central reference points in guiding and evaluating the African experience. The narrative of the Asian Green Revolution as a state-driven, market-mediated and smallholder-based development with scientific–industrial technology as a necessary but not sufficient condition for growth is especially pertinent.

Against this backdrop, the purpose of the present chapter is to analyse and discuss the drivers behind changes in staple food production. The role of three key processes, namely commercial drivers, farm technology and the agrarian policies of the state, will be evaluated and discussed on the basis of data on maize for the period 2002–2008. This is done on the basis of a model of production and changes in production, which draws on data from a panel of 1805 maize-growing smallholder households in eight African countries.

Theoretical Overview and Previous Research

Following decades of neglect, smallholder-based agriculture has for the past few years been promoted as the foundation for a broad-based development effort in the regional context of sub-Saharan Africa. Evidenced by a range of national, regional and global initiatives, such strategies have focused on promoting access to technology and inputs aimed at raising productivity within the smallholder sector. The most publicized case of such recent initiatives at the national level is probably Malawi's Agricultural Input Support Programme, which to some extent has revived the pre-structural adjustment programme (SAP) focus on widespread fertilizer and seed subsidies. In some cases, renewed interest in smallholder fortunes has also translated into policies geared towards enhancing commercial incentives on the demand side. On the whole, however, food markets characterized by uncertainty, depressed prices, atomism and prohibitive transaction costs are identified as major causes of farmer reluctance regarding input adoption and, by extension, failure to improve productivity and food security (Jayne *et al*., 2006; Poulton and Dorward, 2008). The considerable variability in production levels that exists among smallholders, even within the same villages, underscores the crucial role of farm inputs as a source of yield differentials, with the use of chemical fertilizer and improved seed specifically being an important explanation of such discrepancies (see Chapter 4, this volume; see also Sanchez *et al*., 1997; Holmén, 2005a,b).

The prospects of agriculture relieving the food security situation in sub-Saharan Africa and the presumed importance of agriculture within development in general is therefore increasingly centred on the interaction of markets with technological advances in stimulating agricultural development and promoting food security and poverty reduction (Crawford *et al.*, 2003; Barrett, 2008).

Despite recent political efforts to stimulate either or both aspects of the smallholder production balance, the role of the state in general and more specifically the slant of the agrarian policies it chooses to pursue is debated among academics and policy makers alike. Although the long-standing debate on African agriculture appears to have closed, at least temporarily, in favour of the African smallholders, the policy prescriptions unfolding from this realization vary widely (Lipton, 2005; Haggblade *et al.*, 2007; World Bank, 2007).

The role of the staple crop sector, for instance, is to some extent still debated. Proponents of African smallholder-based agriculture point to the historically positive relationship between productivity increases within the smallholder staple crop sector and broader economic growth (Tiffen, 2003). A substantial increase in the productivity of staple food agriculture over time enables investments in more diversified production, including high-value crops, and in economic activities outside the farm (Lipton, 2005; Haggblade *et al.*, 2007). While productive off-farm incomes tend to benefit the already well off, increased farm incomes, especially within the staple crop sector, accrue largely to the poorer segments of the economy (Haggblade *et al.*, 2007).

Arguably, the state and the development community at large, despite recent efforts pointing in an agriculture friendly direction have a lot of catching up to do. Indeed, political commitments made in Maputo in 2003 to devote 10% of public expenditure to agriculture, should be seen against a backdrop of falling agricultural spending from 7% in 1980 to 4% in 2004 (World Bank, 2007).

The historical role of the state as a provider of both input and output markets has, in practice, been challenged by the experience of structural adjustment. None the less, arguments related to the potentially vital role of the state as a substitute for private markets continue to be advanced in the academic literature. State involvement may in certain contexts be justified to counter prohibitive transactions costs and lacking economies of scale, which in turn create disincentives to private trade (Dorward *et al.*, 2004; Dyer, 2004). Some commentators, moreover, suggest that the dismantlement of public procurement systems under structural adjustment has impacted negatively on smallholders' access to staple crop markets, despite the notoriously inefficient operation of such organizations (Holmén, 2005b).

Arguments which have sought to resurrect the state as an active player in smallholder-based agrarian policy have recently turned into policy practice in a number of African countries through state intervention in input markets, for instance in Malawi, Rwanda and Zambia. In other cases, the partial revival of state-run marketing boards has been experimented with, for instance in Zambia.

The following chapter will consider the relative importance and interaction of the three drivers outlined above, namely technological advances, commercialization and state involvement as explanations of smallholder production dynamics in the maize sector. Data from a panel of 1805 farm households from eight African countries: Ethiopia, Ghana, Kenya, Malawi, Mozambique, Nigeria, Tanzania and Zambia will be used to substantiate the discussion.

Data Collection and Modelling Strategy

The panel data constitutes a subset of a larger household and village cross-sectional sample consisting of roughly 4000 households for two survey rounds, carried out in 2002 and 2008 respectively. The panel of maize growers comprises of 1805 households interviewed in both 2002 and 2008 and for whom retrospective data for the year of household formation (what is referred to as the reference year) is also available. Household-level data is complemented by village- as well as country-level data. The data collection and sampling strategies have been detailed in the Introduction to this volume.

Despite the constraints of the survey methodology identified in the Introduction to this volume, the data can be used as a basis for indicating structural changes at the local level. In this sense they constitute a reliable gauge of processes and changes in farmer behaviour, which in turn can be used to draw comparisons across the set of countries as well as identify changes over time.

In general, longitudinal panel data on production patterns, income sources and income diversification are exceptional in the African context and, as such, the data present a rare opportunity for analysing changes in production over time.

The modelling strategy departs from the overarching purpose of the paper – i.e. to capture the drivers of production changes, while it is adapted to the multi-level and longitudinal data used. Both countries and villages have been sampled purposively (in the case of villages within purposively sampled regions), so that the most advantageous treatment of villages and countries in the context of our modelling strategy is as random effects.[2]

The retrospective nature of many questions related to the reference year (t_0) and the focus on attaining robust rather than detailed but less reliable information has consequences for data structure and the way that data can be treated and analysed. In contrast to most longitudinal data analysis, which presumes that data refer to points in time and consists of scale variables, our data refers to time periods (t_0 to t_1 (p_1), t_1 to t_2 (p_2) and t_0 to t_2 ($p_1 + p_2$) respectively) and mainly to ordinal scale variables relating the difference between two points in time.

The data, hence, is typically ordinal with respect to time; for example, was production higher, lower or the same in the reference year compared to currently? An ordinal data structure therefore departs from a variable, reflecting the situation in the reference year (t_0). On the basis of this, three dummies are

[2] For more details on fixed and random models consult any textbook, e.g. Frees's lucid treatment Frees, E.W. (2004) *Longitudinal and Panel Data: Analysis and Applications in the Social Sciences*. Cambridge University Press, Cambridge, UK.

used to describe each time period, for example: (i) used fertilizer at the beginning of the period (t_0); (ii) used less or no fertilizer at t_1; and (iii) used more fertilizer at t_1.

In matrix notation, a regression equation can be written as:

$$y_i = \beta_i \mathbf{X}_i + u_i \qquad (5.1)^3$$

where y_i is a vector of the response variable, $\mathbf{X}_i$ is a matrix of explanatory variables and u_i is vector of the residuals. The intercept, which in the traditional form is known as α, is here β_0. A well-formulated regression model should fulfil two requirements: the residuals should be approximately zero mean normal with same variance and should be uncorrelated with the explanatory variables in $\mathbf{X}$.[4]

These requirements often create problems in social science applications. The complexity of phenomena may create an *omitted variable bias*, in which the model fails to include relevant variables in **X**, including their effects instead in the residual (u), distorting the distribution of the latter. In our case such unobservable characteristics would include factors like farmer skills and farm characteristics or the agro-ecological potential of the farm. Since we lack indicators for these variables we cannot directly estimate their influence.

Furthermore, although the advantages of panel data in terms of handling endogeneity are obvious when compared with cross-sectional data (Hildebrand, 1960; quoted in Mundlak, 2001), aspects of endogeneity still need to be considered. The interactive effects between the dependent and independent variables may occur at different intervals from when the data has been collected, potentially causing endogeneity. An example in this context would be production decisions based on changes in short-term price incentives, which may shift several times during the period covered by the panel.

Given the characteristics of the data described above, an ideal strategy would be to model a dependent variable y for the period p_2 as a function of a vector **X** for the period p_1, since causal attribution from **X** to y in this case would appear unproblematic, given the time lag:

$$y_{p2} = \beta \mathbf{X}_{p1} + u \qquad (5.2)$$

In this case, all the independent variables in **X** would be exogenously determined, but the possibility of an omitted variable bias in the residual still remains, implying a potential bias also in the estimation of the regression coefficients (β). Using the instrumental variable approach to deal with endogeneity does so at the cost of making tenuous assumptions about the causal relations between the

[3] While in traditional notation and for two independent variables it is written: $y_i = \alpha + \beta_1 x_1 + \beta_2 x_2 + u_i$.

[4] If it is not we have a case of endogeneity, as the term is defined in statistics. As Frees has pointed out, this definition deviates from that of economists, who use the term in another sense. Frees, E.W. (2004) *Longitudinal and Panel Data: Analysis and Applications in the Social Sciences*. Cambridge University Press, Cambridge, UK. Here we use the statistical definition.

instrumental and dependent variables. Given the complexity of smallholder decision making, and the difficulties in establishing clear-cut causality in many real-life situations, we attempt to control for, rather than eliminate, endogeneity in the model, using an approach inspired by Hausman and others (Hausman, 1976).

In developing the formulas to follow, we first break down the residual (u_i) into components, consisting of random factors (u) and *latent* variables (λ).[5]

We can model any household property (P) with time-dependent data as follows:

$$P_{it} = \theta(\mathbf{X}_i, A_{it}, (\lambda_{it}, u_{it})) \tag{5.3}$$

P_{it} can stand for any property of the individual household i at time t, in our case production of maize. The symbol θ denotes the functional form of the relationship, e.g. log–log. $\mathbf{X}_i$ is a vector of exogenous variables. A_{it} is a vector containing two variables: (i) *household age* at time t, broken out of $\mathbf{X}$ because the model will be adapted to panel data; and (ii) *descendant household*, indicating if the household has been partitioned during the period 2002–2008 (p_2). The effects of partitioning hence are controlled for in the modelling.

Bracketed at the end of Eqn 5.3 are the unobserved variables, i.e. latent ones (λ) and the residual (u). The symbol λ_{it} stands for a vector of *unobservable or latent characteristics of the individual household, or of village- or country-level variables*. Finally, u_{it} is a residual of random factors.

The latent variables (λ_{it}) can be classified into three groups: (i) those that are constant over time (λ_{1i}); (ii) those that are time or age dependent (λ_{2it}); and (iii) those that are variable over time but not dependent on age (λ_{3it}). The aggregate effects of time-dependent latent variables are captured through the age variable, whereas the other two types of variables in an ordinary regression are impossible to distinguish from the residual. As suggested above, the endogeneity aspects of such latent variables will be dealt with through an indirect technique.

The use of panel data enables estimation of the determinants of the difference in production between 2008 and 2002. The use of a reduced form model, inspired by Glewwe and Hall (1998), in this case enables us to identify the drivers behind changes in production through the combination of two separate models for the two panel rounds, for 2002 and 2008 respectively, and a model for the change in production between 2002 and 2008 (i.e. the reduced form).[6]

$$\ln(P_{i02}) = \beta_{c02} + \mathbf{X}_i\,\beta_{02} + \alpha_{c02}A_{i02} + \lambda_{1i} + \lambda_{3i02} + u_{i02} \tag{5.4}$$

$$\ln(P_{i08}) = \beta_{c08} + \mathbf{X}_i\,\beta_{08} + \alpha_{c08}A_{i08} + \lambda_{1i} + \lambda_{3i08} + u_{i08} \tag{5.5}$$

[5] In statistics *latent variables* refer to variables which have not or cannot be directly measured, at best they can estimated through indicators or manifest variables, an approach we will be using below.

[6] Note that λ_{2it} disappears from the equation since it is equal to $\alpha_{ct}A_{it}$.

Where
P_{it} = production of maize in the ith household at time t
$\mathbf{X}_i$ = a vector of exogenous variables for the ith household
A_{it} = a vector containing two variables: (i) household age at time t, and (ii) the descendant household dummy
λ_{it} = unobserved latent variables of type 1: constant over time and type 3: age-independent ones
u_{it} = residual

In the above equations, the first term denotes the constants for the two years, corresponding to (β_1) in Eqn 5.1, while the two penultimate terms contain two types of latent variables (λ_{1i} and λ_{3it}). The third term contains a constant α_{c02}, which is the regression coefficient for the age-related variables.

Let $\Delta\beta$ denote the changes in the vector β between 2008 and 2002 (i.e. $\beta_{08} - \beta_{02}$) and $\Delta\alpha$ similarly the changes in the time- and age-dependent latent variables (λ_{2it}). Subtract Eqn 5.4 from Eqn 5.5 to obtain:

$$\begin{aligned}\ln\left(\frac{P_{i08}}{P_{i02}}\right) &= \beta_{c08} - \beta_{c02} + (\beta_{08} - \beta_{02})\mathbf{X}_i + \alpha_{c08}A_{i08} \\ &\quad -\alpha_{c02}A_{i02} + (\lambda_{3i08} - \lambda_{3i02}) + (u_{i08} - u_{i02}) \\ &= \beta_{c08} - \beta_{c02} + (\beta_{08} - \beta_{02})\mathbf{X}_i + \alpha_{c08}A_{i02} + \alpha_{c08}6 - \alpha_{c02}A_{i02} \\ &\quad + (\lambda_{3i08} - \lambda_{3i02}) + (u_{i08} - u_{i02}) \\ &= \beta_{c08} - \beta_{c02} + (\beta_{08} - \beta_{02})\mathbf{X}_i + (\alpha_{c08}A_{i02} - \alpha_{c02}A_{i02}) + \alpha_{c08}6 \\ &\quad + (\lambda_{3i08} - \lambda_{3i02}) + (u_{i08} - u_{i02})\end{aligned}$$

or

$$\ln\left(\frac{P_{i08}}{P_{i02}}\right) = \Delta\beta_c + \Delta\beta\mathbf{X}_i + \Delta\alpha A_{i02} + \alpha_{c08}6 + (\lambda_{3i08} - \lambda_{3i02}) + (u_{i08} - u_{i02}) \quad (5.6)$$

where $\Delta\beta_c = \beta_{c08} - \beta_{c02}$ is a constant, $\Delta\beta = \beta_{08} - \beta_{02}$ denotes the changes in the vector β between 2008 and 2002, $\Delta\alpha = \alpha_{08} - \alpha_{02}$ denotes the changes in the time- and age-dependent latent variables (λ_{2it}), $A_{i08} - A_{i02}$ is the length of panel wave (6 years in our case) and $\lambda_{3i08} - \lambda_{3i02}$ is the difference between the latent variables of type 3 between 2008 and 2002.

The penultimate term is the difference between the time- and age-independent latent variables at t_{08} and t_{02} respectively. In the estimation of Eqn 5.6, this difference would not be possible to distinguish from the difference between the residuals ($u_{i08} - u_{i02}$).

Note, furthermore, that the latent variable λ_{1i} disappears, since $\lambda_{1i} - \lambda_{1i} = 0$. According to the reduced form model in its original formulation, as suggested by Glewwe and Hall (1998), constant latent factors in this way are eliminated from the equation. This is a weak part of the original model, since constant latent variables may reflect conditions for the activation of other drivers, which in this way would not be represented in the model. The random errors u_{i02} and u_{i08} are now differentiated, which, according to Glewwe and Hall (1998), may reduce the correlation between them and the observed variables.

The same applies to the latent variables that vary over time (λ_{3i}), which, however, also remain a potential source of bias in the model.

Multi-level data opens up a possibility of controlling for this latter source of bias, however. To the extent that the time-independent latent variables (λ_{3i}) are meso- or macro-level, their effect can be controlled for by village and country dummies, respectively. Total village-level effects can be estimated through a set of village dummies, which means that, even if the effect of individual factors cannot be distinguished, their aggregate effect can. The same applies to country-level factors. This again means that the gross effect of λ_{3it} type of variables turn from being unobservable to being possible to estimate, in part at least.

The following discussion therefore will be based on a model consisting of three equations: (i) total production of maize in 2002, logged (y_{it1}); (ii) total production of maize in 2008, logged (y_{it2}); and (iii) the change in total production of maize in 2008 over 2002, logged (y_{ip2}). In order to control for the effects of multi-collinearity, the model will be developed in steps. In Step 1 we control for the variables in A_{it}, i.e. household age, logged and descendants[7] and lastly for *area under maize, logged*. By taking the logged value of area, the β coefficients in the models of production for t_1 and t_2 will directly reflect scale effects on production, while for the third model they reflect elasticity of production with respect to area, with values over unity reflecting intensification – i.e. increased production stemming from increased yields. Values below unity, by contrast, reflect extensification, i.e. expanded area with lower mean yields. The interpretation of dummy variables is also straightforward, since the β values indicate the percentage difference in logged production between households having the value of 1 for the dummy, as compared to those having the value 0.

Following on from this first step, we add variables block-wise, starting in Step 2 with technological drivers of intensification (seed fertilizer technology and ploughing); in Step 3 with indicators of commercialization, continuing in Step 4 with estimates of the effects of macro-level policies. The social distributional profile of increased production of maize will be investigated in Step 5, while in Step 6 macro-level variables are removed and replaced by country dummies. In Step 7, finally, endogeneity is checked by introducing the residuals from two models dealing with fertilizer use and market participation. In the following tables, we will be using the conventional *, ** and *** to denote test results significant at 5, 1 and 0.1% level. Results at 5% should be interpreted with care, since they have a propensity to fluctuate between different runs, while results at lower levels of significance tend to be stable.

Descriptive Statistics

Before constructing the model, we describe and discuss the variables involved. Starting with the dependent variables, these are the natural logarithm of production of maize in 2002 and 2008 and the change in production between

[7] This is a dummy variable taking the value of 1 for descendants and 0 otherwise.

these two years. Over the period, production increased by 23% on average (see Appendix 1, Table 5A.1). The averages for the panel thus point to a considerable dynamism.

The panel is biased compared to the two statistically representative cross sections, which we reported on in Chapter 4. The production figures for the cross sections were 1584 and 1219 kg for 2002 and 2008, respectively, suggesting a decline in production in the cross-sectional data, which contrasts to the relative dynamism of the panel households. Since the aim here is not to produce representative point estimates but to search for the drivers of increased production, this bias should not affect the results, however.

In the models we use logged figures of production and change in production. Similarly we take the log of the few independent variables that are scales, i.e. area and household age.

Like production, area under maize has increased over the period and, on average, panel households have 12% more logged area under maize today than in 2002. This suggests that, to some extent, production increases for the panel are due both to increased area and to higher yields, i.e. to a mixed pattern of extensified and intensified production.

There is some evidence of increased use of industrial and scientific inputs like improved/hybrid seed or chemical fertilizer. We use fertilizer as a proxy for these inputs and, as is clear from Table 5A.1, use, disuse and adoption varies only marginally over time. However, here the length of the periods should be considered: over the longer period from the reference year to 2002, 14% started using fertilizer, while over the shorter period from 2002 to 2008 12% did.

Much the same could be said about oxen or, less frequently, tractor ploughing: usage figures change modestly over the period and there is evidence of both adoption and disuse. From the reference year to 2002, 4% adopted ploughing, while over the shorter period from 2002 to 2008, 6% did. Adoption of ploughing explains part of the dynamism of maize production, although this may in part be related to re-stocking in Zambia following earlier outbreaks of cattle disease. Nineteen per cent have adopted ploughing in Zambia, but the technique is spreading also in Mozambique, Tanzania, Nigeria and Kenya.

Commercialization, both in the sense of entering the market and through increasing sales volumes, is connected to more recent dynamism: 36% of the panel have started to sell or increased their sale of maize since 2002, compared to 28% of the households who did so during the longer period between the reference year and 2002. Thus there is evidence of a substantially increased market engagement over the last 5-year period, which also proves to be an important driver of production increases.

The macro-variables are included to reflect the influence of government policy. Three indicators are used: (i) government expenditure on agriculture and rural development as a percentage of total government expenditure; (ii) import of maize as share of total domestic production; and (iii) change in GDP per capita between t_1 and t_2 at constant 2000 USD values. In the models for 2002 and 2008 we use lagged data, to allow time for government policies to impact on farmers' production and production decisions. For the model dealing with change in production over the period 2002–2008, we use the change in these lagged figures.[8]

The distributional consequences of production changes are captured through two variables, i.e. gender and elite membership. As a proxy for elite membership we use relative rank in the village distribution of landownership in 2002, considering the 10% biggest landowners as belonging to the elite.[9]

Finally, like in all panel data models, we need to control for household age, or more precisely for the year in which the farm came under its present management. As Table 5A.1 shows, the average here is 1982, which means that the average household age in 2002 was 20 years. Since 3% of the panel household consists of descendant households, where the household head interviewed in 2002 has deceased, the average household age in the panel in 2008 is slightly lower than would have been the case if such partitioning of the original household had not occurred. To check for any possible effects of generational shifts, we use a dummy variable for the descendants.

The Results from Modelling

The first model in the attempt to model maize production and its drivers deals with the period from the reference year to 2002 and treats logged production as a function of a series of independent variables (c.f. Eqn 5.1 above). All independent variables refer either to the reference year or to the period until 2002. The first type of variable should be straightforward in terms of causal interpretation, while for the latter type, endogeneity cannot be excluded. We return below to our strategy for controlling the effects of endogeneity.

The second model is a similar function, referring to the period from the reference year to 2008. Besides covering a longer time period, the model reassesses the role of the independent variables as explanations of production. To the extent that the second model replicates the results of the first, this would be an indicator of robustness and reliability of the models. Moreover, changes in regression coefficients, as we will see, may also be significant in terms of our hypotheses.

[8] The figures on public expenditure refer to the situation in 1999 and 2005 respectively. The 2005 data in the case of Nigeria refer to 2003, Zambia to 2004, Ghana to 2004 and Malawi to 2006. The figure on imports as a share of domestic maize production is the 5-year average for the period 1995–1999 and 2001–2005 respectively. The figures for GDP per capita refer to 2001 and 2007 respectively, at constant 2000 USD values.

[9] In a previous publication Larsson used the wealth group classification made in 2002 as a proxy for membership of the village elite: Larsson, R. (2005) Crisis and potential in smallholder food production – evidence from micro level. In: Djurfeldt G., Holmén, H., Jirström, M. and Larsson, R. (eds) (2005) *The African Food Crisis: Lessons from the Asian Green Revolution.* CAB International, Wallingford, UK. In this earlier round we asked interviewers to group households into five categories, with the middle one denoting average wealth. When repeating this exercise in the latest survey and comparing the results, it is evident that this method is far from reliable, which is why we have chosen a new indicator.

The third model, finally, resembles what is technically known as a reduced form model, i.e. the dependent variable is logged change in production from 2008 to 2002. Again we use the panel data so that the independent variables refer either to a state in 2002 (alternatively to the reference year) or to the period from 2002 to 2008. The above comments on endogeneity apply here too. Table 5.1 presents an overview of the variables involved and the reasons for including the variables in the analysis. Model development is done stepwise with variables entered in blocks.

Step 1 begins by introducing the control variables, i.e. household age and a dummy for descendant households. In the first block we also include the most important determinant of production, i.e. area. Since production and area refer to the same point in time for the first two models, causal attribution is tricky. In this case, data are cross-sectional rather than longitudinal and do not permit conclusions on elasticity, although they allow for comparisons among the households. Regression coefficients above unity would point to increasing returns to area, while coefficients below unity would indicate the reverse (given homogeneity of land, which cannot be taken for granted, however).

Table 5.2 details the results of the three variables introduced in the first step. Logged farm age in the first model shows the familiar and expected curvilinear (i.e. logarithmic) relation to production, increasing with age but less so as age increases.[10] Similarly the control for descendants shows an effect in the second model but not in the third one. The effect of being a descendant probably accounts for the non-significance of age in the second model.

In the first model, we include descendant households to establish that their production volumes did not differ significantly from other households in 2002. The positive effect of a generational transfer on production is evidenced by a regression coefficient of 0.48 in the second model, indicating that descendants currently have 48% higher logged production than others (significant at the 1% level). This, too, is an expected and well-known phenomenon in studies of farm economics and rural sociology, as the ambitions of a younger generation in the process of an inter-generational shift are generally higher than those of their parents. The extent to which such ambitions translate into investments that enhance the productivity of the farm units depends on the smallholder business climate, however. The positive sign of the regression coefficient for descendants in the second model can thus be taken to indicate that the business climate in maize production has been somewhat positive during the period from 2002 to 2008. This conclusion is supported by other results, to be reported below.

The regression coefficient for area in model three, i.e. for change in production between 2002 and 2008, is only 0.42 and significant at the 0.1% level. With the time factor taken care of, causal interpretation is more straightforward here, suggesting that production increases during the period have been mainly land extensive. Most farmers interviewed in 2002 said they would be able to put

[10] This effects is sometimes referred to as the Chayanov effect, after the Russian agricultural economist who was first to document it: Chayanov, A.V. (1966) *A.V. Chayanov and the Theory of Peasant Economy*. Richard D. Irwin, Homewood, Illinois.

Table 5.1. Overview of variables and associated hypotheses.

Variable group	Variable	Hypothesis or reason for inclusion
Controls	Years since farm established, logged	Production is expected to be curvilinear with respect to age (Chayanov effect)
	Descendant household, dummy	Descendants are expected to invest in higher production than their predecessors
Area	Area under maize, ha logged	Growth in production is expected to be largely extensive (regression factor in third model <1)
Weather	Drought in E. and S. Africa 2002, village dummy	Control for and estimate the effects of drought
	Floods in W. Africa in 2008, village dummy	Control for and estimate the effects of flood
Fertilizer	Used fertilizer at the start of the period, dummy	Estimate the effects of seed fertilizer technology on production and change in production
	Decreased or stopped using fertilizer during the period, dummy	
	Increased or started using fertilizer during the period, dummy	
Ploughing	Used ploughing at the start of the period, dummy	Estimate the effects of ploughing on production and change in production
	Stopped using ploughing during the period, dummy	
	Started using ploughing during the period, dummy	
Commercialization	Sold maize at the start of the period, dummy	Estimate the effects of commercialization on production and change in production
	Decreased or stopped selling maize during the period, dummy	
	Increased or started selling maize during the period, dummy	
Policy	Government expenditure on agriculture, lagged and logged	Increased government expenditure is expected to stimulate production or growth in production
	Import of maize, per cent of total production, lagged and logged	Increased import dependence is expected to be a disincentive to domestic production
	GDP per capita in 2001 and 2007 and change over those years, constant USD	To estimate the elasticity of maize production to economic growth (third equation)
Socio-economic characteristics	Proxy for elite membership (belonging to the 10% biggest landownership in village in 2002)	Smallholder-friendly development would decrease the importance of elite membership

Continued

Table 5.1. Continued.

Variable group	Variable	Hypothesis or reason for inclusion
	Gender of farm manager in 2002	Gender discrimination is expected to negatively affect production
Country	Country dummies (Kenya reference country)	To control for between-country differences and to estimate between-country variance
Village	Village dummies	To check for deviant villages and to estimate between-village variance

Table 5.2. Model of maize production and change of production, Step 1.

	Period 1 (p_1) t_0 to t_1			Period 1+2 (p_{1+2}) t_0 to t_2			Period 2 (p_2) t_1 to t_2		
	B	Std error	Sig.	B	Std error	Sig.	B	Std error	Sig.
Constant	6.60	0.09	***	6.83	0.09	***	0.61	0.09	***
Years since farm established, logged	0.07	0.03	*	−0.05	0.03		−0.17	0.03	***
Descendant households	0.09	0.16		0.48	0.15	**	0.15	0.17	
Area under maize, logged	0.83	0.03	***	0.78	0.02	***	0.42	0.03	***
Drought in E. and S. Africa 2002	−0.25	0.05	***						
No. of cases	1317			1442			1515		
R^2	0.44			0.44			0.16		
Missing cases (%)	27			20			16		

more land under cultivation. Many farmers appear to have done exactly so, possibly in response to increasing demand and prices during the last few years. While doing so they seem to have drawn on marginal and less productive land.

In 2002 widespread drought in eastern and southern Africa affected farmers in several of our case study countries. This effect can be traced by the regression coefficient in the first model, which suggests a mean harvest reduction of 25%.

The proportion of variance explained is quite high for the first two models ($R^2 = 0.44$ in both cases), mostly due to area, which is highly correlated with production. For change in production, as is to be expected, R^2 is lower (0.16). The percentage of missing cases is tolerable (27, 20 and 16% respectively).

Step 2 introduces indicators of farm technology, i.e. seed fertilizer technology and ploughing. We would have liked to include indicators of the use of pre-industrial technologies such as crop rotation, intercropping, manuring, fallowing, etc. Our data on such technologies indicate lots of noise, however, resulting in more or less randomly distributed answers, which prevent their use.

The strong association of both seed fertilizer and ploughing with higher production levels is brought out in Table 5.3. Those who used fertilizer throughout had significantly higher logged production (52% and 41%, respectively, in the first two models) than those who did not. Those who started or increased their use of fertilizer during the first period (p_1) had significantly higher logged production (38%) than those who did not; the percentage is not significantly lower (30%) for the longer period (p_{1+2}). The introduction or increase of fertilizer use in the third model is not statistically significant, however, suggesting

Table 5.3. Model of maize production and change of production, Step 2.

	Period 1 (p_1) t_0 to t_1			Period 1 + 2 (p_{1+2}) t_0 to t_2			Period 2 (p_2) t_1 to t_2		
	B	Std error	Sig.	B	Std error	Sig.	B	Std error	Sig.
Constant	6.26	0.10	***	6.44	0.09	***	0.57	0.10	***
Controls									
Years since farm established, logged	0.06	0.03	*	−0.06	0.03	*	−0.17	0.03	***
Descendant households	0.15	0.16		0.34	0.14	*	0.12	0.17	
Area									
Area under maize, logged	0.80	0.03	***	0.70	0.02	***	0.42	0.03	***
Weather									
Drought	−0.29	0.05	***						
Fertilizer									
Used fertilizer at the start of the period	0.52	0.06	***	0.41	0.06	***	−0.02	0.07	
Decreased or stopped using fertilizer during the period	−0.55	0.07	***	−0.61	0.08	***	−0.02	0.09	
Started or increased using fertilizer during the period	0.38	0.08	***	0.30	0.08	***	0.07	0.09	
Ploughing									
Used ploughing at the start of the period	0.19	0.07	**	0.59	0.06	***	0.17	0.08	*
Stopped using ploughing during the period	−0.10	0.10		−0.42	0.09	***	−0.39	0.14	**
Started using ploughing during the period	0.43	0.12	***	0.58	0.10	***	0.40	0.11	***
Model info									
No. of cases	1317			1442			1515		
R^2	0.49			0.51			0.17		
Missing cases (%)	27			20			16		

that the relative dynamism of the latter period has been less driven by seed fertilizer technology than earlier.

The use of ploughing at the start of the periods[11] implies significantly higher production in all three models.[12] None the less, the significance of the β value for introducing the plough for the third model, as well as the higher value of the regression coefficient in the second, hints at a recent dynamism, where plough farms have increased their production more than others. This too seems to be a country effect: many Zambian farmers whose livestock were affected by disease in the early years of the new millennium have now recovered their stock and reintroduced the plough. When we introduce country dummies at a later step, these regression coefficients lose significance, which supports this interpretation.

The β values for area are largely unaffected by the introduction of the technology variables, which points to a robust model and few problems of multi-collinearity.

Step 3 introduces the commercialization variables, as outlined in Table 5.2. These variables overall have the strongest influence on production dynamics.

The results in Table 5.4 indicate that increases in production are connected to having sold maize at the start of the periods as well as to having entered the market or increased market participation during the periods (the differences between households who have sold maize throughout and new sellers are not statistically significant though). Those who have entered the market or increased their market engagement have a 59% higher logged production in the first period. The tendency is also pronounced with respect to the third model, where the β value for households who have entered the market or increased their sale of maize is 0.59.

In Step 4 (Table 5.5), macro-level policy indicators are introduced in the model. Although generally increased levels of public spending on agriculture reflects the altered policy atmosphere following the Maputo declaration, there is little evidence in our data that political commitment has translated into field-level effects. Indeed, the regression coefficient for share of agricultural spending in the second model is negative, while there is no statistically significant relation between change in production and re-ordered political priorities in the third model.

The effect of trade policies seems to have shifted over the period, however. The relationship between maize imports during the latter half of the 1990s and domestic production is statistically significant and negative during the first period, suggesting that cheap imports may have undermined domestic production. During the second period, these disincentive effects disappear. In turn this may be connected to the effects of new global trade regimes associated with the World Trade Organization and the protective tariffs introduced following the collapse of the Doha round, as well as with rising world market prices for staples and improving commercial incentives for domestic producers from 2007 onwards.

[11] This largely points to Ethiopia, so that country's effects are mirrored here. When, in a later step, country dummies are introduced so that Ethiopia is controlled for, the real effects of ploughing are much more visible.
[12] Although only significant at 5% level in the reduced form model.

Table 5.4. Model of maize production and change of production, Step 3.

	Period 1 (p_1) t_0 to t_1			Period 1+2 (p_{1+2}) t_0 to t_2			Period 2 (p_2) t_1 to t_2		
	B	Std error	Sig.	B	Std error	Sig.	B	Std error	Sig.
Constant	5.91	0.10	***	6.02	0.09	***	0.50	0.10	***
Controls									
Years since farm established, logged	0.03	0.03	*	−0.09	0.03	**	−0.14	0.03	***
Descendant households	0.12	0.15		0.25	0.13	*	0.06	0.15	
Area									
Area under maize, logged	0.68	0.03	***	0.60	0.02	***	0.28	0.03	***
Weather									
Drought in 2002	−0.23	0.05	***						
Fertilizer									
Used fertilizer at the start of the period	0.54	0.06	***	0.37	0.05	***	−0.07	0.06	
Decreased or stopped using fertilizer during the period	−0.46	0.07	***	−0.47	0.08	***	0.03	0.08	
Started or increased using fertilizer during the period	0.31	0.07	***	0.21	0.07	**	0.07	0.08	
Ploughing									
Used ploughing at the start of the period	0.15	0.06	*	0.49	0.06	***	0.16	0.07	*
Stopped using ploughing during the period	−0.08	0.09		−0.32	0.08	***	−0.38	0.13	**
Started using ploughing during the period	0.40	0.12	***	0.45	0.09	***	0.33	0.10	**
Commercialization									
Sold maize at beginning of period	0.38	0.08	***	0.51	0.06	***	−0.10	0.07	
Stopped or decreased selling maize during the period	−0.01	0.09		−0.23	0.06	***	−0.52	0.09	***
Started or increased selling maize during the period	0.59	0.07	***	0.72	0.05	***	0.59	0.06	***
Model info									
No. of cases	1317			1442			1515		
R^2	0.56			0.61			0.31		
Missing cases (%)	27			20			16		

A similar effect is witnessed in the effects of economic growth in general: while the level of economic development as operationalized by gross domestic product (GDP) per capita had no significant association with production in the first two models, more recent trends suggest change. In the third model,

Table 5.5. Model of maize production and change of production, Step 4.

	Period 1 (p_1) t_0 to t_1			Period 1 + 2 (p_{1+2}) t_0 to t_2			Period 2 (p_2) t_1 to t_2		
	B	Std error	Sig.	B	Std error	Sig.	B	Std error	Sig.
Constant	5.18	0.46	***	5.63	0.69	***	0.16	0.13	
Controls									
Years since farm established, logged	0.00	0.03		–0.06	0.03	*	–0.14	0.03	***
Descendant households	0.09	0.14		0.22	0.12		0.09	0.15	
Area									
Area under maize, logged	0.69	0.03	***	0.65	0.02	***	0.30	0.03	***
Weather									
Drought	–0.19	0.05	***						
Fertilizer									
Used fertilizer at the start of the period	0.57	0.06	***	0.31	0.05	***	0.00	0.06	
Decreased or stopped using fertilizer during the period	–0.39	0.07	***	–0.46	0.07	***	–0.01	0.08	
Started or increased using fertilizer during the period	0.32	0.07	***	0.23	0.07	***	0.15	0.09	
Ploughing									
Used ploughing at the start of the period	0.15	0.07	*	0.49	0.06	***	0.08	0.08	
Stopped using ploughing during the period	0.00	0.09		–0.41	0.08	***	–0.33	0.13	*
Started using ploughing during the period	0.43	0.12	***	0.27	0.09	**	0.26	0.11	*
Commercialization									
Sold maize at beginning of period	0.25	0.08	**	0.49	0.06	***	–0.17	0.08	*
Stopped or decreased selling maize during the period	0.07	0.09		–0.30	0.06	***	–0.54	0.09	***
Started or increased selling maize during the period	0.51	0.07	***	0.69	0.05	***	0.56	0.07	***
Macro-level variables									
Share of state budget for agriculture at the end of period, logged	0.19	0.07	**	–0.24	0.07	***	–0.06	0.04	
Import of maize as per cent of total production at the end of period, logged	–0.09	0.02	***	0.02	0.02		0.01	0.03	
GDP per capita 2001 and 2007 and change 2007 over 2001, logged	0.11	0.07		0.15	0.10		2.00	0.43	***
Model info									
No. of cases	1317			1442			1515		
R^2	0.57			0.63			0.31		
Missing cases (%)	27			20			16		

growth in GDP per capita during the period 2001–2007 had a strong and statistically significant effect on change of production. The regression coefficient is 2.00, indicating that over this period, and other things being equal, a 1% growth in GDP per capita translated in to 2% growth in maize production. Adding a 95% confidence interval to this, we can say with 95% probability, and other things equal, the effect of a one unit increase in GDP per capita corresponds to between 1.15% and 2.85% increased growth in maize production.

That there can be strong macroeconomic linkages between the agricultural sector and economic growth, especially in the early steps of development, has long been recognized (Mellor, 1966, 1995). A growing body of literature pointing at the interdependence of the agricultural and non-agricultural sectors shows that agriculture can stimulate, as well as be stimulated by, overall economic growth (Haggblade *et al.*, 2007). The Asian experience in this respect has often been one of agriculture and industry working in tandem. Growth of employment in industry and services implied increasing demand for food among net consumers and stimulated domestic agricultural production.

Although production increases are connected to macro-level developments, such developments are the reflection less of state priorities than of global processes and of domestic economic growth in general. In this sense, a state-driven Green Revolution is not (yet?) traceable in our case study countries.

The Asian model of agricultural development emphasizes the smallholder base. For our case we assess this by looking at the distributional profile of the farmers behind the production increases. We look at both the village elite versus other farmers and men versus women. This is done in Step 5.

Interesting results emerge for the elite, as suggested by Table 5.6. In the first model, elite status is related to higher production, as earlier claimed by Larsson (2005), but this does not hold for the longer period, where the regression coefficient is not significant. This may signal a change, which is also brought out by the coefficient in the third model, which is negative, although only significant at the 5% level. This may suggest that recent dynamism in maize markets may have brought new groups of smallholders into commercial production. Such a conclusion is reinforced by the strong effects we have already seen for market entry (c.f. Table 5.5). Thus there is some, but admittedly not very strong, evidence of smallholder-driven development. This development is not state-driven, as suggested by the Asian model, but driven by the market.

The discrimination against women is evident only in the second model, being non-significant in the first and third models. This is in line with our general hypothesis that discrimination against women lies primarily in access to land. Assuming that generational transfer privileges sons rather than wives and daughters, we would predict a negative and statistically significant regression coefficient for gender, which we also get when controlling for descendants in the second model.

Yet another step in the development of the models is the introduction of country dummies. When doing so we remove the macro-level variables, which

Table 5.6. Model of maize production and change of production, Step 5.

	Period 1 (p_1) t_0 to t_1			Period 1 + 2 (p_{1+2}) t_0 to t_2			Period 2 (p_2) t_1 to t_2		
	B	Std error	Sig.	B	Std error	Sig.	B	Std error	Sig.
Constant	5.25	0.46	***	5.62	0.68	***	0.17	0.13	
Controls									
Years since farm established, logged	0.00	0.03		−0.05	0.03		−0.13	0.03	***
Descendant households	0.08	0.14		0.24	0.12		0.09	0.15	
Area under maize, logged	0.67	0.03	***	0.64	0.02	***	0.30	0.03	***
Weather									
Drought	−0.18	0.05	***						
Fertilizer									
Used fertilizer at the start of the period	0.58	0.06	***	0.34	0.05	***	0.01	0.06	
Decreased or stopped using fertilizer during the period	−0.39	0.07	***	−0.48	0.07	***	−0.01	0.08	
Started or increased using fertilizer during the period	0.33	0.07	***	0.24	0.07	***	0.14	0.09	
Ploughing									
Used ploughing at the start of the period	0.15	0.07	*	0.47	0.06	***	0.07	0.08	
Stopped using ploughing during the period	0.02	0.09		−0.38	0.08	***	−0.32	0.13	*
Started using ploughing during the period	0.45	0.12	***	0.27	0.09	**	0.25	0.11	*
Commercialization									
Sold maize at beginning of period	0.24	0.08	**	0.48	0.06	***	−0.16	0.08	*
Stopped or decreased selling maize during the period	0.07	0.09		−0.31	0.06	***	−0.54	0.09	***
Started or increased selling maize during the period	0.49	0.07	***	0.68	0.05	***	0.57	0.07	***
Macro-level variables									
Share of state budget for agriculture at the end of period, logged	0.19	0.07	**	−0.24	0.07	***	−0.06	0.04	
Import of maize as per cent of total production at the end of period, logged	−0.09	0.02	***	0.03	0.02		0.00	0.03	
GDP per capita 2001 and 2007 and change 2007 over 2001, logged	0.10	0.07		0.16	0.10		1.93	0.44	***

Continued

Table 5.6. Continued.

	Period 1 (p_1) t_0 to t_1			Period 1 + 2 (p_{1+2}) t_0 to t_2			Period 2 (p_2) t_1 to t_2		
	B	Std error	Sig.	B	Std error	Sig.	B	Std error	Sig.
Distributional dimensions									
Elite membership in 2002	0.19	0.09	*	0.04	0.08		−0.15	0.08	*
Gender of farm manager in 2002	−0.10	0.05		−0.21	0.05	***	−0.01	0.06	
Model info									
No. of cases	1317			1442			1515		
R^2	0.57			0.63			0.31		
Missing cases (%)	27			20			16		

are highly collinear with these dummies. We choose Kenya as the reference country, with the β values for the country dummies pointing to the difference between a given country and Kenya. During the first period, three countries stand out negatively. Controlling for technology and commercialization and other variables, Mozambique, Ethiopia and Zambia (albeit only significant at the 5% level in the latter case) had lower production of maize than Kenya (see Table 5.7).

Nigeria had significantly higher production in the first period, as had Malawi and Tanzania (the latter significant only at 5% level). There is considerable fluidity though, as indicated by the pattern for the longer period, where Ethiopia and Malawi (the latter only significant at 5% level) stand out negatively, while Ghana, Nigeria and Zambia divert positively from Kenya. In the case of Malawi, tendencies in our sample villages deviate from the debated but politically important national maize surpluses of 2007/8, which may be due to a bias in our sample of villages or in the data. In the third model, finally, Zambia, Mozambique (significant at 0.1% level) and Ghana (5% level) feature positively when compared with Kenya.[13]

Robustness of model

The robustness of the model is checked for through the stepwise development of the models. If β values for a variable introduced in an earlier step change as a consequence of the introduction of other variables, it implies questionable robustness. There are very few statistically significant changes in β coefficients over the different steps in all three models, however.

[13] Such country-level variation is also confirmed by the individual reports prepared by the national teams for the respective countries.

Table 5.7. Model of maize production and change of production, Step 6.

	Period 1 (p_1) t_0 to t_1			Period 1 + 2 (p_{1+2})t_0 to t_2			Period 2 (p_2) t_1 to t_2		
	B	Std error	Sig.	B	Std error	Sig.	B	Std error	Sig.
Constant	5.71	0.13	***	5.94	0.12	***	0.18	0.13	
Controls									
Years since farm established, logged	0.03	0.03		–0.01	0.03		–0.07	0.03	*
Descendant households	0.09	0.14		0.09	0.12		0.04	0.15	
Area									
Area under maize, logged	0.62	0.03	***	0.61	0.03	***	0.34	0.03	***
Weather									
Drought	–0.14	0.05	**						
Floods in West Africa 2008				–0.30	0.13	*			
Fertilizer									
Used fertilizer at the start of the period	0.57	0.06	***	0.36	0.06	***	–0.02	0.07	
Decreased or stopped using fertilizer during the period	–0.40	0.07	***	–0.51	0.07	***	–0.11	0.08	
Started or increased using fertilizer during the period	0.37	0.07	***	0.26	0.07	***	0.10	0.09	
Ploughing									
Used ploughing at the start of the period	0.53	0.08	***	0.53	0.08	***	0.16	0.09	
Stopped using ploughing during the period	–0.17	0.10		–0.45	0.09	***	–0.40	0.13	**
Started using ploughing during the period	0.47	0.11	***	0.31	0.09	***	0.10	0.11	
Commercialization									
Sold maize at beginning of period	0.37	0.08	***	0.36	0.06	***	–0.19	0.08	*
Stopped or decreased selling maize during the period	–0.02	0.09		–0.29	0.06	***	–0.49	0.09	***
Started or increased selling maize during the period	0.55	0.07	***	0.62	0.05	***	0.54	0.06	***
Macro-level variables									
Share of state budget for agriculture at the end of period									
Import of maize as percent of total production at the end of period									

Continued

Table 5.7. Continued.

	Period 1 (p_1) t_0 to t_1			Period 1 + 2 (p_{1+2}) t_0 to t_2			Period 2 (p_2) t_1 to t_2		
	B	Std error	Sig.	B	Std error	Sig.	B	Std error	Sig.
GDP per capita change, logged, previous 5 years									
Distributional dimensions									
Elite membership in 2002	0.18	0.09		0.06	0.08		−0.16	0.08	*
Gender of farm manager in 2002	−0.14	0.05	**	−0.21	0.05	***	−0.04	0.06	
Country dummies (Kenya reference cat.)									
Ethiopia	−0.42	0.12	***	−0.36	0.11	**	0.01	0.13	
Ghana	0.18	0.11		0.76	0.16	***	0.26	0.12	*
Malawi	0.24	0.09	**	−0.21	0.09	*	−0.07	0.09	
Nigeria	0.42	0.10	***	0.41	0.12	***	0.11	0.10	
Tanzania	0.22	0.09	*	0.20	0.09	*	0.18	0.10	
Zambia	−0.23	0.09	*	0.44	0.09	***	0.68	0.09	***
Mozambique	−0.77	0.14	***	0.06	0.10		0.50	0.11	***
Model info									
No. of cases	1317			1442			1515		
R^2	0.60			0.65			0.35		
Missing cases (%)	27			20			16		

Leaving aside changes in constants (β_1) and requiring statistical significance at 1% or lower, we find a few but explicable changes. In the first period, while the β value for change in maize area did not appreciably change from step 1 to step 2, i.e. after introducing the technological variables, the corresponding β values decrease from 0.80 in the second step to 0.68 in the third step and after the introduction of the commercialization variables (significant at 1% level). While this points to mild endogeneity from the latter group of variables, it also signals how market stimuli may encourage land-extensive growth (i.e. not intensification as some would have believed and hoped for). A similar pattern is also visible over the longer period, i.e. in the second and third models.

It should also be noted that the β value for used plough at the start of the period increases from 0.19 to 0.53 in step 6, i.e. after the introduction of the country dummies. In other words, when controlling for Ethiopia in particular, the model allows us to see the inherent potential of plough-based agriculture.

Another step on which we avoid details is the introduction of village dummies, partly in order to control for deviant villages but, more importantly, in order to estimate the total between-village effect. Country dummies can similarly be used to estimate the between-country effect (see Table 5.8).

As can be seen below, the largest share of variance is at the farm and household level. Between-village effects vary from 0.08 in the second model to 0.12

Table 5.8. Proportion of total variance explained by country, village and farm household effects.

Variance components	t_0 to t_1	t_0 to t_2	t_1 to t_2
Between-country effects	0.03	0.03	0.04
Between-village effects	0.12	0.08	0.12
Between-household effects	0.56	0.62	0.30
Residual	0.29	0.27	0.53
Total	1.00	1.00	1.00
Total variance in y	2011.46	2326.76	2014.00
Proportion of the above explained by:			
Proportion of between-country variance explained by macro-level variables	0.95	0.98	0.60

in the first and the third ones. As we see the between-country variance constitutes a relatively small share of total variance. Another implication of the small variance among countries is that the detected patterns seem to be quite general and with some likelihood apply to large parts of the African maize and cassava belt.

From the lower portion of the table, finally, it is clear that the manifest variables, i.e. the macro-level variables, chosen do explain sizeable parts of the between-country variance. Consequently, these results substantiate the contention that trade and agricultural policies, as well as general economic growth, produce much of the variance in production of maize and its development. This is notwithstanding the parallel conclusion that the recent changes in policy priorities have not yet had much impact on the ground. This again would point to the great potentials of a more vigorous agricultural policy.

Thus our models appear robust and also point to generalizable patterns in the case study countries.

Checking for endogeneity

The final stage in the development of the models in this chapter is to check for the effects of endogeneity on the results. As a reminder of the basic logic: we control for latent variables at village and country level through dummies, and for time- and age-dependent latent variables (through the controls for age of household and for descendant households). These controls notwithstanding we may be unable to track the influence of latent variables that vary over time.

We are not using an instrumental variables (IV)[14] approach to check for endogeneity, since we are critical of this approach as problematic assumptions may be hidden in the choice of instrumental variables.

[14] We are grateful to Christopher Udry and his critical comments on our draft chapter. Two pieces of criticism are well taken, i.e. on errors in the notation and on some sloppy interpretations of regression results, e.g. on ploughing. We are more doubtful, however, on the third piece of criticism, which is of our way of dealing with endogeneity.

So far we have modelled the log production of maize (in 2002 and 2008 and the change between those two years) as a function of a vector of independent variables (*X*). In the seventh step we add a linear function (*G*) to the (*X*) matrix. *G* is a function of a residual containing three types of latent variables (λ_1, λ_2 and λ_3) and a set of random disturbances (*u*):

$$\ln(P) = X\eta + G(\lambda_1;\lambda_2;\lambda_3;u) \tag{5.7}$$

As argued earlier, we can control for two of the three types of latent variables but not for those that are time-variant but uncorrelated with household age (λ_{3it}). This again means that λ_1 can be eliminated from the model (as in Eqn 5.8 below) by letting $y = \left(\frac{P_{i08}}{P_{i02}}\right)\left(\text{or } \ln(y) = \ln\left(\frac{P_{i08}}{P_{i02}}\right)\right)$, so that

$$\ln(y) = X\beta + H(\lambda_2;\lambda_3;u) \tag{5.8}$$

Thus, we model two of the explanatory variables in Eqn 5.7 above, i.e. fertilizer use (X_f) and market participation (X_m) as a function of a vector of explanatory variables (*Z*) and residuals (e_1 and e_2). See Tables 5A.1 and 5A.2 in Appendix 1.

$$x_f = Z_1\gamma_1 + e_1 \tag{5.9}$$

and

$$x_m = Z_2\gamma_2 + e_2 \tag{5.10}$$

where x_f and x_m are (possibly) endogenous variables in the model for the period P_2 in the maize production model.

Next we introduce estimates of the two residuals above into Eqn 5.8, giving:

$$\ln(y) = X\beta + \theta_1\hat{e}_1 + \theta_2\hat{e}_2 + H(\lambda_2;\lambda_3;u) \tag{5.11}$$

By testing whether the parameters for θ_1 and θ_2 are 0, we can check for possible endogeneity. That is, we test whether there is correlation between the dependent variable (ln(*y*)) and the errors (e_1 and e_2).

We estimate the above model for ln(*y*), change in logarithm of production between 2002 and 2008, using the same set of independent variables (*X*) as in Eqn 5.8 and the estimates of the residuals (e_1 and e_2) from Eqns 5.9 and 5.10. The residual in Eqn 5.11 is defined as a function (*H*) of a residual containing two types of latent variables ($\lambda_2;\lambda_3$) and a set of random disturbances (*u*). When estimating Eqn 5.5 we get regression coefficients (β, θ_1 and θ_2), from which estimates $\hat{\theta}_1$ and $\hat{\theta}_2$ are used for testing endogeneity.

More explicitly then, by means of Eqn 5.11 we test the hypothesis that besides the indirect influence of *Z* on change in production (ln(*y*)) via x_f and x_m respectively, there is a direct causal link between *Z* and ln(*y*). This boils down to a test of the significance $\hat{\theta}_1$ and $\hat{\theta}_2$ from zero. If one or both tests are positive, it indicates that there is endogeneity in the main model, while if they are negative there is no such evidence.

When adding the residuals mentioned to the model developed in Step 6, none of the residuals is statistically significant (details withheld). On the basis of this, we conclude that the model presented does not suffer from any biasing endogeneity.

A new era for African agriculture, or merely an episode?

We have used the experience of 1805 smallholders in eight African countries across the maize and cassava belt to identify and evaluate the role of three main types of production drivers, comparing two periods – one lasting on average between 1982 and 2002 and the other capturing the past 6 years (2002–2008). Our interest has been focused mainly on the latter period, since political interest in the welfare of smallholders across the subcontinent, both as sources of domestic food production and, more generally, as the core sector of broad-based development has received increased attention in the last 5 years. Looking at the three tenets of our original model: state drivenness, technology and market mediation, production increases are primarily connected to commercial drivers and, although to a lesser extent, to inputs and use of technology. In the latter case, public interventions, as found, for instance, through input subsidy programmes in Malawi and Zambia, may have played some part in democratizing access to chemical fertilizer and improved seeds, for instance. None the less the use of such technology is equally widespread in countries such as Kenya, which have not experimented with such programmes. The effects of this kind of state involvement therefore remain the subject of debate. Improved commercial incentives, although they are the strongest drivers of production increases, appear, as yet, to be largely disconnected from public efforts to improve smallholder market participation and more connected to economic growth in the non-farm sector.

Thus the much-publicized – and in many cases real – state commitment to rural development since 2003 has failed to make its mark on smallholder realities and, in this sense, production increases arising from improved commercial incentives are occurring despite, rather than because of, state involvement. Although the images of heavy-handed marketing boards conjured up by references to the pre-SAP era are clearly undesirable, the state has an important role to play as an enhancer of smallholder access to markets and provider of technology. The role of the state as a facilitator of both input and especially output markets needs to be revamped to suit the realities emerging from globalized markets for staple foods, growing regional trade and the price volatility connected to unpredictable weather conditions and growing populations.

When the research underlying this work was started in the early years of the millennium, it was done with the expectation that important changes were underway in African agriculture and that a long recession might be nearing its end. Initially we found little evidence of this (see Holmén, 2005a,b; Larsson, 2005). This analysis, however, has pointed to two changing drivers: firstly, lower import dependency and hence less competition from producers overseas, in turn connected to changing trade regimes, less protection and export subsidies in OECD countries, the collapse of the Doha round and subsequent imposition of protective tariffs by African countries, etc. Secondly, quite dynamic growth in the non-farm sector since the early years of the millennium has, as we have demonstrated, imputed a new dynamism into the maize sector. This driver has, of late, decelerated due to the global financial crisis. If it does not regain speed in the coming years, the new era may turn into an interlude.

References

Barrett, C.B. (2008) Smallholder market participation: concepts and evidence from eastern and southern Africa. *Food Policy* 33, 299–317.

Chayanov, A.V. (1966) *A.V. Chayanov and the Theory of Peasant Economy*. Richard D. Irwin, Homewood, Illinois.

Crawford, E., Kelly, V., Jayne, T.S. and Howard, J. (2003) Input use and market development in sub-Saharan Africa: an overview. *Food Policy* 28, 277–292.

Dorward, A., Kydd, J., Morrison, J. and Urey, I. (2004) A policy agenda for pro-poor agricultural growth. *World Development* 32, 73–89.

Dyer, G. (2004) Redistributive land reform: no April rose. The poverty of Berry and Cline and GKI on the inverse relationship. *Journal of Agrarian Change* 4, 45–72.

Frees, E.W. (2004) *Longitudinal and Panel Data: Analysis and Applications in the Social Sciences*. Cambridge University Press, Cambridge, UK.

Glewwe, P. and Hall, G. (1998) Are some groups more vulnerable to macroeconomic shock than others? Hypothesis tests based on panel data from Peru. *Journal of Development Economics* 56, 181–206.

Haggblade, S., Hazell, P. and Reardon, T. (2007) *Transforming the Rural Nonfarm Economy: Opportunities and Threats in the Developing World*. John Hopkins University Press, Baltimore, Maryland.

Hausman, J.A. (1976) Specification tests in econometrics. *Econometrica* 46, 1251–1271.

Hildebrand, J.R. (1960) Some difficulties with empirical results from whole-farm Cobb-Douglas-type production functions. *Journal of Farm Economics* 42, 897–904.

Holmén, H. (2005a) Spurts in production – Africa's limping Green Revolution. In: Djurfeldt, G., Holmén, H., Jirström, M. and Larsson, R. (eds) *African Food Crisis – the Relevance of the Asian Green Revolution*. CAB International, Wallingford, UK, pp. 65–85.

Holmén, H. (2005b) The state and agricultural intensification in sub-Saharan Africa. In: Djurfeldt, G., Holmén, H., Jirström, M. and Larsson, R. (eds) *African Food Crisis – the Relevance of the Asian Green Revolution*. CAB International, Wallingford, UK, pp. 87–112.

Jayne, T.S., Zulu, B. and Nijhoff, J.J. (2006) Stabilizing food markets in eastern and southern Africa. *Food Policy* 31, 328–341.

Larsson, R. (2005) Crisis and potential in smallholder food production – evidence from micro level. In: Djurfeldt, G., Holmén, H., Jirström, M. and Larsson, R. (eds) *African Food Crisis – the Relevance of the Asian Green Revolution*. CAB International, Wallingford, UK, pp. 113–134.

Lipton, M. (2005) *The Family Farm in a Globalizing World: the Role of Crop Science in Alleviating Poverty*. IFPRI, International Food Policy Research Institute, Washington, DC.

Mellor, J.W. (1966) *The Economics of Agricultural Development*. Cornell University Press, Ithaca, New York.
Mellor, J.W. (1995) *Agriculture on the Road to Industrialization*. Johns Hopkins University Press, Baltimore, Maryland.
Mundlak, Y. (2001) Production and supply. In: Gardner, B.L. and Rausser, G.C. (eds) *Handbook of Agricultural Economics*. Elsevier, Amsterdam, pp. 3–86.
Poulton, C. and Dorward, A.R. (2008) Getting agricultural moving: role of the state in increasing staple food crop producitivity with special reference to coordination, input subisidies, credit and price stabilisation. Paper prepared for AGRA Policy Workshop, Nairobi, Kenya, 23–25 June, 2008.
Sanchez, P.A., Shephard, K., Soule, M., Place, F., Buresh, R. and Izac, A.M. (eds) (1997) *Repleneshing Soil Fertility in Africa*. Soil Science Society of America, Madison, Wisconsin.
Tiffen, M. (2003) Transition in sub-Saharan Africa: agriculture, urbanization and income growth. *World Development* 31, 1343–1366.
World Bank (2007) *World Development Report 2008: Agriculture for Development*. The World Bank, Washington, DC.

Appendix 1: Descriptive Statistics

Table 5A.1. Descriptive statistics for variables in Steps 1 to 7.

	n	Mean	Std error
Dependent variables			
Maize production in 2002 (2005), kg, logged	1685	6.22	0.03
Maize production in 2008, kg, logged	1757	6.47	0.03
Change in maize production 2008 over 2002, logged	1652	0.23	0.03
Controls			
Age of household 2002, logged	1730	2.75	0.02
Descendant household, dummy	1799	0.03	0.00
Area			
Area under maize 2002, ha, logged	1784	−0.40	0.02
Area under maize 2008, ha, logged	1774	−0.28	0.03
Change in maize area 2008 over 2002, logged	1759	0.12	0.02
Weather			
Poor rainfall in 2002, dummy	1661	0.45	0.01
Floods in West Africa 2008, dummy	1799	0.14	0.01
Fertilizer			
Fertilizer use in reference year, dummy	1588	0.56	0.01
Stopped using fertilizer between ref. year and 2002, dummy	1799	0.13	0.01
Started using fertilizer between ref. year and 2002, dummy	1799	0.14	0.01
Stopped using fertilizer between ref. year and 2008, dummy	1799	0.10	0.01
Started using fertilizer between ref. year and 2008	1799	0.12	0.01
Used fertilizer on maize in 2002, dummy	1779	0.56	0.01
Stopped using fertilizer between 2002 and 2008, dummy	1799	0.13	0.01
Started using fertilizer between 2002 and 2008, dummy	1799	0.12	0.01
Ploughing			
Used plough in ref. year, dummy	1761	0.27	0.01
Stopped using plough between ref. year and 2002	1788	0.09	0.01
Started using plough between ref. year and 2002	1788	0.04	0.00

Continued

Table 5A.1. Continued.

	n	Mean	Std error
Stopped using plough between ref. year and 2008	1787	0.09	0.01
Started using plough between ref. year and 2008	1787	0.06	0.01
Used plough in 2002, dummy	1788	0.23	0.01
Stopped using plough between 2002 and 2008	1787	0.04	0.00
Started using plough between 2002 and 2008	1787	0.07	0.01
Commercialization			
Sold maize in reference year, dummy	1737	0.49	0.01
Stopped or decreased sale of maize between ref. year and 2002, dummy	1740	0.32	0.01
Started or increased sale of maize between ref. year and 2002, dummy	1740	0.28	0.01
Stopped or decreased sale of maize between ref. year and 2008, dummy	1799	0.48	0.01
Started or increased sale of maize between ref. year and 2008, dummy	1799	0.53	0.01
Sold or intended to sell maize 2002, dummy	1799	0.39	0.01
Given up or reduced sale of maize between 2002 and 2008, dummy	1799	0.23	0.01
Started to sell or increased sale of maize between 2002 and 2008, dummy	1799	0.37	0.01
Macro-level variables			
Government expenditure on agriculture and rural development, 2002 (lagged by 3 years), logged	1798	1.35	0.02
Government expenditure on agriculture and rural development, 2005, logged, Nigeria 2003, Zambia 2004, Ghana 2004, Malawi 2006	1799	1.96	0.01
Change in budget allocations to agriculture logged, 2008 over 2002	1798	0.61	0.02
Import of maize as share of total domestic production 1995–1999, logged	1798	1.13	0.05
Import of maize as per cent of total production 2000–2005, logged	1799	1.19	0.04
Change in import dependence, 2008 over 2002, logged	1798	0.06	0.03
GDP per capita 2002, constant 2000 USD (Mozambique, 2005)	1799	5.54	0.01
GDP per capita 2007, constant 2000 USD, logged	1799	5.74	0.01
Change in GDP per capita, 2007 over 2002, logged	1799	0.19	0.00
Distributional dimensions			
Proxy for elite membership in reference year	1772	0.07	0.01
Sex of farm manager	1792	0.26	0.01
Country dummies			
Ethiopia dummy	1799	0.08	0.01
Ghana dummy	1799	0.08	0.01
Malawi dummy	1799	0.17	0.01
Mozambique dummy	1799	0.13	0.01
Nigeria dummy	1799	0.11	0.01
Tanzania dummy	1799	0.12	0.01
Zambia dummy	1799	0.16	0.01

Continued

Table 5A.1. Continued.

	n	Mean	Std error
Endogeneity controls			
Standardized residual, absolute value, for model of market participation in 2008	1501	−0.70	0.02
Standardized residual, absolute value, for model of fertilizer use in 2008	1696	−0.85	0.01

Table 5A.2. Descriptive statistics for variables in appendix models A2 and A3.

	n	Mean	Std error
Fertilizer use in 2008, dummy	2327	0.56	0.01
Years since farm established, logged	3305	2.70	0.02
Descendant household, dummy	4875	0.02	0.00
Additional land available 2002, dummy	3357	0.85	0.01
Family labour resources increased since 2002	2109	0.44	0.01
Increased cattle ownership since 2002, dummy	4875	0.10	0.00
Farm management feminized since 2002, dummy	4875	0.07	0.00
Used fertilizer on maize in 2002, dummy	2900	0.49	0.01
Started to sell or increased sale of maize since 2002	1824	0.37	0.01
Change in country-level nominal producer price of maize since 2002, logged	3409	0.11	0.01
Started or increased sale of other food crops since 2002, dummy	4875	0.30	0.01
Started receiving extension services since 2002, dummy	1805	0.21	0.01
Government expenditure on agriculture and rural development, 2008, (lagged by 3 years Nigeria 2003, Zambia 2004, Ghana 2004, Malawi 2006), logged	3807	1.94	0.01
Import of maize as share of total domestic production 2001–2005, logged	3807	0.91	0.02
Change in GDP per capita 2007 over 2001, logged	3807	0.21	0.00
Valid N (listwise)	1230		

Appendix 2: Fertilizer and Seed Technology Use

Table 5A.3. Binary logistic regression of fertilizer use, 2008.

	B	Std error	Sig.	Exp(B)
Years since farm established, logged	−0.12	0.08		0.89
Descendant households	0.11	0.39		1.12
Additional land available, dummy	0.45	0.18	**	1.57
Family labour resources increased since 2002	0.28	0.14	**	1.32

Continued

Table 5A.3. Continued.

	B	Std error	Sig.	Exp(B)
Increased cattle ownership between 2002 and 2008, dummy	–0.12	0.16		0.88
Farm management feminized since 2002, dummy	–0.02	0.18		0.98
Used fertilizer on maize in 2002, dummy	1.24	0.14	***	3.46
Started to sell or increased sale of maize between 2002 and 2008, dummy	0.51	0.15	***	1.66
Change in country-level mean nominal producer price of maize, 2002–2008, logged	0.58	0.22	**	1.78
Increased or started selling of other food crops since 2002, dummy	0.67	0.14	***	1.95
Started receiving extension services between 2002 and 2008, dummy	–0.06	0.17		0.94
Government expenditure to agriculture and rural development, 2005, logged and lagged	–0.46	0.15	**	0.63
Import of maize as share of total domestic production 1995–1999, logged	–0.33	0.06	***	0.72
Constant	0.45	0.48		1.57
No. of cases	1230			
Nagelkerke's R^2	0.28			
Missing cases (%)	32			

GDP indicator omitted due to multi-collinearity with other macro-level variables. Data on government expenditure refer to the following years for Nigeria 2003, Zambia 2004, Ghana 2004, Malawi 2006.

Appendix 3: Model of Market Participation

Table 5A.4. Binary logistic regression of sale of maize in 2008.

	B	Std error	Sig.	Exp(B)
Years since farm established, logged	–0.08	0.07		0.93
Descendant households	0.80	0.37	**	2.23
Additional land available, dummy	0.62	0.16	***	1.86
Maize area increased between 2002 and 2008, dummy	0.68	0.14	***	1.97
Yields of maize increased since 2002, dummy	0.90	0.13	***	2.47
Used fertilizer on maize in 2002, dummy	0.68	0.14	***	1.97
Started using fertilizer since 2002	0.18	0.21		1.20
Sold maize in 2002 makes model autoregressive	1.42	0.14	***	4.15
Increased or started selling other food crops since 2002, dummy	0.63	0.12	***	1.88
Change in country-level mean nominal producer price of maize, 2002–2008, logged	–0.25	0.20		0.78
Import of maize as share of total domestic production 1995–1999, logged	–0.41	0.06	***	0.66

Continued

Table 5A.4. Continued.

	B	Std error	Sig.	Exp(B)
Change in GDP per capita 2007 over 2001, logged	0.34	1.13		1.40
Proxy for elite membership in reference year	0.45	0.24		1.56
Non-farm income, some income from non-farm sector, dummy	–0.19	0.13		0.83
Earning income from sale of non-staples, dummy	0.12	0.15		1.13
Constant	–1.77	0.46		0.17
No. of cases		1601		
Nagelkerke's R^2		0.37		
Missing cases (%)		11		

6 Maize Remittances, Market Participation and Consumption among Smallholders in Africa

Agnes Andersson

Department of Human Geography, Lund University, Lund, Sweden

Following decades of neglect, smallholder-based agriculture has recently been promoted as the foundation for a broad-based development effort in the regional context of sub-Saharan Africa. A range of initiatives have focused on promoting access to technology and inputs to raise productivity within the smallholder sector. Despite such renewed policy interest, commercial incentives on the demand side have failed to increase productivity. Food markets characterized by uncertainty, depressed prices, atomism and prohibitive transaction costs have been identified as major causes of farmer reluctance regarding input adoption and, by implication, failure to improve productivity and food security (Jayne *et al.*, 2006b; Poulton *et al.*, 2006). As such, the potential of smallholder-based agriculture to relieve the food security situation in sub-Saharan Africa is increasingly tied to the ability of the market to stimulate production increases. By implication, weak or malfunctioning markets are considered to be explanations for lacking agricultural development (Von Braun and Kennedy, 1994). The literature contains ample examples of underdeveloped markets and low productivity. Figures show that maize production, although growing in real terms, has failed to keep pace with population growth during the past four decades, which has led to suggestions by some commentators that the African population is best fed through increased imports of maize (Wood, 2002) or, as advocated by more recent commentaries, through concentrating investment in the domestic large-scale commercial sector (Collier, 2008). Attention has also been directed towards non-agricultural sources of rural livelihoods and tendencies towards what has become known as de-agrarianization – that is a situation in which smallholder households are gradually squeezed out of agriculture (Bryceson, 1997, 1999; Ashley and Maxwell, 2001; Ellis, 2005).

Rapid urbanization and the increasing reliance of rural households on urban incomes has been one of the foci of such studies, suggesting the declining importance of agriculture for rural livelihoods. These studies may, however, be underestimating the importance of (agrarian) household-level linkages from rural to urban areas. Meanwhile, proponents of improved commercial incentives for smallholders tend to focus on transactions occurring within the formal realm of the market, while aspects affecting rural livelihoods may also be found within informal exchange relations. More generally, empirical research on participation in staple markets, when compared with technology-adoption studies, remains scant and therefore deserves attention (Barrett, 2008).

Recent household data from households in Ethiopia, Ghana, Kenya, Malawi, Mozambique, Nigeria, Tanzania, Uganda and Zambia suggest that smallholders are engaged in in-kind remittances, which, in effect, bypass ordinary market channels, widening not only the subsistence responsibilities of farm households, both spatially and numerically, but also possibly signalling the malfunctioning of markets. This chapter discusses how such remittances should be perceived. Should in-kind remittances be interpreted as a sign of the failure of thin and uncoordinated grain markets to meet the food demands of both rural and urban dwellers? In contrast, if they are viewed instead as an expression of multi-spatial livelihoods, remittances may be a method of countering shortcomings of the market as a provider of grain for rural net food purchasers and urban household members. Although these rural households remit a number of staples and other food crops, this paper focuses on maize.

Maize, in addition to being Africa's largest contemporary staple crop, occupies a special historical role, i.e. smallholder production geared towards the market both during colonial times and post-colonially (Ranger, 1985; Bates, 1989), and for this reason is a crucial crop to study in the context of agricultural market development in Africa in general. Maize markets have also been the subject of renewed policy interest, while grabbing academic attention for the past decade, especially in the wake of rising global food prices. Maize, as a relatively non-perishable, weight-efficient crop, can be transported and remitted over long distances.

The purpose of this chapter is twofold. First, it intends to shed light on the phenomenon of in-kind remittances of staple crops. Whether remittances should be perceived as signs of market failure is analysed through evaluating the connection between remittances and the commercial behaviour of the remitting households. Secondly, the chapter assesses the reciprocal and livelihood implications for the remitters to assess the role of remittances as an expression of multi-spatial livelihoods. The data used is part of a larger survey, conducted in 2008 and presented in detail in Chapter 4, this volume, also covering villages in which remittances did not occur. The discussion draws on the data collected in villages where maize remittances were occurring – 91 villages in total. The sample population in these villages was 3388 households, out of which 2857 were maize producers. In general, the analysis will treat the subset of maize producers. The households were surveyed in 2008.

Methodology

This study draws on parts of a larger survey of farm households in nine countries in the African maize and cassava belt. The areas sampled can be characterized as typical of environments in which a majority of the smallholder population in sub-Saharan Africa lives. The households are assumed to be statistically representative of farmers in the region, being above average in terms of ecology and market access, but exclude the most vibrant local rural economies. Questions on in-kind remittances were developed following a 3-week pilot study on maize circulation in Kenya in July of 2006 (see Andersson and Wambugu, 2007). The pilot study detailed data on the flows of maize among household members and what the maize was used for once it left the villages. For the larger survey, questions have been standardized to suit the questionnaire format. The focus in the questionnaire has been on documenting the size and direction of remittance flows, of both maize and other staple products.

Data has been collected on a number of socio-economic characteristics of the household, production patterns and the various uses of maize. Information on the annual amount of maize sent to relatives, as well as the amount of maize collected by relatives, was gathered. In addition, data on the destinations of maize remittances was gathered. In the presentation below, these amounts have been grouped together and are collectively discussed as in-kind remittances of maize.

Maize Production, Commercialization and Rural Livelihoods

Poor market participation is commonly attributed to a number of characteristics of African maize markets as well as production environments in general. Smallholder populations suffer from a range of poverty traps, aggravated by the functioning of markets but also related to the inability to participate in the market. The instability of maize prices, both seasonally and from one year to the next, leads to disinvestment in maize production among sellers, as the risks associated with such fluctuations discourage the use of inputs for production of marketable surplus. Lacking consumer confidence in the market and seasonally inflated prices, meanwhile, prompts the poor to continue growing maize, even at very low levels of productivity, to meet their own consumption needs (see Dorward *et al.*, 2008 for a discussion on Malawi, for instance). Farmers and consumers fail to benefit from large variations in cereal production at the regional level as a result of poor infrastructure and limited coordination, which prevents the emergence of local and regional markets.

Prompted by differential land access as well as divergences in productivity, as much as 70% of food grains are estimated to be non-tradable (World Bank, 2007), while more than half (55%) of the rural small-scale farm population is forced to buy some of its maize requirements – in Kenya, for example (Jayne *et al.*, 2006b). Given the price volatility, insularity and unreliability of the maize market in the lean season especially, self-provisioning of food staples remains an attractive option, even for relatively wealthy households (see e.g. Jayne *et al.*, 2006a for a discussion of Kenya). Although rural smallholder access to food has been the subject of numerous studies, the extent to which urban households self-provision is less well known. The staple-based links between rural and urban areas occurring among family members merit attention, since these largely invisible linkages, if undocumented, may under-report the urban dependence on rural food, with implications for food security of both groups.

Production systems and income composition

In many respects, the national samples reflect the patterns described above. Land sizes are small, and although maize dominates the crop portfolios, cropping patterns are relatively diverse in terms of staples' production. Female-headed households are in the minority, varying from 8% of the maize-producing households (n = 2857) in Nigeria to 23% in Malawi. Land sizes vary considerably from 1.2 ha to 5.6 ha, with the largest land sizes, by far, found in Nigeria. Due to the difficulties in estimating land sizes, such figures should, however, be treated with caution.

Although a variety of staples are grown and dependence upon the various staples as sources of food and income varies among the countries, maize is by far the most commonly grown crop in all the countries, with a few exceptions. In Ethiopia, sorghum is grown by 93% of the maize-growing households, whereas in Uganda around 85% of these also grow sweet potatoes and cassava. At the national level, the percentage of maize growers varies from 60% in Ghana to 100% in Kenya (see Table 6.1).

Although maize is the most commonly grown crop, production levels vary considerably among the countries (see Table 6.2). To some extent such variation is reflected in differences in technology use and input intensity in maize production. Mozambique, where fertilizer is used by only 1% of the maize-growing households, has the lowest production figure in the sample, while Zambia, where 78% of the maize producers used fertilizer, exhibited the highest production figures. In terms of fertilizer expenditure, Nigeria and Zambia were found at the top of the sample. Although 70% of the maize producers in Kenya were using fertilizer, Kenya is found in the middle range in terms of production. None the less, the relatively low production figures may also be explained by the post-electoral violence at the time of data collection.

In total, 41% of the maize producers used chemical fertilizer. Although data was collected on fertilizer expenditure, such figures are difficult to convert into quantities of fertilizer use. In some cases, such as Malawi and Zambia, higher expenditure could reflect that the household lacks access to cheaper, subsidized

Table 6.1. Percentage of maize-growing households in national samples. (From: own survey data, 2008.)

Country	Maize-growing households (%)	n
Ethiopia	84	243
Ghana	60	569
Kenya	100	180
Malawi	98	398
Nigeria	99	374
Tanzania	85	400
Uganda	80	398
Zambia	96	423
Mozambique	76	403
Total	84	3388

Table 6.2. Mean maize production 2008. (From: own survey data, 2008.)

	Maize production (kg)		
	Mean	Median	Producers (n)
Ethiopia	1052	600	205
Ghana	664	400	340
Kenya	853	360	180
Malawi	820	600	391
Nigeria	2657	2000	369
Tanzania	1028	567	341
Uganda	984	500	318
Zambia	2562	1200	406
Mozambique	492	350	307
Total	1313	600	2857

fertilizer. Low expenditure may mask the use of relatively larger amounts of subsidized fertilizer. Moreover, expenditure differences may reflect local variations in fertilizer prices rather than amounts used.

The share of farmers using hybrid or improved maize varieties spans from 5% in Mozambique to 86% in Kenya. In general, the use of improved maize varieties is more frequent than the application of fertilizer, with 52% of the sample using improved maize varieties. In most cases, the use of fertilizer and improved maize varieties go hand in hand, although in the Kenyan case, improved maize varieties are being grown by a larger share of the sample than fertilizer use would suggest. In the case of Nigeria, by contrast, the use of fertilizer is more widespread than the use of improved maize varieties.

In terms of income sources, staple sales are by far the most important source of cash income for households in Ethiopia and Uganda, where they constitute 53% and 44% of average household income, respectively. Staple sales are the least important source of cash income in Kenya, where they comprise 17% of average household cash income. On average, 586 kg of maize were sold by the households. Again, however, this figure varies considerably among the countries and is directly related to production, with sales being the highest in absolute terms in Zambia, followed by Nigeria, where 1285 and 1136 kg of maize were sold in the 2007/8 season respectively. Mozambique had the lowest level of sales, with only 80 kg being sold.

Sale of other food crops constitutes 18% of cash income for the maize sample, with households in Malawi, Zambia, Nigeria and Ghana sourcing around a quarter of their cash income from the sale of other food crops. Sale of non-food cash crops, by contrast, only constitutes 11% of average household income among the maize growers, although the importance of cash crops varies among the countries. Generally speaking, cash remittances from non-resident household members are relatively minor sources of cash income, with less than 5% of average household income consisting of cash remittances. Despite reports of de-agrarianization tendencies (see e.g. Bryceson, 1999), only 13% of average cash income was derived from small-scale micro-business.

Divergent patterns of income sources may explain at least some of the differences in the use of modern farm technology. Although beyond the scope of this chapter, investments in farm technology could be expected to be more prevalent where maize takes on the dual role of cash and food crop.

Maize remittances

The share of maize remitters varies from 22% in Tanzania to 70% in Nigeria and 61% in Ghana. Forty-two per cent of the maize producers (2857 households) in the sample reported remitting maize to their relatives. The amounts remitted, however, like the amounts sold, are the highest in Nigeria and Zambia, which also represent higher production figures among the countries (see Table 6.3).

Remittances of other staple foods were also common, with cassava being remitted by 527 of the maize remitters in the sample. Among the maize-remitting households, on average a total of 307 kg of staple crops were remitted to relatives following the last harvest. Nigeria and Uganda stand out in terms of amounts remitted, with average staple remittances totalling 544 and 418 kg, respectively.

In-kind remittances and market participation

The focus on agriculture as a source of comprehensive economic growth for sub-Saharan Africa is often placed on commercialization, and for this reason the connection between in-kind uses of maize and sale of maize is especially interesting. Whether individual farmers' commercial incentives are affected by obligations to remit maize and how such incentives vary with the local market structure is, in this sense, relevant for a wider discussion of farmer market behaviour.

Table 6.3. Mean and median maize production (2008) and maize remittances by country. (From: own survey data, 2008.)

	Maize production (kg)			Maize remittances (kg)		
	Mean	Median	Producers (n)	Mean	Median	Remitters (n)
Ethiopia	1051	600	205	71	35	47
Ghana	663	400	340	69	50	207
Kenya	852	360	180	134	90	69
Malawi	820	600	391	127	100	154
Nigeria	2657	2000	369	229	150	258
Tanzania	1027	567	341	131	77	74
Uganda	983	500	318	130	75	144
Zambia	2562	1200	406	318	200	131
Mozambique	491	350	307	57	50	122
Total	1312	600	2857	152	100	1206

Fafchamps and Minten's (2001) characterization of the African grain trade as a 'flea market economy' points to a set of constraints which may be preventing market engagement outside the local village economy. Such limitations, moreover, may also explain why in-kind transfers in social networks are an understandable alternative to formal commercial transactions.

At the receiving end, urban households' reliance on informally accessed food may be indicative of the failure of urban food markets to meet urban demand at a price affordable to consumers (see Andersson, 2002). In-kind transfers of food from rural to urban areas are touched upon in general terms and in more detail in the historical literature on food provisioning in African cities (Bryceson, 1987; Guyer, 1987). Recent interest in urban multi-spatial livelihoods (e.g. Foeken and Uwuour, 2008) has focused on the importance of rural farming practised by urban household members. Bryceson (1993) notes the role of family networks in providing urban dwellers with maize in Tanzania, where such provisions account for around 15% of total consumption needs. The understanding of maize remittances as a component of urban livelihoods remains impressionistic, however, while the connection between remittances and the commercial behaviour of the remitters has not been previously considered in a cross-country context.

When the sample is divided by whether households remit maize or not, the differences between the two groups in terms of production and market participation is striking. Contrary to the notion that remittances are indicative of poorly developed markets, amounts and percentages sold are higher among the remitters than the non-remitters (see Table 6.4).

When divided by production quartiles[1] and separated according to whether households remit maize or not, the picture becomes less straightforward: the lower-production quartiles stand out in terms of the percentage of total production devoted to remittances (see Table 6.5). Although production levels are higher

Table 6.4. Remitting and non-remitting households: average production, amounts remitted and sold, percentage sold.

	n	Production (kg)	Amount remitted (kg)	Amount sold (kg)	Maize sold (%)
Remitters	1206	1727	152	824	35
Non-remitters	1651	1010	0	413	21
Difference	–	717***	–	411***	14***
Total sample	2857	1313	64	587	27

Total sample refers to the remitters and the non-remitters together.
***The differences in production, amounts and shares sold are significant between the two groups (remitters and non-remitters) at the 0.001 level.

[1] The size of the production quartiles varies slightly. This is the result of all households with the same production being placed in the same quartile when the sample is ranked according to production.

Table 6.5. Remitting and non-remitting households divided by maize production quartiles.

	Total production (kg)	Sig.[a]	Amount remitted (kg)	Remitted (%)	Amount sold (kg)	Sig.[a]	Sold (%)	Sig.[a]	Sellers (%)	Sig.[a]	Maize buyers (%)	Sig.[a]
Remitters												
Q1 (N 156)	154	***	47	40	21	**	12	**	26	***	47	
Q 2 (N 251)	417	**	68	17	130	***	30	***	64	***	33	*
Q 3 (N 355)	884		106	12	294		33	*	72	**	23	**
Q 4 (N 444)	3694		273	9	1922		47		88		13	
Non-remitters												
Q 1 (N 507)	131	***	0	0	12	**	6	**	13	***	56	
Q 2 (N 425)	397	**	0	0	64	***	16	***	36	***	41	*
Q 3 (N 426)	907		0	0	265		28	*	63	**	33	**
Q 4 (N 293)	3570		0	0	1828		45		86		16	

Differences in means were tested between the remitting and non-remitting households in the same production quartiles for total production, amount sold, share sold and share of sellers. ***Significant at the 0.1% level; **at the 1% level; *at the 5% level.

[a]Sig. represents the significance of the difference between remitters and non remitters within the same quartile.

among the remitting households in all the production quartiles, differences in production levels between remitters and non-remitters are only statistically significant in the bottom half of the sample. Remitters, as a whole, are over-represented in the higher-production quartiles, especially in the uppermost quartile, where 60% of the households reported remitting maize. In statistical terms, this over-representation in the upper half of the sample is strongly significant. Substantial amounts of total production are remitted, varying from 9% to 40%, with the share remitted being inversely proportional to total production. In absolute terms, remittances varied from 47 kg in the first production quartile to 273 kg in the fourth production quartile. Among the remitting households in the first quartile, an extremely small amount of maize is sold. Around half of these households reported buying maize at some time during the year.

Commercialization in terms of market participation (percentage of sellers) is higher for remitters in the first two production quartiles. The amounts sold are higher for the remitting households, but such differences are statistically significant only in the case of the first two quartiles, although differences in percentages sold are also statistically significant among households in the third production quartile. The amounts sold are generally small, although comprising a sizeable part of total production in relative terms. The remitted amounts are larger than sales in the first quartile but lower than sales for the second quartile (see Table 6.5).

In the two lower quartiles, the remitters sell around twice the amount of maize that the non-remitters do, whereas sales volumes are almost identical in the two uppermost production quartiles (see Table 6.5). None the less both remitters and non-remitters, in many cases, resort to buying maize at some stage of the year, although generally speaking the percentage of maize buyers is lower among the remitters. The differences in the percentage of maize buyers among remitters and non-remitters in the different quartiles are statistically representative only in the third and second quartiles.

The correlation between amounts remitted and amounts sold is relatively weak (Pearson's r = 0.246, significant at the 0.01 level) but positive. If remittances, as such, were undermining market participation, this correlation would be expected to be negative. Since differences in commercialization, when measured in terms of amounts of sales, are statistically significant for the bottom half of the sample, this suggests that market participation is not *directly* undermined by remittances among these households.

The positive correlation between remittances and sales indicates that remittances are likely to be a reflection of a culture of gift-giving and multi-spatial support networks that extend beyond the village boundaries, rather than a sign *primarily* of malfunctioning markets. The functioning of a multi-spatial unit of consumption appears to be especially pronounced among households found in the lower-production quartiles.

Although remittances, as such, do not appear to be directly undermining commercial participation among the remitting households, remittances may, to some extent, be indicative of poorly functioning markets for maize, in the sense that recipients may need to rely on remittances rather than the market for part of their food requirements. The limited recent research that exists on the subject of informal provisioning concerns the Soviet Union and China, where the role of informal practices in circumventing markets characterized by shortages is highlighted (see Ledeneva, 2008).

Both the amount remitted and the amount sold are correlated to total household production (as suggested by Table 6.5).[2] The direction of causality is difficult to determine, as decisions related to both sales and remittances can be made post- rather than pre-harvest. Since remittances are connected to higher production levels and also positively correlated with sales, this does suggest that households extend their consumption obligations relative to their production rather than in response to poor market conditions.

Despite being connected to higher levels of production, remitters are over-represented among those households whose share of total cash income is predominantly non-staple based (see Table 6.6), underscoring the role of maize as a crop grown mainly for own consumption. This over-representation towards non-staple-based incomes is strongly statistically significant for the remitters as a group. When compared with the non-remitters, the cash incomes of the remitters are also more staple-based. This over-representation relative to the non-remitters again suggests that remittances and market participation are not mutually exclusive. In turn, this underlines the notion that remittances should be viewed as an expression of household subsistence obligations found within the context of kinship networks.

Table 6.6. Household composition of cash income by percentage of staple sales. (From: own survey data, 2008.)

	Remitters		Non-remitters		Total	
	n	%	n	%	n	%
Staple sales constitute up to 25% of total cash income	595	51	961	61	1556	57
Staple sales constitute between 25.1 and 50% of total cash income	228	20	185	12	413	15
Staple sales constitute between 50.1 and 75% of total cash income	148	13	156	10	304	11
Staple sales constitute more than 75% of total cash income	195	17	263	17	458	17
Total	1166	100	1565	100	2731	100

A Chi Square test of the differences in the expected and actual distribution of remitters and non-remitters among the different types of income composition is significant at the 0.001 level. The distribution of the standard residual among the types of income composition shows that the greatest deviation is in the first and second categories, where staple sales constitute up to 50% of cash income. The number of cases is smaller in this table, since it only covers those households who had cash income.

[2] For the sample as a whole (n = 2857) the correlation between production and amount remitted is lower (Pearson's r = 0.351, significant at the 0.01 level) than that between production and sale (Pearson's r = 0.910, significant at the 0.01 level).

The largely non-staple-based livelihoods found within the sample may, to some extent, mirror the poor commercial environments facing staple crop producers and lacking incentives for productivity growth within the smallholder staple crop sector, especially in certain national markets. At the receiving end, moreover, remittances may represent a mechanism for dealing with food shortages, price shocks and volatility characteristic of such markets. Although remittances may occur immediately following harvests, they may none the less counteract local shortages through smoothing out differences in cropping seasons among villages.[3] To the extent that the latter is the case, remittances should be interpreted as driven primarily by a culture of gift-giving, rather than market failure.

In-kind transfers may also be connected to patterns of remittances, with roots in the migrant labour histories of a number of African countries (see Stichter, 1982 on Kenya, for instance) and the operation of labour systems based on a circulation of both labour and remittances – in cash and in-kind – between rural and urban areas. While the literature on the African family generally suggests that the historically long-standing operation of extra household support systems is being undermined by the pressures of economic decline on the one hand (Devereux, 1999) and a growing culture of individualism on the other (Bank and Qambata, 1999; Ferguson, 1999), such systems need to be considered in a situation in which households have few other resources to bank upon in the event of misfortune. The role of the wider family network in cushioning the impact of economic hardship for individual households also needs to be considered as an enabling aspect of maize remittances. An implicit degree of reciprocity, which may not materialize as incoming remittances of food or cash as such, may none the less be a source of security in the event of adversity. Given the vicissitudinous nature of localized maize markets, investing in these relations by way of maize remittances may be an alternative to maize sale, with a secondary effect of deflecting maize from the market.

The body of literature on the interaction between market exchange and reciprocal exchange (such as the classical works of Polanyi, 1957 and Sahlins, 1965, 1972 and, more recently, Bloch and Parry, 1991) that has developed mainly within economic anthropology offers interesting insights into the relationship between gift and commodity exchanges. In this tradition, Seppälä (1998), on the basis of detailed field work in Tanzania, studies the interplay between these various spheres of exchange as a source of diversification and accumulation among individual households. Ledeneva's (2008) work on how informal networks are used to counteract shortages and substitute for market relations in the former Soviet Union suggests that in-kind uses of maize can be seen as a complement to the market.

Poor commercial incentives for maize could, in this situation, encourage in-kind uses rather than outright sale, since the in-kind value may be higher. Using maize as payment would also suggest an economy based, in part, on

[3] This was suggested by the qualitative data collected in Kenya and reported on by Andersson and Wambugu (2007).

in-kind transactions. A well-known phenomenon in the literature, using maize as payment for labour is perhaps the most obvious in-kind use of maize, especially given the dominant role of maize in local dietary patterns. Sixteen per cent (477) of the households, out of which 299 (63%) also remitted maize, were using maize to pay for agricultural labour. The use of maize to pay for labour is largely concentrated in Nigeria, however, where 48% of the households reported using maize for this purpose. The average amounts of maize used for labour payments were 719 kg in Nigeria, compared with 80 kg in Kenya at the bottom of the sample, where only five farmers were using maize for this purpose. The mean payment for those who used maize to pay for labour was 492 kg, compared with 152 kg used for maize remittances (Table 6.3). Although sizeable amounts of maize were used, labour payment in maize was restricted in terms of the number of individuals involved and connected to practices within national labour markets. The use of maize for in-kind payments when compared with remittances is limited, suggesting that remittances should not be viewed as one of a number of expressions of an economy relying on maize as an alternative currency. Rather, remittances appear to be indicative of households operating beyond the boundaries of the village, at least in terms of consumption.

Own consumption

Although remittances are neither detrimental to market participation nor signs primarily of an economy based, in part, on in-kind transactions of maize, the co-resident household may be compromising its own food security as a result of its subsistence obligations towards non-resident relatives and family members. The connection between price volatility and seasonal food insecurity is well documented in the literature, and as suggested by Barrett (2008), in general, autarky in terms of food security remains the preserve of those households with sufficient resources to avoid the market. Although autarky is undesirable as an alternative to an efficiently working market that connects producers with consumers, the vantage point from the individual household may be different, with producers preferring or needing to self-provision to the largest extent possible. In this respect, remitting households are foregoing their own possibility of disengaging from the market, at least partially, especially considering that many remitting households also bought maize.

A comparison with the consumption needs of the resident household is necessary to consider whether remittances should be viewed as a spatial extension of rural subsistence obligations. Table 6.7 details average production in relation to the consumption units[4] of the household. The household unit is based on co-residence for both remitting and non-remitting households. Hence, as can be seen in Table 6.7, remitting and non-remitting households

[4] A consumption unit takes into consideration the age composition of the household. The definition of a consumption unit can be found in Chapter 2, this volume.

Table 6.7. Food security of remitting and non-remitting households divided by maize production quartiles. (From: own survey data, 2008.)

	Household CU	Production per CU	Production per consumption unit excluding remittances	Sig.	Number of meals eaten during lean season
Remitters					
Q1	4.9	41	31	**	2.3
Q2	5.0	115	97		2.2
Q3	5.4	226	198	**	2.2
Q4	7.8	581	537	***	2.3
Non-remitters					
Q1	4.6		40	**	2.2
Q2	4.9		105		2.2
Q3	5.3		220	**	2.2
Q4	6.8		659	***	2.3

Where CU is the consumption unit.
Differences were tested between the remitting and non-remitting households in the same production quartiles for production per consumption unit, which in the case of the remitters included remittances.
***Significant at the 0.1% level; **at the 1% level.

contain almost identical numbers of consumption units and the production per consumption unit is very similar, regardless of whether the household remits maize or not. When the amounts remitted are taken into account, however, the production per consumption unit is lower among remitters in all production quartiles.[5] These differences are significant for the first, third and fourth quartile but not significant for the second quartile. The tendency for production per consumption unit to level among the two groups when remittances are taken into consideration is the more important of the two patterns though. Remitters, hence, do appear to be forfeiting their own consumption needs. As suggested by the number of meals eaten during the lean season, however, households in all the production quartiles fail to eat three meals per day, and lacking food security is a striking feature of the sampled households in general.

From a developmental point of view, in-kind maize remittances which bypass ordinary market channels may hold important food security implications for family members outside the village, constrained by either land or income. None the less, although not directly undermining commercial incentives for the remitters, remittances could perhaps be considered an informal means of enabling a degree of disengagement from a market characterized by seasonality and insularity. Survey data, as well as a wealth of evidence from

[5] Remittances have been deducted from total household production of maize to calculate the amount of maize that remains within the household. This amount has then been divided by the number of consumption units in the household to enable comparison with the households that do not remit maize.

numerous African countries (Stephens and Barrett, 2006; World Bank, 2007), show that smallholders, in general, sell maize immediately following the harvest to cover the most pressing cash needs of the household, only to be forced to later buy maize for their own consumption at prices elevated by short supplies. Even delinking partially from the market may be coveted by receiving households, as a means of attaining a measure of food security outside the forces of the market, even though such disengagement may carry latent expectations of reciprocity.

Direction of remittances and reciprocity

Numerous studies of the role of urban remittances to rural livelihoods attest to the traditional connection between rural and urban areas (see e.g. Baker and Pedersen, 1995; Baker, 1997) throughout the African subcontinent.

In most countries, rapid urbanization has characterized the period since the mid-1990s, while some African countries, most notably Kenya and Zambia (see e.g. Foeken and Uwuour, 2008), have experienced rising urban poverty rates, and recession has levelled the differences in rural and urban poverty levels. Measuring changes in rural and urban poverty levels for all the countries covered by the sample is difficult, due to the lack of time-series data, but Table 6.8 shows how the share of population between rural and urban areas has shifted since the mid-1990s. Mozambique, Ghana and Nigeria have experienced the largest

Table 6.8. Percentage of population by country, 1995 and 2007. (Adapted from: World Bank, 2009.)

	1995 (%)	2007 (%)	Change 1995 to 2007 (%)
Ethiopia rural	86	83	
Ethiopia urban	14	17	3
Ghana rural	60	51	
Ghana urban	40	49	9
Kenya rural	81	79	
Kenya urban	19	21	2
Malawi rural	87	82	
Malawi urban	13	18	5
Mozambique rural	74	64	
Mozambique urban	26	36	10
Nigeria rural	61	52	
Nigeria urban	39	48	9
Tanzania rural	80	75	
Tanzania urban	21	25	5
Uganda rural	88	87	
Uganda urban	12	13	1
Zambia rural	63	65	
Zambia urban	37	35	–2

relative changes in population composition since 1995, although in the case of Mozambique this may be related to mobility caused by the civil war. In the case of Zambia, the rural population has increased more rapidly than the urban population, and return migration has been occurring, as has also been noted elsewhere (World Bank, 2007).

To some extent, the generally rapid urbanization is reflected in the direction of remittances, with 33% of the remitters reporting that they sent staples to relatives in towns outside the district, while 23% remitted maize to the capital city. Urban destinations, in many cases, overlap with rural destinations, as household remittances were directed to a multitude of destinations.

While the connection between rural and urban areas has been emphasized in the literature since the mid-1990s, less interest has been placed on rural-to-rural linkages. In this context, the direction of remittances underscores the results from the Kenyan pilot study, which suggested a support system directed towards rural relatives, especially in surrounding rural areas. Indeed, 47% of the remitters in the 2008 cross-country study reported that staples were being sent or collected by relatives in neighbouring villages, and an additional 32% to relatives in other rural areas.[6] Whether such remittances are indirect payments in return for family labour or constitute important sources of food for the recipients is impossible to distinguish on the basis of the collected data. Given the large number of destinations, the amount remitted to each receiver is likely to be relatively small, but the spatial reach and subsistence burden of the co-resident household is underestimated if remittances are not taken into consideration. Although remittances may, to some extent, exist as a result of poor production potential in surrounding rural areas and be used to even out production differences between deficit and relatively surplus areas, they may also be symbolic aspects of gift-giving. Cultural aspects of remittances and the existence of multi-spatial household constellations may explain why households spread their remittances thinly among a number of destinations and relatives.

Moreover, the nature of remittances is likely to vary with the position of the recipient, in the sense that remittances to relatives or family members that are relatively well off should be viewed as gifts rather than subsistence obligations. In the context of underdeveloped milling structures in urban areas, remittances of maize meal may, as pointed out by A. Isinika (Cape Town, 2009, personal communication), also serve to counter the practical difficulties of processing maize among urban households.

Although female-headed households are under-represented (13%) among the remitters, compared with 21% for the non-remitters, de facto female heads could be expected to remit maize to absentee husbands working in urban areas. Only five remitting households were listed as de facto female headed. Remittances therefore appear to be occurring to relatives who are not part of

[6] The respondents could indicate a number of destinations for the remittances (neighbouring villages, other rural areas, towns in the same district, towns outside district, major urban areas and capital cities).

the household, as defined on the basis of co-residence in the context of data collection. Only 16% of remitting households received cash remittances, compared with 15% for those households that were not remitting maize. Moreover, as noted earlier, cash remittance in the sample constitutes only 5% of total cash income.

Hence, reciprocity in terms of cash was a very minor phenomenon and only small differences were noted between remitting and non-remitting households. Whether remitting households receive in-kind contributions in terms of labour or agricultural inputs is not possible to determine from the data, however. On the basis of cash remittances, it does appear as if food remittances should be conceived primarily as a support mechanism for less-food-secure relatives rather than a direct reciprocal arrangement which enhances the maize-remitting households' access to cash. Expectations of future reciprocity may underpin the relationship between the remitter and the recipient, however, although such latent reciprocity was not possible to capture using the survey format.

Conclusions

Failure to consider in-kind food remittances may lead to two sources of underestimation of the importance of smallholder-based agriculture as a source of subsistence-based food security, one urban and one rural. Firstly, the role of subsistence agriculture to urban livelihoods may be underestimated since such remittances are not directly visible in market transactions. Secondly, the subsistence obligations of the rural household may be underestimated, since the consumption needs of non-resident household members are not considered in surveys related to smallholder production and food security. In this way, the size of the household in terms of consumption is underestimated, in the sense that the productive capacity of the household is stretched both spatially and numerically. Rural and urban areas are interconnected not only through the more visible market relations but also at the household level.

Remittances are flowing from rural areas, often from households that operate under precarious production conditions; the withdrawal of such remittances from even less productive rural environments and even more vulnerable households may be devastating. Meanwhile, in-kind remittances of maize may, among receivers, constitute crucial sources of supplementary food. At the rural end, such underestimation points to the importance of raising productivity within the staple sector, generally as a means of increasing the subsistence base of both resident and non-resident household members. At the urban end, the existence of food remittances points to the necessity of improving the commercial incentives for maize production and the importance of eliminating physical barriers to the free movement of maize across the national space, easing the vulnerability of the urban poor to price volatility in staple foods.

References

Andersson, A. (2002) *The Bright Lights Grow Fainter – Migration, Livelihoods and a Small Town in Zimbabwe*. Almkvist och Wiksell, Stockholm, Sweden.

Andersson, A. and Wambugu, S. (2007) Maize circulation within family based networks: a case study from Lugari District, Kenya. Department of Sociology and Department of Social and Economic Geography. Lund University, Lund and Department of Geography, Kenyatta University, Nairobi.

Ashley, C. and Maxwell, S. (2001) Rethinking rural development. *Development Policy Review* 19, 395–425.

Baker, J. (ed.) (1997) *Rural–Urban Dynamics in Francophone Africa*. Nordiska Afrikainstitutet, Uppsala, Sweden.

Baker, J. and Pedersen, P.O. (eds) (1995) *The Rural–Urban Interface in Africa: Expansion and Adaption*. Institute of African Studies (Nordiska Afrikainstitutet) in cooperation with Centre for Development Research, Uppsala, Sweden.

Bank, L. and Qambata, L. (1999) *No Visible Means of Subsistence: Rural Livelihoods, Gender and Social Change in Mooiplaas, Eastern Cape 1950–1998*. Afrika Studiecentrum Working Paper. Africa Studiecentrum, Leiden, the Netherlands.

Barrett, C.B. (2008) Smallholder market participation: concepts and evidence from eastern and southern Africa. *Food Policy* 33, 299–317.

Bates, R. (1989) *Beyond the Miracle of the Market: the Political Economy of Agrarian Development in Kenya*. Cambridge University Press, New York.

Bloch, M. and Parry, J. (eds) (1991) *Money and the Morality of Exchange*. Cambridge University Press, New York.

Bryceson, D.F. (1987) A century of food supply in Dar es Salaam. In: Guyer, J. (ed.) *Feeding African Cities*. Manchester University Press, Manchester, UK.

Bryceson, D.F. (1993) *Liberalizing Tanzania's Food Trade*. UNRISD in association with James Currey, London.

Bryceson, D.F. (1997) De-agrarianisation in sub-Saharan Africa: acknowledging the Inevitable. In: Bryceson, D.F. and Jamal, V. (eds) *Farewell to Farms. De-agrarianisation and Employment in Africa*. Ashgate Publishing Ltd, Aldershot, UK.

Bryceson, D.F. (1999) African rural labour, income diversification & livelihood approaches: a long-term development perspective. *Review of African Political Economy* 26, 171–189.

Collier, P. (2008) The politics of hunger. *Foreign Affairs* 87, 67–80.

Devereux, S. (1999) Making less last longer – informal safety nets in Malawi. IDS Discussion Paper. Sussex University, Brighton, UK.

Dorward, A., Chirwa, E., Kelly, V., Jayne, T.S., Slater, R. and Boughton, D. (2008) Evaluation of the 2006/2007 Agricultural Input Supply Programme, Malawi, Final Report, March 2008. Imperial College, London, Wadonda Consult, Michigan State University, Overseas Development Institute, Ministry of Agriculture and Food Security, Malawi.

Ellis, F. (2005) *Small Farms, Livelihood Diversification and Rural–Urban Transitions, Strategic Issues in Sub-Saharan Africa*. The future of small farms, research workshop, jointly organized by the International Food Policy Research Institute (IFPRI) and the Overseas Development Institute (ODI) and Imperial College, London.

Fafchamps, G. and Minten, B. (2001) Property rights in a flea market economy. *Economic Development and Social Change* 49, 229–268.

Ferguson, J. (1999) *Expectations of Modernity, Myths and Meanings of Urban Life on the Zambian Copperbelt*. University of California Press, Berkeley.

Foeken, D. and Uwuour, S. (2008) Farming as a livelihood source for the urban poor of Nakuru, Kenya. *Geoforum* 39, 1978–1990.

Guyer, J. (ed.) (1987) *Feeding African Cities*. Manchester University Press, Manchester, UK.

Jayne, T.S., Mather, D. and Mghenyi, E. (2006a) *Smallholder Farming under Increasingly Difficult Circumstances: Policy and Public Investment Priorities for Africa*. MSU International Development Working Paper. Department of Agricultural Economics, Michigan State University, East Lansing, Michigan.

Jayne, T.S., Zulu, B. and Nijhoff, J.J. (2006b) Stabilizing food markets in eastern and southern Africa. *Food Policy* 31, 328–341.

Ledeneva, A. (2008) Blat and Guanxi: informal practices in Russia and China. *Comparative Studies in Society and History* 50, 118–144.

Polanyi, K. (1957) *Trade and Market in the Early Empires*. The Free Press, Glencoe, UK.

Poulton, C., Kydd, J. and Dorward, A. (2006) Overcoming market constraints on pro-poor agricultural growth in sub-Saharan Africa. *Development Policy Review* 24, 243–277.

Ranger, T.O. (1985) *Peasant Consciousness and Guerilla War in Zimbabwe*. James Currey, London.

Sahlins, M. (1965) The sociology of primitive exchange. In: Banton, M. (ed.) *The Relevance of Models for Social Anthropology*. Tavistock, ASA Monographs, London.

Sahlins, M. (1972) *Stone Age Economics*. Aldline Press, Chicago, Illinois.

Seppälä, P. (1998) *Diversification and Accumulation in Rural Tanzania*. Nordiska Afrikainstitutet, Uppsala, Sweden.

Stephens, E.C. and Barrett, C.B. (2006) *Incomplete Credit Markets and Commodity Marketing Behaviour*. Working Paper, Department of Applied Economics and Management. Cornell University, Ithaca, New York.

Stichter, S. (1982) *Migrant Labour in Kenya: Capitalism and African Response, 1895–1975*. Longman, Harlow, UK.

Von Braun, J. and Kennedy, E. (eds) (1994) *Agricultural Commercialization, Economics Development and Nutrition*. International Food Policy Research Institute (IFPRI), The Johns Hopkins University Press, Baltimore, Maryland.

Wood, A. (2002) *Could Africa Be More Like America?* Department for International Development, Government of the United Kingdom, London.

World Bank (2007) *World Development Report 2008: Agriculture for Development*. World Bank, Washington, DC.

World Bank (2009) *World Development Indicators*. World Bank, Washington, DC.

7 Meeting the Financial Needs of Smallholder Farmers in Ethiopia

WOLDAY AMHA

Association of Ethiopian Microfinance Institutions (AEMFI), Addis Ababa, Ethiopia

Ethiopia is one of Africa's largest countries, with about 77 million people. The agricultural sector in Ethiopia contributes about 44% of the national gross domestic product (GDP), 86% of the total export earnings and 80% of employment. Moreover, about 90% of the poor depend on agriculture for their livelihood (World Bank, 2009). Ethiopia's economy has grown very rapidly during the last 4 years. GDP grew, on average, by 11.6% between 2003–2004 and 2007–2008. Agriculture has grown by 13% per year, on average, since 2003–2004, followed closely by the service and industry sectors (Loening *et al.*, 2009). The high growth rate in the agricultural sector, which is the key to the entire growth of the economy, is partly explained by the commitment of the government, reflected in the allocation of a relatively higher budget for the sector. About 25% of government expenditure in the country is for rural infrastructure and agriculture and rural development, one of the highest shares in the world (World Bank, 2009).[1] Moreover, meeting the targeted economic growth and Millennium Development Goals (MDGs) in Ethiopia entirely depends on the performance of the agricultural sector. In order to attain the targeted 11% annual growth in the sector and meet MDGs, financial systems, extension services, rural infrastructure, marketing and distribution systems need to be addressed. In spite of the impressive growth in agriculture from a low base, the sector continues to be of a subsistence nature and fails to produce enough agricultural production which can ensure food security.

Area cultivated and yield of cereals in Ethiopia have shown an increasing trend. Although cereal yields are relatively higher than the average in eastern Africa, they

[1] The country has also registered a remarkable growth in expanding physical and social infrastructure in the last 7 years. The paved road network has increased by 43%; power-generation capacity has nearly doubled; primary school enrollment has increased from 5.2 million to 13 million (World Bank, 2009).

are still low – less than a fifth of the level in Egypt, less than a third of that in China and Vietnam (Alemayehu Seyoum, 2009). Increasing agricultural production and productivity and ensuring food security are the key objectives of the development policies/strategies and programmes of Ethiopia. However, farmers have limited internal capacity or resources to procure additional farm inputs, such as improved seeds, fertilizers, chemicals, etc., and make farm-related long-term investments. Delivering financial services, such as credit, savings, money transfer, etc., to the smallholder farmers has been identified as an important instrument capable of breaking the vicious cycle of poverty and ensuring food security, as it would increase household production, productivity, employment, income, consumption and empowerment of disadvantaged groups. Improving financial access to the poor would also facilitate economic growth by easing liquidity constraints and providing capital to start up new production-related activities or adopt new technologies. Access to financial services provides new opportunities, builds the confidence and self-esteem of the smallholder farmers and empowers the disadvantaged groups, such as women. However, access to finance alone is not a panacea for poverty and related development challenges. Improving access to finance will have a sound impact when it is complemented with other development interventions, such as the diffusion of modern technology through extension programmes, development of input and output markets and building networks and linkages, increasing access to both producers and buyers in both domestic and international markets, and building the capacity of smallholder farmers and their organizations.

Statement of the Problem

Improving financial access to smallholder farmers has been one of the most prominent instruments in the development programmes and strategies used by the Ethiopian government and development partners. Over the past 40 years, huge financial resources have been injected in the form of credit to support agricultural production, increase productivity and create employment in rural areas. However, delivering loans to smallholder farmers during Derg and pre-Derg[2] has been characterized by poor loan repayment and unsustainable subsidies and financial service providers. Over the last decade, finance providers such as the deposit-taking microfinance institutions (MFIs) and financial cooperatives have been exerting commendable efforts in Ethiopia in the provision of sustainable financial services to smallholder farmers. Despite the continued hard work and effort of finance providers, governments, donors and other development partners to expand outreach in delivering financial services to smallholder farmers, there is still a huge unmet demand for such services. Such unmet loan demand is seen not only in the production sphere but also in the processing, marketing and other areas. Thus, there is a need to revisit the

[2] Derg refers to the era of a socialist-oriented military regime in Ethiopia. Derg represents an Amharic word for the provisional Military Administative Council. The regime lasted for 17 years (1974–1992). Pre-Derg refers the era before 1974.

entire approach of the delivery system in order to satisfy this unmet demand. The effort to expand outreach and efficiency in delivering financial services to smallholder farmers in Ethiopia should address the following issues:

1. Like many African countries, smallholder farmers in Ethiopia are located in dispersed areas due to low population densities as well as a difficult terrain, which increases transaction costs of finance providers.
2. Smallholder farmers often demand relatively small loans and savings accounts, which again increase the per unit transaction costs of financial providers.
3. Smallholder farmers are heterogeneous, with varying skill and cultural background.
4. Getting information to make an accurate assessment of willingness and capacity of smallholder farmers to repay loans often takes time and money.
5. The weak institutional capacity of finance providers affects their outreach, efficiency and sustainability.
6. Providing financial services to smallholder farmers is perceived as being less sound or risky because of the covariate risks tied with agricultural production and marketing risks, seasonality and absence of formal insurance mechanisms to mitigate risks.
7. Smallholder farmers have little acceptable collateral, due to either lack of assets or unclear property rights or proper registry system for movable assets they possess.
8. Poor communication systems and physical infrastructure increase transaction costs.
9. Inadequate meso-level infrastructure, regulation and supervision, as well as inadequate contract enforcement mechanisms, limit the provision of financial services to smallholder farmers (Wolday Amha, 2008a).

As a result of the above challenges, credit to smallholder farmers in Ethiopia is characterized by high lending costs and high demand, resulting in relatively high interest rates being charged to borrowers compared to the formal banking sector.

Although delivering financial services to the smallholder farmers in Ethiopia, particularly in remote areas, is very challenging, lessons and innovative practices on how to advance the delivery of financial services in sustainable ways are emerging. Towards the end of the 1990s, new and innovative approaches for delivering financial services to smallholder farmers in the rest of the world have been implemented by deposit-taking MFIs, financial cooperatives, banks and non-governmental organizations (NGOs). Practitioners in Ethiopia have identified and implemented the essential requirements needed to establish financial systems and developed innovative financial products and services that match the needs of smallholder farmers. Moreover, the renewed emphasis of governments and donors on increasing agricultural production, particularly after the recent worldwide increase in prices of agricultural products, has also put agricultural development and rural finance back in the spotlight of the development agenda. To this end, the macro, meso, micro and sectoral policies and development programmes are giving due focus to the provision of sustainable finance to smallholder farmers.

In the Ethiopian context, although commercial banks can provide financial services (if they will) to the smallholder farmers, it is the deposit-taking MFIs and financial cooperatives that provide financial services in sustainable ways. Unlike many countries, the delivery of microfinance or rural finance services in Ethiopia is strictly regulated. The government issued its first microfinance legislation in 1996 (Proclamation 40/1996), with the aim of providing microfinance services to the poor households through deposit-taking MFIs. The government has created an environment conducive to the development of cooperatives. The Proclamation on Cooperatives 147/1998 states that:

> it has become necessary to establish cooperative societies which are formed of individuals on voluntary basis and who have similar needs for creating savings and mutual assistance among themselves by pooling their resources, knowledge and property; (...) has become necessary to enable cooperative societies to actively participate in the free market system.
>
> (FDRE, 1998:942)

The main objective of the MFIs and cooperatives in Ethiopia is to deliver micro-loans, micro-savings, micro-insurance, money transfer, leasing, etc., to a large number of productive yet resource-poor people in a cost-effective and sustainable way. At the end of the day, the interventions of MFIs and financial cooperatives or Savings and Credit Cooperatives (SACCOs) in the country should bring measurable impacts on the well-being of millions of households. However, in spite of the government's efforts to deliver financial services through MFIs and cooperatives, particularly financial cooperatives, providing efficient, sustainable and widely accessible financial services for a large number of smallholders is still a key challenge to implement development policies and programmes in Ethiopia.

This study attempts to inform policy makers, development partners and other key stakeholders on how to develop appropriate strategies, regulation and legislative framework and meso-level infrastructure and establish sustainable finance providers, which play a critical role in expanding the delivery of financial services to smallholder farmers in Ethiopia. It also provides useful information to finance providers in designing a range of financial services that meet the needs of various categories of smallholder farmers.

Objectives

The specific objectives of the study include:

1. Assess the policies, strategies and regulatory framework and the meso-level players that affect the delivery of financial services to smallholder farmers in Ethiopia.
2. Review the financial landscape or 'who is who' in the delivery of financial services to smallholder farmers.

3. Examine the developments in accessing financial services (trends), financial products and determinants of taking loans and accessing savings services using the *Afrint I* and *Afrint II* surveys.
4. Propose areas of interventions to address the critical challenges of expanding financial access to smallholder farmers.

Method of Data Collection and Analysis

The study used both qualitative and quantitative information collected from secondary and primary sources. The secondary data were collected from the Association of Ethiopian Microfinance Institutions (AEMFIs), the Federal Cooperative Agency (FCA), the National Bank of Ethiopia (NBE) and other institutions. Primary data were collected mainly through panel data *Afrint I* in 2002 and *Afrint II* in 2008, conducted by Lund University, Sweden. A total sample of 480 households was administered in Assebot, Bekoji, Bako and Yetmen. Each sample area contains two kebeles or peasant associations (PA), one of which is the PA that was sampled in the 2002 *Afrint* survey. Each kebele that was surveyed in 2002 contains 80 sample households. The newly sampled kebeles in 2008 include 40 sample households. On top of the multiple regression and logistic models, simple quantitative statistics, such as percentile, mean and frequency, were used to analyse the data.

Providing Financial Services to Smallholder Farmers: a Conceptual Framework

Improving financial access helps smallholder farmers to improve production and productivity through investment in irrigation, production equipment and inputs and in postharvest handling, processing and marketing. This will have a direct implication on the implementation of the various development policies and programmes, such as the food security and poverty reduction programmes. It also helps the rural households to increase their income and rural livelihoods by creating opportunities to engage in non-farm enterprises such as handicraft, trade, etc. Access to credit, savings, insurance, money transfer, etc. helps smallholder farmers to manage seasonal liquidity shortages, develop a savings culture, finance the emergency needs of the household and mitigate risks and shocks. According to a World Bank report (2005), substantial agricultural development – and development related to processing and marketing facilities in rural areas – and real increases in incomes of rural families have happened almost nowhere without access to financial services.

Finance is one of the key elements in addressing development issues in Ethiopia. It is even considered to play a leading role in guiding development interventions in the country. Whatever development strategies or programmes (poverty reduction strategy, rural development strategy, industrial development strategy, food security strategy, etc.) we propose for Ethiopia, there is a need for finance and financial systems to implement the development programmes. The provision of finance contains two basic elements: (i) capital, the funds that are being provided; and (ii) financial system, the process of providing them and the institutions involved in this process. If the objective is to deliver financial services to smallholder farmers and promote agricultural investments in Ethiopia, there is a need to have both the capital and well-functioning financial systems and institutions in place. In order to increase outreach, efficiency and sustainability, there is a need for interventions/support at macro, meso and micro/individual financial institutions and client levels.

The conceptual framework is expected to provide the road map on how to provide a broad range of financial services, expanding outreach and ensuring sustainability of finance providers based on a market-based paradigm. The new framework, as illustrated in Fig. 7.1, shows that improving financial access to smallholder farmers has the potential to make a significant difference in agricultural production, food security, poverty reduction and economic growth, which requires: (i) enabling policies, strategies, legislative and regulatory framework and infrastructure development; (ii) supportive meso-level infrastructure and technical service providers; (iii) efficient and sustainable financial systems or institutions; and (iv) the clients, who are the targets in building inclusive financial systems.

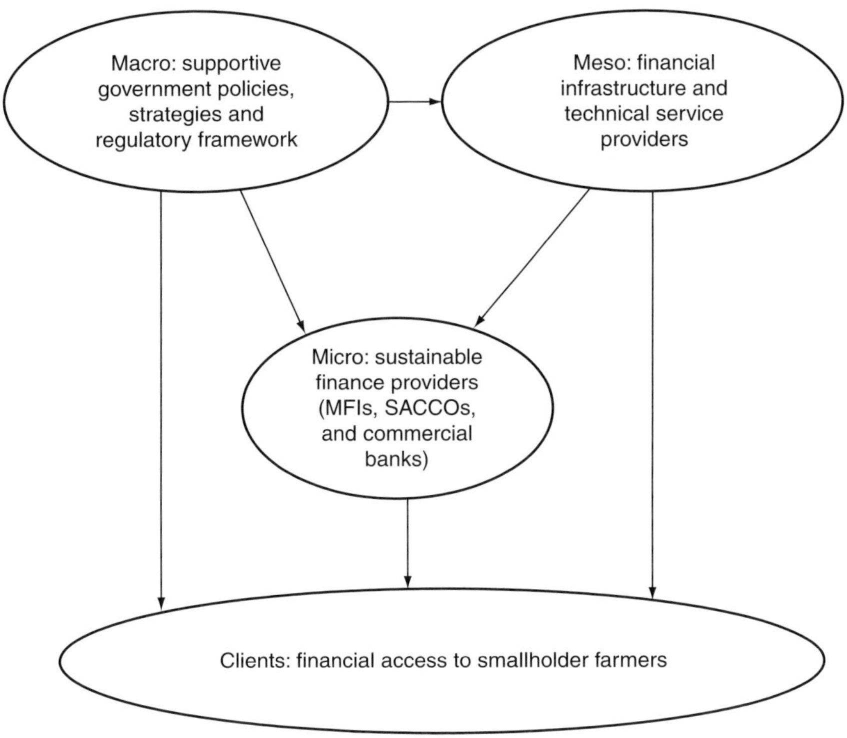

Fig. 7.1. Conceptual framework to expand the delivery of financial services to smallholder farmers.

The government is the key player at the macro level, which creates the enabling strategies, legal and regulatory framework and builds the infrastructure. The government is expected to be involved in implementing an inclusive finance strategy, deregulation of interest rate, liberalization of the financial sector, establishment of a legal system that enforces contracts, protects property and land-use rights and ensuring due legal process. In the Ethiopian context, the key macro-level players include the NBE or the Central Bank, which regulates banks, insurance companies and deposit-taking MFIs, and the Federal Cooperative Agency, which regulates and promotes the financial cooperatives. In addition to the above institutions, the Ministry of Finance and Economic Development plays a critical role in the development of the entire financial sector in Ethiopia.

On top of creating an enabling macro-policy and regulatory framework, creating sustainable finance providers in rural areas requires a well-functioning financial infrastructure (the 'architecture', along with a web of other technical service providers) offering a range services which reduce transaction costs at meso level. According to Helms (2007), an efficient meso-level infrastructure is critical for the functioning of the financial system as a whole and especially for expanding access to financial services for the poor. It extends from financial infrastructure to systems that promote transparency of financial institutions, technical service providers that offer training and consulting services, and professional associations and networks (WWB/AMAF, 2009).The key players at the meso level are training providers, networks such as the Association of Ethiopian Microfinance Institutions (AEMFI), federation of financial cooperatives, auditors, accountants, IT service providers, credit reference bureaus, domestic rating agencies, payment system and professional certification institutes, wholesaling institutions and other service providers, such as Business Development Service (BDS) providers, to enhance the capacity of clients. In the Ethiopian context, the meso-level support focuses on reducing transaction costs, improving sector information and transparency, increasing access to refinancing for finance providers, and enhancing skills and other capacity-building support across the sector.

The micro-level players include those inclusive finance providers which are involved in delivering a range of financial services to the unbanked through formal and informal channels (which include banks, deposit-taking MFIs, financial cooperatives, NGOs, government projects, multipurpose cooperatives, woreda or district administration, moneylenders, Rotating Savings and Credit Associations (ROSCAs) or Ikkub, etc) in a sustainable manner. However, since some of the finance providers are not operationally and financially sustainable, there is a need to support them through institutional capacity building (including systems development), promoting market-based interest rates, encouraging competition and benchmarking, providing continuous training, promoting transparency, developing client-centred financial products, etc.

The target groups or the clients of the inclusive finance providers are those who are excluded from the formal financial sector. They are typically the smallholder farmers, self-employed, low-income entrepreneurs, women, men, youth, micro-, small and medium enterprise operators, marketers and agro processors, those who are employed in low-salary jobs and in the informal sector in both rural and urban areas.

The Delivery of Financial Services to Smallholder Farmers: Literature Review

The flow and impact of credit and other financial services to rural households has not been properly documented. This has contributed to the lack of understanding on how financial services and credit policies to the smallholder farmers and their enterprises have affected agricultural production and productivity, and the dissemination of agricultural technologies. Although there were some unsuccessful and unsustainable credit programmes implemented by formal banks and government/donor projects in Ethiopia and elsewhere, there are successful cases, which demonstrate that credit is a powerful instrument to improve the livelihoods of smallholder farmers. Many studies have indicated that credit is one of the critical inputs required to increase agricultural production and productivity and income, smooth consumption, improve household welfare (including education, health, etc.), build household assets and mitigate risks.

The study of Khandker and Rushidur Faruquee (1999) reveals that the Agricultural Development Bank of Pakistan (ADBP) contributes to household welfare, and its impact is higher for smallholders than for medium and large holders in agriculture, but large holders nevertheless receive the bulk of ADBP finance, which was not cost-effective. Khandker (1998) observed that micro-credit programmes are as cost effective as other programmes, such as the food for work, in benefitting the poor. Binswanger and Khandker (1995) estimated the impact of formal credit using district-level data from India and found that formal credit increased rural income and productivity and that rural benefits exceeded the cost of the formal system by at least 13%. Pitt and Khandker (1998) examined the impact of credit from the Grameen Bank and other targeted credit programmes in Bangladesh on a variety of individual and household-level outcomes, including school enrolment, labour supply, asset holding, fertility and contraceptive use. They found credit to be a significant determinant of many household outcomes, and programme credit has a significant effect on the well-being of poor households in Bangladesh.

On the other hand, the result of the study by Kochar (1997) revealed that the demand for credit, particularly in productive areas, is low and the role that credit can play in enhancing agricultural development is limited. The evidence from the study of Swain (2002) in Puri (a relatively poor region in India) indicates that there is relatively high demand for credit, suggesting a need for further development of the credit programmes, which requires not just increased outreach of credit to farmers but also well-designed credit facilities that benefit the disadvantaged and not just the rich and large landowners. The study concluded that credit policies still have an important role to play in agricultural development. On the other hand, many government- and donor-funded programmes in the past have failed not only in delivering credit to target households but also in promoting a viable credit delivery system. High covariate risk of agricultural production (Binswanger and Rosenzweig, 1986), the asymmetric information and lack of enforcement of loan contract (Hoff and Stiglitz, 1990), government imprudent interference in credit markets and rent-seeking as a result of credit rationing (Braverman and Guasch, 1998) are some of the factors alleged for the poor performance of the government-directed credit schemes.

The study of Diagne and Zeller (2001) indicates that formal lenders in Malawi – such as rural banks, savings and credit cooperatives and special credit programmes supported by the government and NGOs – prefer to give loans to households with diversified asset portfolios and therefore diversified income sources. This is presumably done to increase and stabilize repayment rates. The level of interest rates charged on loans seems not to be an important factor for households in deciding in which microfinance to participate. Non-price attributes of credit institutions and their services play a larger role. Those attributes include types of loan provided and the restrictions on their use, as well as the types of non-financial services provided by the programmes, such as training in management of microenterprises. The findings of the same study reveal that improving access to credit is not always a potent means of alleviating poverty. The study found no significant impact of access to credit (provided for productive activities) on the per capita income, food security and nutritional status of credit programme members. The study concludes that the contribution of rural microfinance institutions to the income of smallholders can be limited or outright negative if the design of the institutions and their services does not take into account the constraints on and demands of clients. Developing attractive credit requires both identifying farm and non-farm enterprises and technologies that are profitable under the conditions experienced by subsistence-oriented farmers and responding to the numerous constraints of resource-poor rural households. The results suggest that a strategy of expanding financial institutions in rural, drought-prone areas, with inadequate market and other infrastructure may – at least in a below average rainfall year – have no significant positive welfare effects.

According to the results of the Tegemeo household survey in Kenya, households that received credit for maize production had a higher productivity, averaging 7.65 bags per acre as compared to 6.5 bags per acre among households that did not receive credit (Kibaara, 2007). A farm household facing binding credit constraint is more likely to misallocate its resources and under-invest than its unconstrained peers. Thus, availability of finance and its accessibility crucially affect production start-up and subsequent performance of farmers (Hussien and Ohlmer, 2006).

The study of Freeman *et al.* (1998a) looked at both the supply of credit from financial institutions in Ethiopia, Nigeria and Uganda, particularly the credit-allocation policies of formal banks, and at the demand for livestock credit in the sample households. The results of the study revealed that although the formal banks covered in the study, namely Agricultural and Industrial Development Bank of Ethiopia, the Nigerian Agricultural and Co-operative Bank and the Uganda Commercial Bank, had the aim of increasing the flow of institutional credit to large numbers of smallholder livestock producers, very few livestock producers obtained formal credit in these countries. Smallholder producers were often screened out of the formal credit markets because of the criteria banks used to screen loan applicants. For example, the Uganda Commercial Bank required potential borrowers to show that they have the infrastructure for keeping livestock before the bank approves the loan. The study found that farmers who had a larger proportion of cross-bred cows used more inputs – especially feed – and tended to have more profitable dairy operations than those who had fewer cows. The study identified that the loan products of the three banks were not sufficiently flexible to address the various loan needs of livestock producers. It also recommended revisiting the loan delivery systems, loan policies and loan term structures.

The findings of Jabbar *et al.* (2002) revealed that borrowers used their loans mainly to acquire hybrid cows, so that the primary impact of credit was to increase milk production through increased dairy herd size. Moreover, the study reveals that the sex of household head, education, dairy training, prevalence of outstanding loans and the number of improved cattle on the farm had significant influence on both borrowing and liquidity status. In Ethiopia and Kenya, an additional cross-bred cow on a credit-constrained farm contributes about twice as much as milk output (in litres) per farm as it would on a non-credit-constrained farm. The study of Freeman *et al.* (1998a) indicated that a 1% increase in credit to purchase cross-bred dairy cows leads to a 0.6% increase in milk productivity on credit-constrained farms and a 0.4% increase on non-credit-constrained farms in Ethiopia. In Kenya, a 1% increase in credit for investment in cross-bred dairy cows leads to a 1.6% increase in milk productivity on credit-constrained farms and a 0.9% increase on non-credit-constrained farms. This suggests that credit should be targeted at credit-constrained farms (depending on their household characteristics) to achieve the greatest impact.

The findings of Freeman *et al.* (1998b) on credit and uptake of improved technologies in Ethiopia showed that the bulk of the credit to smallholder dairy farmers was used to purchase cross-bred cows and that borrowing farmers with liquidity constraint had significantly larger cattle herds than non-borrowing farmers, suggesting that credit was used mainly for acquiring cattle. Very little credit was used for the purchase of variable inputs such as improved feed or veterinary services. While the adoption of technologies is closely related to the investment decision in cross-bred cows, lack of credit for the purchase of variable inputs is a major constraint to increasing yields and ultimately the profitability of investments in improved dairy technologies. One clear implication from the study is that improving access to adequate credit to farmers whose activities are constrained by liquidity will accelerate the uptake of dairy technologies in Ethiopia.

A similar study conducted in Kenya (Oluoch-Kosura and Ackello-Ogutu, 1998) showed that, with respect to milk production per animal, liquidity-constrained farmers produced significantly less than non-liquidity-constrained farmers. There were strong indications that credit had an important role to play in overcoming liquidity constraint and in the use of improved technology and subsequently increased yield. Moreover, non-credit-constrained farmers, on average, achieved almost ten times the level of dairy gross margins obtained by the credit-constrained farmers. This implies that improving access to credit will lead to greater incentives to adopt improved dairy technologies and hence achievement of higher output and net returns.

The study of Bbuza *et al.* (1998) in Uganda showed a high proportion of liquidity-constrained farmers among borrowers, suggesting that many farmers received inadequate credit. A large number of farmers who received credit expressed a desire to receive more loans to finance their dairy operations. While the study showed that smallholder dairying was a profitable activity, the lack of any significant differences in performance between liquidity-constrained and non-constrained farmers suggested that many factors other than farmers' liquidity position were important in explaining differences in performance. For example, donor intervention in the study area, which provided in-calf heifers and supplementary feed, was important in explaining some of the observed differences.

A comparative study by Freeman *et al.* (1998c) concluded that the total flow of institutional credit from various sources to smallholder dairy producers in Ethiopia was too small to make an impact on dairy production because credit policies and the credit delivery system discriminated against the smallholder livestock farmers. In contrast, Kenya's dairy cooperatives were the most important sources of credit for smallholder producers. The survey results showed that 67% of borrowers in Kenya obtained loans from cooperatives, while the corresponding proportion in Ethiopia was less than 30%. The findings suggest that the functioning and effectiveness of credit delivery systems in different countries is perhaps one of the most important determinants of smallholder farmers' credit-constraint conditions because they largely determine their access to additional liquidity.

Macro-, Meso- and Micro-level Players Influencing Financial Access to Smallholder Farmers in Ethiopia

Creating employment, increasing production and productivity, and improving input and output markets of smallholder farmers requires a package of interventions, with finance as a key component. Increasing outreach (growth), efficiency and sustainability of financial institutions providing financial services to the smallholder farmers and their enterprises requires deliberate interventions at the macro level (including the regional- and sectoral-level policies and infrastructure development), meso level (focusing on creating the financial infrastructure and building the capacity of technical service providers) and micro level (expanding the outreach, efficiencies and sustainability of finance providers) and building the capacity of smallholder farmers themselves to utilize and manage the financial resources in productive and value-adding activities.

Macro-level policies, strategies and regulations

An enabling policy, legal and regulatory environment is important to expand financial access to smallholder farmers and attain operational and financial sustainability of finance providers. The macro and sectoral policies, the commitment of governments at various levels, the performance of the macroeconomy (the state of the real economy in terms of improving efficiency in agriculture, industry, commerce, etc.), the performance of the financial sector, the efficiency of the legal and regulatory system, fiscal and monetary system, infrastructural development and the political system affect the level of outreach and performance of finance providers for smallholder farmers and their enterprises. The federal and regional governments in Ethiopia are aware of the role of finance in development, which is reflected in all the development strategies and programmes and sectoral policies. The federal government of Ethiopia has shown its commitment to promoting the delivery of financial services to rural households in various ways. The government considers access to finance as an important tool to fight poverty and ensure food security. This is reflected in the poverty reduction programme, food security strategy, rural development programme, industrial policy, etc.

Stable macroeconomic conditions and relatively low inflation support the growth of finance providers. However, macroeconomic stability and growth, although important, is not, by itself, a sufficient condition for the growth of sustainable finance providers and expanding outreach. There are poorly performing finance providers in countries with stable macroeconomic conditions. It is also possible for finance providers to operate effectively even when the economy is not that stable. Dedebit Credit and Savings Institution (DECSI) in Tigray region is a good example. Although the Tigray regional state, where DECSI operates, is one of the poorest regions in Ethiopia (with a relatively stagnant economy and seriously affected by war and drought), DECSI's performance in providing loan and other financial services to smallholder farmers in the last 12 years has been remarkable by all standards. This is partly attributed to the political commitment as well as a high degree of social cohesion and traditional social structures that facilitated the enforcement of contracts, increased outreach and improved its efficiency.

Key policies promoting rural finance

The Rural Development Strategy of the Ethiopian Government (2002) stated that rural finance is a vital tool to implement the Agricultural Led Industrialization Strategy, Agriculture Sector Development and other sectoral development programmes. Access to finance plays a critical role in increasing agricultural productivity, production, investment and employment and improving agricultural marketing (input and output markets). The strategy reiterates that an efficient rural financial intermediation and functional financial system is the basis to transfer resources from agriculture to other sectors of the economy. According to the Rural Development Strategy, there are three key financial institutions, namely banks, MFIs and cooperatives, which can support rural development in Ethiopia. Although banks have a limited role in providing financial services directly to smallholder farmers and their enterprises, they can play a useful role in providing finance to cooperatives and microfinance institutions. The Plan for Accelerated and Sustainable Development to End Poverty (PASDEP) document states that the NBE should foster the role of MFIs in intermediating financial assets in the rural areas. To this end, the NBE is expected to encourage commercial banks to on-lend to microfinance institutions.

One way to encourage commercial banks is by strengthening the regulatory framework and capacity of NBE's Microfinance Supervision Department.

On the other hand, MFIs in Ethiopia have been very successful in filling this gap by using innovative lending methodologies to large number of smallholder farmers. However, the MFIs need support from the government and other development partners to meet the huge demand for loans and reach millions of rural households. MFIs should link their activities with cooperatives and play a very important role in implementing the warehouse receipt system, developing crop and livestock insurance schemes and diversifying their products to reduce covariate risks. Governments at various levels should also support the activities of MFIs by creating an enabling policy and regulatory environment and providing all-round support at grass roots level. The strategy stresses that governments, at various levels, should not subsidize and interfere in the activities of finance providers.

The Rural Development Strategy indicates that cooperatives should play a useful role in the delivery of financial services to members and non-members by linking their activities with formal banks and MFIs. Cooperatives can even establish their own cooperative banks and other specialized banks to meet the financial needs of cooperatives at various levels. However, credit provided through multipurpose cooperatives should not erode the repayment culture of rural households, which has been the case during the era of Derg. The financial institutions identified in the Rural Development Strategy for delivering financial services are the commodity-based or multipurpose cooperatives. However, the strategy did not mention the role that financial cooperatives or SACCOs can play in sustainable delivery of financial services to the unbanked. Since sustainable delivery of financial services requires specialized financial institutions engaged in banking activities, commodity-based cooperatives may not be the appropriate institutions to deliver such services in a sustainable way. Efforts should be made to support the development of SACCOs at various levels (primary, union and federation). At the end of the day, since SACCOs are independent and self-governed private sector enterprises, they have to develop their own strategies to establish a federation or apex and cooperative banks.

The PASDEP indicates that access to finance is one of the main constraints in promoting private sector development, agricultural development, micro- and small enterprise development and investment in general. About one-third of households need to travel for 20 or more kilometres to reach the nearest financial service-providing centre. The proportion with financial services within the distance of 5 km is 77% in urban areas and only 17% in the rural areas. Despite the emergence of many financial providers, the PASDEP document indicates that only 6% of smallholder farmers in Ethiopia have access to their services. In the process of implementing PASDEP, it is stipulated that steps will be taken to promote the provision of credit in the farm input retailing system, service cooperatives and expansion of rural microfinance institutions.

The agricultural strategy, according to PASDEP, will revolve around a major effort to support the intensification of marketable products –for both domestic and export markets – and by both small and large farmers. Elements of the strategy include the shift to high-value crops, promoting niche high-value export crops, a focus on selected high-potential areas, facilitating commercialization of agriculture, supporting the development of large-scale commercial agriculture where it is feasible, and improving the integration of farmers with markets – both locally and globally. The majority of the strategies should be implemented by the private sector (which includes millions of smallholder farmers), but given the early stages of the transition to market-led agricultural development, a range of public investments and services are needed to help jump-start the process. The instruments to achieve these under PASDEP include: constructing farm-to-market roads, development of agricultural credit markets, introducing a specialized extension service or differentiated agricultural zones and types of commercial agriculture, developing national business plans and tailored packages for specialized export crops (such as spices, cut flowers, fruits and vegetables), enlarging the irrigated land through multipurpose dams, taking measures to improve land tenure security and to make land available, where feasible, for large-scale commercial farming and reforms that improve the availability of fertilizer and seeds.

An enabling regulatory environment

The various policies and the regulatory framework in Ethiopia designed to guide and monitor the activities of finance providers have direct impacts on outreach (expanding the delivery of financial services to millions of smallholder farmers) and viability (operational and financial sustainability). Ensuring the safety of clients and building healthy finance providers for the development of the financial sector appears to require prudential regulation and supervision compatible with the objective condition of Ethiopia. Actually, finance providers, such as deposit-taking MFIs, providing financial services (including savings mobilization from the public, particularly from the poor), with numerous repeated loans, physically attempting to provide their services to clients, quick repayment, using group-lending methodology, a highly decentralized system and with high operating cost per loan or deposit amount and management orientation towards poverty reduction (not always profit) do have specific risk profiles different from those of conventional banks. The high-risk profiles of MFIs will then increase the importance of prudential regulation, strict supervision and effective governance. It should also be noted that, unlike many countries, the delivery of financial services through deposit-taking MFIs in Ethiopia is part of the financial sector.

In order to clearly separate between charity (handout) and finance, the policy makers in Ethiopia introduced a regulatory environment that has a direct impact on building sustainable MFIs and reaching millions of poor households. The need for prudential regulation and supervision has also brought the activities of the MFIs under Ethiopia's monetary and financial policy framework. Until recently (2009), Proclamation No. 40/1996 'a proclamation to provide for the licensing and supervision of the business of microfinance institutions' was the major law that was used to regulate and supervise MFIs. The NBE is empowered to license, supervise and regulate the delivery of financial services to the rural and urban poor through MFIs. The 19 directives of the NBE still serve as the basis for prudential regulation, influencing good governance of

MFIs and prudent lending. The law allows MFIs to mobilize public savings from day one of their registration under the NBE. With the exception of banks, cooperatives and MFIs, the proclamation prohibits NGOs, government institutions and others from delivering loans and other financial services in urban and rural areas (Wolday Amha, 2008b). The Proclamation No. 84/1994, Licensing and Supervision of Banking Business, precludes foreign nationals from undertaking banking business, including microfinance activities. The microfinance law issued in 1996 (Proclamation No. 40/1996) was revised and replaced by Proclamation No. 626/2009. Although there isn't major change in the new microfinance law, it is relatively very strong in institutionalizing financial discipline, prudent lending and transparency of MFIs.

The cooperative movement in Ethiopia is guided by Proclamation 147/1998, a law that has relatively addressed most of the critical issues for promoting member-owned, need-based and sustainable cooperatives. However, Proclamation 147/1998 does not address all the critical issues of promoting financial cooperatives. Although regulation and supervision are very critical to sustainability and growth of financial cooperatives, the existing cooperative law does not provide the necessary guidance to regulate financial cooperatives as part of the financial sector. If significant resources are to be channelled through financial cooperatives, there is a dire need to issue a separate law for financial cooperatives and a separate regulatory framework to supervise and monitor their activities. The financial cooperative law and the regulatory framework will help the cooperatives to carry out the fiduciary responsibilities and protect the deposits of members and government and donors (who inject funds though various programmes, such as the food security programme). This will have a positive impact on the sustainable delivery of financial services, particularly for poor and remote households. Developing a separate law directly contributes to the establishment of financial cooperatives which are stable and efficient. Issuing a separate law and designing and implementing a regulatory framework for financial cooperatives will also provide member protection of their deposits against excessive risks that may arise from fraud, failure (insolvency) or opportunistic behaviour on the part of the financial cooperatives. Moreover, although the Federal Cooperative Agency can play a key role in promoting financial cooperatives, establishing a strong and independent federation of SACCOs can contribute in implementing self-regulation of financial cooperatives in Ethiopia. However, although regulation contributes to stable and efficient performance and outreach of the MFIs, financial cooperatives and others, effective regulation and supervision entail significant cost.

Meso-level players

There are five major institutions and/or agencies and programmes which are playing a critical role in supporting financial service providers engaged in inclusive financial services in Ethiopia. These include: (i) the Association of Ethiopian Microfinance Institutions (AEMFI); (ii) the Rural Financial

Intermediation Program (RUFIP); (iii) the Federal and Regional Cooperative Agencies; (iv) the unions of SACCOs; and (v) the credit guarantee schemes of NGOs. Apart from these organizations and a few consultants and auditors, there is very little financial infrastructure to support finance providers at the grass roots level in the country.

The Association of Ethiopian Microfinance Institutions (AEMFI)

The Association of Ethiopian Microfinance Institutions is the representative network of all the 30 licensed deposit-taking MFIs in Ethiopia. In 2006, the association also established a SACCO unit to promote sustainable financial cooperatives in the country. Registered in June 1999 under the Ministry of Justice of the Federal Democratic Republic of Ethiopia, the strategic goal of AEMFI includes, among others, creating an organizational structure that serves as a national industry forum and network for the activities of MFIs. The mission of AEMFI is to see a reduced level of poverty and increased level of consumption and ultimately wealth and capital creation in Ethiopia through the active intervention of efficient and sustainable inclusive finance providers. The main activities of AEMFI include capacity building of MFIs and financial cooperatives through training and technical assistance, creating an enabling policy environment through fact-based advocacy, conducting studies to understand the challenges of the industry, discussing and disseminating the results of its research activities, knowledge management and disseminating best practices through workshops, and monitoring and appraising the performance of MFIs. The Association of Ethiopian Microfinance Institutions has been very active in supporting the industry in fund-raising activities and served as the voice of the industry (Wolday Amha and Tigest Tesfaye, 2009).

Federal and Regional Cooperative Agencies

Proclamation 147/1998 provides the power to supervise and license all cooperatives, including the SACCOS, to the Federal and Regional Cooperative Agencies. Issues such as distribution of surplus, borrowing limits, measures to be taken at the time of the loss of property or fund, protection of creditors, settlement of disputes, revoking of licences, etc. are under the purview of the Federal Cooperative Agency (or its regional bureaus or offices). The Federal Cooperative Agency developed guidelines and by-laws for the promotion of savings and credit cooperatives. These guidelines help cooperatives to introduce some uniformity and standards. However, the basis of these guidelines remains the general law of cooperatives, rather than providing specific rules for SACCOs that accommodate financial norms and standards. Moreover, the cooperative promotion and regulation offices can directly, or by delegation, audit the accounts of cooperatives.

Savings and Credit Cooperatives Unions

Primary financial cooperatives, networked to a union, are expected to have opportunities to conduct joint training, improve supervision and facilitate liquidity exchange. To this end, SACCOs in Ethiopia are organized vertically into 42 unions in 2009, and a federation of SACCOs is still under formation. The financial cooperatives are gradually assuming responsibilities to handle more complex transactions by building their capacity to become sustainable and to be able to provide valued services to primary SACCOs. Promotional activities of the FCA and regional cooperative offices should be separated out so that it can focus on its regulatory function without conflict of interest. The primary and secondary SACCOs or unions need qualified external support from government, NGOs and other partners such as the World Council of Credit Unions and donors experienced in cooperative promotion and regulation.

Rural Financial Intermediation Program (RUFIP)

The Rural Financial Intermediation Program is the largest programme in Ethiopia, with a programme cost of US$95 million to support deposit-taking MFIs and rural SACCOs (RUSACCOs). It is a joint programme of the government of Ethiopia, the African Development Bank (AfDB) and the International Fund for Agricultural Development (IFAD). The programme has four components: (i) institutional development for MFIs and SACCOs; (ii) improved regulation and supervision of MFIs by strengthening NBE and AEMFI; (iii) equity and credit funds for MFIs and RUSACCOs; and (iv) programme coordination and management. RUFIP has been providing loans to MFIs and RUSACCOs with 7-year grace periods and 6% lending interest rate. As per the independent evaluation of the programme, RUFIP is one of the successful programmes of IFAD implemented in Ethiopia. Although the programme is in its final year, IFAD and the government of Ethiopia have agreed to initiate RUFIP II.

Credit Guarantee Schemes for MFIs

Microfinance institutions have access to credit guarantee schemes supported by donors and regional governments. For example, the MFI Wasasa has entered into a guarantee partnership with Awash International Bank. The term of the agreement reflects proper risk sharing between the partners: Wasasa deposits 10% of the loan in the facility of the bank; the guarantee's coverage is 45% while the bank assumes the remaining 45% risk. In 2006, ASCI negotiated with the Swedish International Development Cooperation Agency (Sida) for a 50% partial credit guarantee scheme earmarked for 'missing middle' clients. The guarantee fund amounts to 20 million birr (US$2.2 million). Eshet MFI is also a partner of a 50% guarantee fund with a foreign bank. The German financial cooperation – Kreditanstalt für Wiederaufbau (KfW) – has a €4.5 million (US$6.3 million) guarantee fund for MFIs and micro-, small and medium enterprises' financing – the Microfinance Refinancing Facility (MRF). The MRF will partially guarantee loans (at market rate) from commercial banks to selected MFIs. TERAFFINA, a network of NGOs in the Netherlands supporting MFIs and financial cooperatives, has a credit guarantee scheme for MFIs to access loan capital from commercial banks in Ethiopia. MFIs such as Amhara Credit and Saving Institution (ACSI), DECSI, Oromia Credit and Savings Share

Company (OCSCCO), Addis Credit and Savings Institution (AdCSI) and Omo Microfinance Institutions S.Co. (OMF) have accessed more than 1 billion birr (about US$100 million) from the Commercial Bank of Ethiopia with the credit guarantee scheme of the regional governments.

Other meso-level players

Other meso-level support institutions that often emerge in the context of a more mature microfinance industry are very limited in Ethiopia. These include credit bureaus, specialized consultancy firms, rating agencies, specialized auditors, training providers in universities and private institutes, institutions to certify trainers and other technical providers, organizations specialized in wholesale funding and liquidity-pooling facilities, and IT companies supporting the back-end systems and front-end technology for the sector.

Finance providers at the grass roots level

Ethiopia's rural financial markets are characterized by the coexistence of formal, semi-formal, and informal lenders. The finance providers or lenders vary in the cost of screening, monitoring and contract enforcement. The formal financial providers in Ethiopia include banks, MFIs and cooperatives. Iqqub or Rotating Savings and Credit Associations, iddir, mahiber, etc. are semi-formal financiers. The informal finance providers are the moneylenders, relatives, traders and suppliers, friends, church, etc. Although illegal, as per the regulatory framework in Ethiopia, NGOs, government and donor projects are providing loans to beneficiaries.

Banks

The Ethiopian banking sector consists of one state-owned development bank, the Development Bank of Ethiopia, two state-owned commercial banks (Commercial Bank of Ethiopia and the Construction and Business Bank), and ten private commercial banks. As of September 2008, the private banks had 304 branches and a total paid-up capital of 3.8 billion birr, compared to 264 branches and a paid up capital of 6.7 billion birr of the three public banks. Private banks' participation has increased gradually, and they currently account for 36.2% of commercial banking assets, with the remainder being the share of the two public sector commercial banks. The Commercial Bank of Ethiopia is still the main actor in the financial sector, representing over 63.8% of commercial banking assets. As of 2008, total assets of public banks in Ethiopia (37.6 billion birr) are more than double the private banks (16.4 billion birr), which shows the continued predominance of the public sector. According to the National Bank of Ethiopia's credit bureau, the number of outstanding loans in the formal financial sector amounts to 61,395 loans (March 2007). This clearly demonstrates one of the key deficiencies of the sector: access to credit finance is mainly provided to larger companies, relatively wealthy individuals and government projects. The largest part of the population is excluded from access to credit services. Financial services coverage of formal banks in Ethiopia stands at approximately 134,670 people per branch, which is far below international and African standards (Ghana 54,000; Uganda 130,000; Namibia 11,136). However, this ratio has improved considerably, from 253,000 people per branch in 1995 to 193,000 people per branch in 2006. Bank branches are concentrated in urban areas. More than 52% of all bank branches are located in the eight major towns, where only 6.6% of the population lives; Addis Ababa alone accounts for 37.6% of all bank branches.

Microfinance institutions (MFIs)

Although the development of deposit-taking MFIs started very recently (1997), the industry showed a remarkable growth in terms of outreach. Since the issuance of the microfinance law (1996), 30 MFIs have been registered by NBE to deliver financial services to the poor. A large number of these MFIs have achieved significant progress in terms of both outreach and sustainability. As of December 2008, the 27 microfinance institutions registered under the NBE had an active loan portfolio of about 4.6 billion birr (US$447 million),[3] delivered to 2.2 million active clients. They also mobilized about 1.6 billion birr (US$165 million) of savings. Moreover, about 40% of the clients of the MFIs are female (Wolday Amha, 2008b).

The average loan size and savings of MFIs in 2006 were about 2069 birr (US$207) and 729 birr (US$74), respectively, which indicates that MFIs target the active poor. The average loan balance of the Ethiopian microfinance industry is five times smaller than that of the global figure and two times smaller than MFIs in Africa. The average loan size of MFIs has increased by 20% in the last 5 years. However, the average savings of clients has not increased as much as expected. This is due to the injection of loan capital through RUFIP, bank loans and others, which partly discouraged MFIs from building savings (Wolday Amha, 2008a).

In spite of the success of MFIs in the last 5 years, MFIs have to deal with seasonality issues, high covariate production such as drought and market risks, low average returns, inadequate information infrastructure, irregular cash flows, difficult terrain, remote and largely illiterate clients, high diversity and sparse population, which increase the transaction costs of providing financial services to smallholder farmers, particularly those residing in remote rural areas. As a result, the very poor or the vulnerable food-insecure households are excluded from the programmes and services of MFIs for the following three reasons: (i) self-exclusion of the poorest households, poor people who think that taking loans will hurt rather than benefit them – they do not take the risk and some of the fears may be about confidence rather than the reality; (ii) the group-lending methodology (self-selection methodology to ensure group liability), which is widely used by Ethiopian MFIs, systematically excludes the poorest without assets such as land, livestock and regular income; and (iii) the policy of MFIs to stay sustainable implies that they should avoid providing loans to high-risk clients at subsidized interest rates. Although MFIs have their own limitation in

[3] US$1 = 9.87 Birr (September 2008).

addressing the poorest and reaching the very remote kebeles, attempts must be made to provide them with incentives and optimally use the existing MFIs within a region to get involved in the delivery of financial services to smallholder farmers, even the very poor.

Financial cooperatives

Cooperatives (both multipurpose and financial cooperatives) are key grass-roots-level organizations which are very critical instruments in implementing the objectives of the various development programmes and strategies such as the Rural Development Strategy, poverty reduction programmes and food security programmes. To this end, the government has been successful in increasing the number of primary cooperatives in a short span of time. Currently there are over 23,000 primary cooperatives and 143 unions in Ethiopia. Out of these, about 6000 are primary SACCOs and 26 are unions of SACCOs. The number of RUSACCOs is estimated to be more than 2000. In spite of the significant increase in the number of primary RUSACCOs and unions in the last 4 years, their capacity is very small. Since RUSACCOs are established within the localities of the communities and are owned by community members, they are the appropriate finance providers to deliver financial services to smallholder farmers. The transaction costs of financial cooperatives are relatively lower because they are managed and supervised by elected members on a volunteer basis. There are several issues that need to be addressed regarding RUSACCOs in Ethiopia as sustainable finance providers in providing financial services to smallholder farmers. These include capacity, governance and leadership, developing client-centred financial products and developing a separate law to regulate, supervise and expand the activities of financial cooperatives. In spite of the challenges, the community-owned models, such as the SACCOs, have high outreach, lower cost and low sustainability.

Non-governmental organizations

The micro-credit projects of NGOs were the foundation for many of the formalized deposit-taking MFIs in Ethiopia. Before the microfinance law of 1996, many NGOs used to provide credit services directly to their 'project beneficiaries', often coupled with non-financial services. However, they suffered from lack of specialized staff, weak loan recovery and no clear path towards sustainability, among other weaknesses. In 1996, the government issued Proclamation No. 40/1996, which prohibited NGOs, other than those incorporated, licensed and regulated 'microfinance institutions', from providing financial services. Following Proclamation 40/1996 and the creation of several MFIs by regional governments, many NGOs started to form their own local microfinance wings as independent, regulated MFIs by putting their Ethiopian employees as shareholders. The rest of the NGOs –especially the local ones – followed a different route in providing financial services to their beneficiaries. Some hired the services of MFIs to handle 'managed funds' (the MFI manages the financial service without owning the loan fund). In most of the cases, the beneficiaries were expected to join the cooperatives to access the resources. Despite the microfinance law, reports from the MFIs to the AEMFI indicate that some local NGOs still directly provide financial services. These NGOs generally use the financial services (in some cases including savings mobilization) to complement non-financial services such as training, family planning, education and gender empowerment. While most of these NGOs work with the very poor and in regions where no MFIs are operating, they contributed to distorting the rural credit markets by providing subsidized and low interest rates, absence of strict follow-up and poor enforcement of repayment of loans. This will have a negative impact on the sustainable finance providers which price their products on the basis of market interest rates. The new law issued in 2009 (Proclamation No. 626/2009) is expected to significantly reduce the intervention of NGOs in directly delivering loans and other financial services in both urban and rural areas.

Government and donor food security programmes

The overall objective of the food security programme in Ethiopia, including the Household Asset Building Program, is to enable up to 8.29 million chronically food-insecure people (essentially Productive Safety Net Program beneficiaries) to attain food security and to improve the food security position of 6.71 million additional people within 3–5 years in the 274 chronically food-insecure woredas. There are two parallel food security programmes implemented by the government and donors. Both programmes use the delivery of credit as a key component to build assets of chronically food-insecure households. The credit component of the government-funded food security programme focuses on providing loans for different household food security packages developed by governments at various levels based on the specific agro-ecology of the areas (e.g. crop, livestock, beehives, poultry, etc.). The beneficiaries of the programme are provided with credit, which they should pay back both the principal and interest. On the basis of the vulnerability of the regions, the federal government started by allocating 150 million birr in 2003–2004 and reached a level of 2 billion birr during 2008–2009 to implement the food security programme. The government-funded package programme (particularly in Tigray and Amhara regions) attempts to design a business plan for each household to ensure household food security. The credit component is expected to finance the business plan of chronically food-insecure households. The household package, particularly in the Tigray and Amhara regions, describes that a household with five members should earn 6000 birr, 12,000 birr and 18,000 birr in the first, second, and third years, respectively. Households with 18,000 birr are expected to graduate from the package. However, the 18,000 birr is being revised to 24,000 birr in Tigray. In Oromia and Southern Nations, Nationalities and Peoples Region (SNNPR), the benchmarks to attain household food security are much lower than in Tigray and Amhara regions (Renate and Wolday Amha, 2009; Wolday Amha, 2009).

The donor-funded US$110 million Food Security Project has the following major components: (i) support to the community through provision of grants to carry out community-initiated income-generating activities or asset-building activities at community or household levels and promotion of community-based child growth promotion initiatives; (ii) institutional capacity building; (iii) food-marketing initiatives; (iv) communication interventions and (v) project administration and monitoring (FDRE and MoARD, 2008). A significant proportion of the project fund has been mainly allocated to income-generating activities or funds for communities. Although the income-generating component of the donor-funded food security project is a grant to the communities, it is provided as a revolving fund to households within a kebele. About US$18,000 are provided to each kebele in any one year. However, subsequent tranches would be possible in subsequent years following demonstrated successful use of each preceding one.

Reviews of the delivery of credit and management of revolving funds in Ethiopia to date have identified some key weaknesses, including: (i) delivery of limited financial products – mostly agricultural production credit; (ii) generally low repayment rates due to poor follow-up of beneficiaries; (iii) unrealistic terms of credit, including subsidized interest rates and limited opportunities for repeater loans; (iv) channelling credit through inappropriate and unsustainable finance providers such as the woreda finance office, kebele administration and multipurpose cooperatives; and (v) distorting the rural credit markets (Renate and Wolday Amha, 2009; Wolday Amha, 2009). Currently, the government of Ethiopia is designing a new 5-year food security programme that is likely to consider the strengths and weaknesses of the existing programme.

Semi-formal finance (iqqub, iddir, mahiber, etc.)

The semi-formal lending institutions such as iqqub (Rotating Savings and Credit Associations), iddir and mahiber are the dominant and sustainable traditional institutions that meet the financial and social needs of the poor. The loan size per borrower from such institutions is, however, low. For instance, 75% of the iddir clients received a maximum of 100 birr each at a time. The average loan size was 260 birr per client (Bezabih *et al.*, 2005). Iqqub, which is popular in both urban and rural areas, is the dominant form of savings and credit association in Ethiopia. It is not a permanent club; it could be continued or dissolved after its members have had a turn each. A member attends an iqqub meeting weekly, bi-weekly or monthly to contribute and/or collect a fixed sum of payments. Compared to iddir, larger loans are provided from iqqub with a relatively longer repayment period. The maximum loan period for iqqub is usually a year, and the mean duration is 8 months. But, due to the collateral requirement of iqqub, only few borrowers have the opportunity to receive a loan from them (Bezabih *et al.*, 2005).

Informal finance providers

Access to institutional credit that contributes to an increase in investment is very limited in Ethiopia. The majority of the poor are therefore left with access to financial services that are limited to informal channels such as moneylenders, iqqub, iddir, friends, relatives, traders, etc. (Bezabeh *et al.*, 2005). The share of informal finance in terms of borrowers and loan size is estimated to reach 69% and 61%, respectively. Among the borrowers from the informal sources, 35% borrowed from friends and relatives, 48% from private lenders, 15% from iddir and 2% from iqqub. Only 3% of them borrowed from both relatives and other informal sources. Moreover, 10% of the borrowers borrowed from multiple informal financial sources. Informal lenders are better equipped with mechanisms for enforcing loan contracts and relatively more flexible loan terms, as a result of which they have high loan recovery rates. However, the interest rates for such loans are very high and the government, through the support of cooperatives and MFIs, is making efforts to curb their roles.

Analysing the Financial Access to Smallholder Farmers Using the *Afrint I* and *Afrint II* Surveys

Improved financial services to smallholder farmers can enhance sustainable agricultural production, processing and marketing and thus play a significant role in promoting agricultural technologies and have a long-term impact on increasing agricultural productivity. Because of the diverse characteristics of the smallholder farmers and their enterprises in Ethiopia, any finance-related intervention should focus on addressing the diverse needs of farmers and developing a flexible approach to delivering credit and other financial services that is responsive to the socio-economic environments, production systems, needs, constraints and priorities of individual households. The main objective of this section is to examine the behaviour of smallholder farmers in accessing financial services, mainly loans and savings from different finance providers, using the *Afrint I* and *Afrint II* surveys.

Access to finance

The survey results in Table 7.1 indicate that about 47.4% of the respondents had access to agricultural input credit in 2008. About 56% of the sample households reported that they took loans from various sources in 2008. About 22.5% indicated that they had problems in repaying their loans, while 77.5% reported that they did not face any problem in repaying loans. About 72.1% reported that access to loans has improved significantly since 2002. About 23.8% of the respondents revealed that they had cash savings every year to meet their future needs. Since the percentage of savers and the amount of savings is relatively low, there is a need to develop flexible and innovative savings products and improve the savings culture.

Finance providers

Table 7.2 shows the finance providers involved in channelling loans to the sample households in 2008. About 19% of the respondents accessed loans from MFIs. Multipurpose cooperatives channelled about 38% of the loan to the sample households. About 15% of the sample households accessed loans from

Table 7.1. Response of sample households in access to finance, 2008. (Adapted from: *Afrint* survey data.)

	No		Yes	
Key indicators of access to finance	Sample respondents	Per cent	Sample respondents	Per cent
Do you at present obtain any form of agricultural input credit?	250	52.6	225	47.4
Are you normally able to save some money every year for future needs?	358	76.2	112	23.8
Did you take loans during the most recent season?	207	43.8	266	56.2
Did you face problems repaying loans?	321	77.5	93	22.5
Compared with 2002, has access to loans increased?	132	27.9	341	72.1

Table 7.2. Major source of loans for sample farmers, 2008. (Adapted from: *Afrint* survey data.)

Finance providers in the study area	Sample respondents	Per cent
Microfinance institutions (MFIs)	51	19.1
Multipurpose cooperatives	102	38.2
SACCOs	41	15.4
Iqqub (Rotating Saving and Credit Associations or ROSCA)	1	0.4
Friends/relatives	52	19.5
Moneylenders	12	4.5
Suppliers (input suppliers)	2	0.7
Others (specify)	6	2.2
Total	267	100.0

financial cooperatives or SACCOs. About 20% of the respondents reported that they accessed credit from friends and relatives. Moneylenders accounted for about 5% of the loans channelled to the sample households. Iqqub and suppliers' credit accounted for less than 1% (each) of the total loans provided to the sample households.

Given the current regulatory framework in Ethiopia, loans to rural or urban households should be provided through banks, cooperatives and deposit-taking MFIs. NGOs, government/donor programmes/projects, woreda offices, kebeles, credit groups, etc. are prohibited, by law, from getting directly involved the delivery of loans and other financial services. However, since the institutional capacity of financial cooperatives, MFIs and banks is very limited, the regional governments still continue to channel credit through unsustainable finance providers such as multipurpose cooperatives, kebele administrations and woreda offices (Wolday Amha, 2009).

It is believed that the sample households obtained loans from limited channels and some of the finance providers are unsustainable. There is a need to develop the right loan products through sustainable finance providers with the necessary institutional set-up to develop financial products and professional skills for their staff in the area of market research, product development and risk management. The finance providers should also shift from the traditional supply-driven approach to demand-driven financial products that involve market research and new product development.

Purpose of loans

Provision of agricultural credit to smallholder farmers by itself is not enough. There is a need to use and manage the credit efficiently to increase agricultural production and productivity. This will partly depend on the capacity of the smallholder farmers and the capacity of finance providers. Production credit provides households with the necessary capital to engage in activities that improve the level and stability of their income by diversifying their income sources (both on-farm and off-farm), adopting new and innovative production technologies (modern farm inputs, irrigation, improved animal breeds, etc.) and acquiring more productive resources (e.g. buying oxen, renting land, hiring labour, etc.). Higher and more stable household income, as a result of creating access to finance, permits access to more and better-quality food for rural households. Table 7.3 shows how the sample households used the loans by specifying the activities they are involved in and inputs purchased with the loans taken from various finance providers.

About 63% of the sample respondents in the survey (Table 7.3) reported that they accessed a loan primarily to buy farm inputs such as fertilizer, improved seeds, chemicals, farm tools, etc. About 10% of the respondents took a loan to buy oxen, while about 13% accessed a loan to buy household consumable items such as food grain, kerosene, edible oil, clothes, etc. About 5% reported that they used the loan to pay school fees and other expenses related with schooling. About 2% of the sample households used the loan to expand existing non-agricultural businesses. About 1.5% took the loan to engage in livestock fattening. About 1% of the respondents accessed the loan to settle other debts. The study of Axel *et al.* (2005) in DECSI in Ethiopia showed that the most important purpose of regular loans was to purchase oxen (29.2%), followed by starting new trade and food-processing businesses (29.3%), farm input (11.2%) and animal fattening (9.3%). Dairy and poultry production accounted for 1.6% of the loans, while purchase of transport animals (donkey, mule, camel, etc.) accounted for 1.9%. On the other hand, the purpose of package loans, under the food security programme of the Tigray region, was mainly for animal fattening (41.9%), followed by purchase of oxen (30.2%), purchase of farm input (12.8%), dairy and/or poultry production (8.1%) and purchase of transport animals

Table 7.3. Main purpose of loans, 2008. (Adapted from: *Afrint* survey data.)

First purpose	Per cent
Purchase of farm inputs (fertilizer, improved seeds, chemicals, farm tools, etc.)	62.9
Purchase of oxen	10.1
For animal fattening (sheep, goat), rearing	1.5
Purchase dairy cows	0.4
Purchase of donkey, mule, camel and other transport animals	0.7
To start new trade business (cereals, coffee, livestock, salt, spices, etc.)	0.7
To expand existing non-agricultural business	1.9
Purchase of household consumable items (food grain, kerosene, oil, clothes, etc.)	13.1
For social ceremony (wedding, tezkar and other festivities)	0.4
For school, health fees	5.2
To hire labour	0.4
For settling other debts	1.1
For paying taxes	
Others	1.5
Total	100

(donkey, mule, camel, etc.). About 55% the clients used loans for agricultural production. Clients use these loans for the purchase of fertilizer, pesticide and chemicals (34.5%), seeds (31%) and oxen (27.5%) (Axel *et al.*, 2005).

Since the loans taken by the sample households were mainly used for productive purposes, it has a direct impact on increasing agricultural production and productivity. However, unless the loans disbursed by the finance providers are collected on a regular basis, it will send a wrong message to the community and distort the credit markets. Once the financial market is contaminated with non-performing loans, it will be difficult to deliver financial services in the community through sustainable finance providers.

Loan period

It takes time to generate sufficient income to pay the loans regularly and put some surplus towards savings. However, if the term of the loan is longer than it takes to generate that income, it puts the farmers at the risk that the income will be spent on other things rather than repayments and there will not be sufficient funds to repay later on, again resulting in the possible sale of the asset. Accessing a loan that meets exactly the life cycle of the asset to be purchased will allow the farmers to take maximum advantage of the opportunities offered

Table 7.4. Loan period (in months) of smallholder farmer, 2008. (Adapted from: *Afrint* survey data.)

Loan period in months	Sample respondents	Per cent
1	9	3.4
2	3	1.1
3	13	4.9
4	5	1.9
5	8	3.0
6	15	5.6
7	7	2.6
8	106	39.6
9	1	0.4
10	8	3.0
11	6	2.2
12	65	24.3
24	21	7.8
Total	268	100

to them. If finance providers fail to identify the appropriate loan periods (which should not be too long or too short), the smallholder farmers may face serious difficulties in repayment. The loan periods reported by sample households in the survey are summarized in Table 7.4.

The loan period of the respondents varied from 1 month to 24 months. About 40% of the sample households took loans which were agreed to be repaid in 8 months. More than 92% of the sample households accessed a loan for less than 12 months. About 8% of the respondents reported that they took long-term loans: 24 months. The average loan period of sample households was 9.5 months. This implies that the sample households in the survey took short-term loans to meet their working capital needs.

Determinants of access to loans, loan size, loan repayment and savings

The survey results indicate that the average loan size of sample households is about 910 birr (about US$91). As indicated earlier, a loan amount made available by a finance provider can be too small to purchase an asset that can make a difference in the life of a household. For example, buying only one goat may not allow a household to pay back the loan and earn sufficient additional income to save so that the goat will not have to be sold. In bad circumstances, a beneficiary may even have to sell the asset itself to pay back the loan. Having the right loan amount for the intended purpose will allow a household to repay the loans on time. Although attempts are made by finance providers to estimate the loan size of individual farmers, there are a number of factors that affect the loan size of smallholder farmers.

The results of the binary logit model (Table 7.5) reveal that the explanatory variables, namely land size, age of the household head, level of education and

Table 7.5. Logit estimation of the probability of taking loans.

Variable	Coefficient	Z	P > | z |
Land size	0.1956161	2.13	0.033
Value of marketable surplus	−2.39e−06	−0.55	0.581
Total cash income	−0.0005063	−1.76	0.078
Household size	0.0356267	0.83	0.408
Sex of head of household	−0.4287699	−1.61	0.106
Age of head of household	−0.0176071	−2.31	0.021
Educational level of farm manager	0.1022764	2.65	0.008
Access to extension	0.3480927	1.48	0.138
Savings	0.0463192	0.19	0.852
Constant	0.4600524	0.86	0.388
Pseudo R^2	0.0527		
Number of observations	466		
LR(Chi^2)(9)	33.69		
Prob > Chi^2	0.0001		

access to extension services, are significant factors influencing the decision of farmers to take loans. The amount of land owned has a positive and significant effect on access to credit, indicating that farmers with more land have a higher probability of taking loans compared to farmers with less land. The results of the logit model also indicate that those farmers with a relatively higher level of education have a higher probability of accessing loans from diverse finance providers. The age of the household head has a negative and significant effect on access to loans, implying that the relatively younger household heads have a higher probability of taking loans. Although significant at the 10% level, the results reveal that those sample farmers who were frequently visited by extension agents had a higher probability of borrowing loans from diverse finance providers.

A double log multiple regression model was run to identify the key independent variables affecting loan size (dependent variable). Size of land, family size, age of head of household, value of marketable surplus and availability of savings were found to have a positive and significant effect on loan size (Table 7.6). The above results are very useful to finance providers in developing loan products for various categories of rural households. For example, clients of MFIs and rural SACCOs in Ethiopia are expected to start with small savings before accessing loans. Developing the culture of saving before taking loans is expected to improve the timely repayment of loans.

According to the logit estimation results in Table 7.7, the probability of timely repayment of loans is determined by the amount of land owned. The smaller the size of land owned by the households, the greater is the probability of default. A higher frequency in accessing extension services results in improving a household's prospects to repay loans on time. Smaller size of the household increases the probability of paying loans on time. The study of Anbes Tenaye (2009) reveals that farm size, educational level of the household head, timeliness of credit, distance of the kebele to the nearest market place, credit

Table 7.6. Estimates of the determinants of loan size, double log regression.

Variable	Coefficient	T	P > | t |
Land	0.247	3.33	0.001
Total cash income	0.051	1.10	0.271
Household size	0.23	2.64	0.009
Age	0.277	2.08	0.038
Value of marketable surplus	0.071	1.68	0.094
Saving	0.159	1.73	0.083
Extension	0.135	1.55	0.122
Education level of farm manager	0.016	1.23	0.218
Sex	0.018	0.93	0.847
Constant	3.84	5.93	0.000
R^2	0.14		
Adjusted R^2	0.12		
Number of observations	468		

Table 7.7. Logit estimation of the probability of smallholder farmers repaying loans.

Variable	Coefficient	Z	P > |z|
Sex	−0.3423014	−1.05	0.293
Age	−0.0154378	−1.49	0.136
Education	0.006027	0.12	0.902
Family size	0.156069	2.77	0.006
Access to extension	−1.26022	−4.56	0.000
Total cash income	−0.0009973	−1.15	0.249
Land size	−0.3701167	−2.65	0.008
Saving	0.1417187	0.40	0.686
Value of marketable surplus	−0.0000202	−1.36	0.173
Constant	0.821201	1.20	0.231
Pseudo R^2			0.1198
Number of observations			407
LR $Chi^2(9)$			51.8
Prob > Chi^2			0.0000

experience and gross farm income were important variables influencing repayment performance of agricultural credit in the logistic model. Moreover, other variables like sex, total livestock unit and amount of credit used are less important variables in influencing repayment performance of agricultural credit.

Saving is very important to smallholder farmers to smooth consumption, manage risks, prepare for investment and increase their economic security by enabling them to accumulate funds slowly over time. There is extensive evidence from the experience of MFIs in Ethiopia that smallholder farmers can save (Wolday Amha, 2008a). Attempts are made here to identify the variables that affect the probability of saving by sample households.

Table 7.8 indicates that the relatively educated farm mangers have a higher probability of saving cash. The probability of saving is determined by size of

Table 7.8. Probability of saving by smallholder farmers, logit estimation.

Variables	Coefficient	Z	P > | z |
Repayment problem	0.0171635	0.05	0.960
Sex	0.5288554	1.37	0.170
Age of head of household	0.0135697	1.33	0.183
Educational level of farm manager	0.1485726	3.41	0.001
Household size	−0.1366805	−2.24	0.025
Access to extension	0.2673648	0.81	0.415
Total cash income	0.0017194	2.55	0.011
Land	0.3298584	3.18	0.001
Value of marketable surplus	9.82e−06	1.34	0.179
Constant	−3.383054	−4.27	0.000
Pseudo R^2			0.141
Number of observations			407
LR $Chi^2(9)$			63.8
Prob > Chi^2			0.000

land owned and cash income. The larger the size of land owned and the higher the cash income of the household, the greater the probability that the sample farmers will tend to save. The results of the estimation also show that households with lower family size have a higher probability of saving compared with those households with larger family size.

Conclusions

Finance, in development theory, is the main lubricant for the engine of growth and development. Finance provides the means through which a country's resources are mobilized and directed to areas of optimal socio-economic benefit. The availability of financial services, such as loans, savings, insurance, money transfer, etc., is a prerequisite to the proper functioning and growth of any sector. A sector that has no access to financial services through which operators can effectively manage their financial resources is doomed to stagnation or may lead to meaningless growth from a long-term development perspective. Moreover, any development policy, strategy or programme that aims at improving the living conditions of smallholder farmers should have a clear financial strategy which stipulates the macro-policies and regulations, meso-level infrastructure and technical service providers and support required to expand the outreach and ensure the sustainability of finance providers and clients at the grass roots levels.

The *Afrint* survey results indicate that land size, age of the household head, level of education and access to extension services are significant factors influencing the probability of borrowing of sample households. Size of land, number of household members, age of head of household, value of marketable surplus and availability of savings were found to have a positive and significant effect on loan size. The probability of timely repayment of loans is determined by the amount of land owned. The smaller the size of land owned by the households, the greater is the probability of default. A higher frequency in accessing extension services results in improving a household's prospects of repaying loans on time. Smaller family size of a household increases the probability of paying loans on time. The probability of saving is determined by size of land owned and cash income. The larger the size of land owned and the higher the cash income of the household, the greater the probability of saving. The results also show that households with lower family size have a higher probability of saving compared with those households with larger family size.

The *Afrint* survey reveals that, although about 72% of the sample households reported that access to loans has improved since 2002, the provision of credit, savings, insurance, remittances and other financial services to smallholder farmers in Ethiopia is still one of the strategic interventions required to promote the adoption of agricultural technologies, improve liquidity management, finance agricultural investments that help smallholder farmers diversify and enlarge their income sources, respond to life-cycle social events and emergencies that arise from illness, death and natural or economic catastrophes. This would require: (i) designing financial products for smallholder farmers by addressing the issue of loan size, the interest charged, the repayment schedule, loan period, etc.; (ii) building sustainable rural finance institutions that address the financial needs of smallholder farmers and their enterprises; and (iii) implementing appropriate macro- and meso-level policies, strategies, and legal and regulatory frameworks to improve financial access to the smallholder farmers.

Developing financial products and innovative lending methodologies that match the needs of smallholder farmers are very critical to improving agricultural production and productivity. Innovative lending methodologies that reduce the lending costs for smallholder farmers should be piloted to increase the demand for loans and expand the frontier of finance. These products will also create additional values if they reduce the transaction costs of accessing financial services. This could be materialized by improving the capacity of the finance providers so that they can identify the needs of smallholder farmers better, improve the quality of their services and/or reduce prices of the financial products. Moreover, focusing on what is of value to the smallholder farmer influences the operational efficiency and profitability of finance providers as well as the satisfaction and retention of clients. Products tailored to the needs of the smallholder farmers will have a greater impact in helping farmers to be effective and efficient in managing their agricultural enterprises. The financial products designed for smallholder farmers should also be tied to their cash flows, which improves their repayment capacity and allows the finance providers to sustain their operations. The whole objective of promoting the delivery of financial services to smallholder farmers should focus on developing sustainable institutions that can create and provide a broad range of microfinance services that will support millions of poor people in their efforts to improve their own and their children's prospects.

The prospect for delivery of effective and sustainable financial services to smallholder farmers in Ethiopia is bright, particularly when the macro- and meso-level supports from various stakeholders are put on the ground and when finance providers are allowed to do what finance is supposed to do. Although there is an enabling policy and regulatory framework to promote inclusive finance in Ethiopia, there are critical issues that need to be addressed at macro level. These include: introducing a national identity card, establishing a national registry system for moveable assets, supporting financial literacy and a consumer protection campaign and taking measures to reduce the effect of inflation in the delivery of financial services to smallholder farmers. Moreover, there is a dire need to address the critical challenges at meso level, which include developing the technology platform of the microfinance industry to address the back-end (MIS) and front-end technology, the formation of credit reference bureaus, opening specialized training institutes, certification of trainers and other technical service providers, establishing a wholesale refinancing facility to meet the huge demand for loan funds by finance providers and promoting national microfinance rating firms.

References

Alemayehu Seyoum, T. (2009) Crop production in Ethiopia: a spatial structural analysis. Paper presented at the 7th International Conference on the Ethiopian Economy. Ethiopian Economic Association (EEA), Addis Ababa, Ethiopia.

Anbes Tenaye (2009) Factors influencing repayment performance of agricultural credit in southern Ethiopia. Paper presented at the 7th International Conference on the Ethiopian Economy. Ethiopian Economic Association (EEA) Addis Ababa, Ethiopia.

Axel, B., Tassew Woldehanna, Gebrehiwot Ageba and Woldeab Teshome (2005) Marginalized groups, credit and empowerment: the case of Dedebit Credit and Saving Institution (DECSI) of Tigray, Ethiopia. Occasional Paper No. 14. Association of Ethiopia Microfinance Institutions (AEMFI), Addis Ababa, Ethiopia.

Bbuza, F.M.B., Dezi Ngambeki and Sabiti, E.N. (1998) Role of credit in the uptake and productivity of improved dairy technologies in Uganda. Livestock policy analysis brief No. 10, Addis Ababa, Ethiopia.

Bezabih Emana, Kejela Gemtessa, Dhunfa Lemessa and Gezahegn Ayele (2005) Informal finance in Ethiopia. Occasional Paper No. 13. Association of Ethiopia Microfinance Institutions (AEMFI), Addis Ababa, Ethiopia.

Binswanger, H.P. and Khandker, R.S. (1995) The impact of formal finance to the rural economy. *Journal of Development Studies* 32, 234–262.

Binswanger, H.P. and Rosenzweig, R.M. (1986) The behavioural and market determinants of production relations. *Journal of Development Studies* 32, 503–539.

Braverman, A. and Guasch, L.J. (1989) Rural credit in LDCs: issues and evidences. *Journal of Economic Development (Korea)* 14, 7–34.

Diagne, A. and Zeller, M. (2001) Access to credit and its impact on welfare in Malawi. Research Report No. 116. IFPRI, Washington, DC.

FDRE (1998) Federal Negarit Gazetta. Cooperative societies. Proclamation 147/1998. Addis Ababa, Ethiopia.

FDRE and MoARD (2008) Annual performance report (2007/2008 or 2000 EC). Food Security Project (cr. 3646 ET IDA, TF 5119, Italy, and TF 52696 CIDA). Addis Ababa, Ethiopia.

Freeman H.A., Jabbar, M.A. and Ehui, S.K. (1998a) The role of credit in the uptake and productivity of improved dairy technologies in sub-Saharan Africa. Livestock policy analysis brief No. 10, Addis Ababa, Ethiopia.

Freeman H.A., Jabbar, M.A. and Ehui, S.K. (1998b) The role of credit in the uptake and productivity of improved dairy technologies in Ethiopia. Livestock policy analysis brief No. 10, Addis Ababa, Ethiopia.

Freeman H.A., Jabbar, M.A. and Ehui, S.K. (1998c) The impact of liquidity and credit on smallholder dairy production: application of a switching regression model. Livestock policy analysis brief No. 10, Addis Ababa, Ethiopia.

Helms, B. (2007) *Access for All*. World Bank, Washington, DC.

Hoff, K. and Stiglitz, E.J. (1990) Imperfect information and rural credit markets: puzzles and policy perspectives. *The World Bank Economic Review* 4, 235–251.

Hussien Hamda Komicha and Ohlmer, B. (2006) Effect of credit constraint on productive efficiency of farm households in southeastern Ethiopia. *Ethiopian Journal of Economics* XV, No. 1.

Jabbar, M.A., Ehui, S.K. and Von Kaufmann, R. (2002) Supply and demand for livestock credit in sub-Saharan Africa. Lessons for designing new credit schemes. ILRI, Addis Ababa, Ethiopia.

Khandker, S.R. (1998) *Fighting Poverty with Microcredit: Experience in Bangladesh*. Oxford University Press, New York.

Khandker, S.R. and Rushidur R. Faruquee (1999) The impact of farm credit in Pakistan. World Bank Policy Research Working Paper No.2653, World Bank, Washington, DC.

Kibaara, B. (2007) Rural financial services in Kenya: what is working and why? Paper presented at the International Conference on Rural Finance Research: Moving results into policies and practice. FAO, Rome.

Kochar, A. (1997) An empirical analysis of rationing constraints in rural credit markets in India. *Journal of Development Economics* 53, 339–371.

Loening, I.J., Durevall, D. and Birru, Y.A. (2009) Inflation dynamics and food prices in an agricultural economy: the case of Ethiopia. Paper presented at the 7th International Conference on the Ethiopian Economy. EEA, Addis Ababa, Ethiopia.

Oluoch-Kosura, W. and Ackello-Ogutu, C. (1998) Role of credit in the uptake and productivity of improved dairy technologies in Kenya. Livestock policy analysis brief No. 10, Addis Ababa, Ethiopia.

Pitt, M.M. and Khandker, S.R. (1998) Household and intra-household impact of Grameen Bank and similar targeted credit program in Bangladesh. *Journal of Political Economy* 106, 958–996.

Renate, K.T. and Wolday Amha (2009) Issue paper under the food security program in Ethiopia. Paper submitted to the task force preparing the second food security program in Ethiopia. World Bank Office-Ethiopia, Addis Ababa, Ethiopia.

Swain, B.R. (2002) Credit rationing in rural India. *Journal of Economic Development* 27.

Wolday Amha (2008a) A decade of microfinance institutions (MFIs) development in Ethiopia: growth, performance, impact and prospect (2008–2017). Occasional Paper No. 21. Association of Ethiopia Microfinance Institutions (AEMFI), Addis Ababa, Ethiopia.

Wolday Amha (2008b) Corporate governance of deposit taking microfinance institutions in Ethiopia. Occasional Paper No. 23. Association of Ethiopia Microfinance Institutions (AEMFI), Addis Ababa, Ethiopia.

Wolday Amha (2009) Assessment of the credit component of government financed household food security package program. Paper submitted to BSF/FAO Office, Addis Ababa, Ethiopia.

Wolday Amha and Tigest Tesfaye (2009) The development of the Association of the Microfinance Institutions (AEMFI) in the last decade. A paper presented at the 10th anniversary of AEMFI. 20–23 May 2009. Addis Ababa, Ethiopia.

World Bank (2005) *Meeting Development Challenges: Renewed Approaches to Rural Finance*. World Bank, Washington, DC.

World Bank (2009) *Ethiopia: Rural Investment Climate Assessment, Diversifying the Rural Economy*. Sustainable Development Network Agriculture and Rural Development Unit Africa Region. World Bank, Washington, DC.

Women World Banking (WWB)/African Microfinance Action Forum (AMAF) (2009) *Diagnostic to Action: Microfinance in Africa*.

8 Agricultural Diversification, Food Self-sufficiency and Food Security in Ghana – the Role of Infrastructure and Institutions

FRED M. DZANKU[1] AND DANIEL SARPONG[2]

[1]*Institute of Statistical, Social and Economic Research, University of Ghana, Legon, Ghana;* [2]*Department of Agricultural Economics and Agribusiness, College of Agriculture and Consumer Sciences, University of Ghana, Legon, Ghana*

Food self-sufficiency has been an important policy objective of many nations, including Ghana. Its importance as a policy priority has diminished over time, as food security became a more appealing policy orientation. Self-sufficiency suggests that a nation produces at least all its food needs, while food security implies the availability and physical access to food by the population, irrespective of whether or not it is produced within the country (Thomson and Metz, 1998). At the household level, economic rationality suggests that resources should be allocated optimally to the production of commodities for which returns are highest. Income generated from trading these commodities could then be used to purchase other food needs. If agricultural diversification is defined as the increasing allocation of household resources to the production of non-staples relative to food staples, then households would diversify, given that the returns to land and labour are higher for the production of non-staples than for food staples (Fafchamps, 1992; von Braun, 1994; Goletti, 1999; Govereh and Jayne, 2003; Joshi *et al.*, 2003; Weinberger and Lumpkin, 2007; Shome, 2009). But it is documented that many farm households, particularly in sub-Saharan Africa (SSA), are subsistent or semi-subsistent producers, which implies an inclination towards self-sufficiency in food production (de Janvry *et al.*, 1991; Finkelshtain and Chalfant, 1991; Fafchamps, 1992; Jayne, 1994; von Braun, 1994, 1995; Govereh and Jayne, 2003; Di Falco and Chavas, 2009).

The theory of comparative advantage, however, assumes that markets exist for the exchange of goods and services the household produces and those it needs to ensure food security. The optimal allocation of household resources is also often based on this assumption. Thus, production and consumption decisions are assumed to be separable (Singh *et al.*, 1986). This assumption may not be realistic because of missing or incomplete markets for some goods and services. High transactions costs in rural Africa make some commodities' market participation prohibitive for some households. Thus, even where markets exist, they fail for some households (de Janvry *et al.*, 1991). Under such circumstances, therefore, household self-sufficiency in food production may be the most reasonable way of achieving food security (Fafchamps, 1992; Minot, 1999) for some households. Transactions cost, a major reason for which markets fail, is largely due to poor or absent infrastructure (both 'hard' and 'soft'),[1] among others (Goletti, 1999). State policy may be warranted to reduce transactions cost as well as to promote diversification (Delgado, 1995; Pingali and Rosegrant, 1995; Delgado and Siamwalla, 1997).

Some synergies, however, have been observed between the self-sufficiency strategy of achieving household food security and diversification into cash crop production (or increased participation in staple crop sales), such that households may first seek to obtain food security insurance through self-sufficiency as a priority (Fafchamps, 1992; von Braun, 1994; Jayne, 1994; Pingali and Rosegrant, 1995; Goletti, 1999; Govereh and Jayne, 2003; Dzanku, 2009; Shome, 2009).

In the context of Ghana, this study addresses two main questions: (i) is the attainment of staple crop self-sufficiency a necessary condition of diversification into non-staple crop production?; and (ii) does the allocation of resources to the production of non-staples hurt or enhance rural household food security? These questions are investigated using panel data collected in 2002 and 2008 from eight Ghanaian villages.

The Policy Context

The national economic development strategy enshrined in the Growth and Poverty Reduction Strategy (GPRS II) aims to achieve accelerated and sustainable shared growth, poverty reduction, gender equity, protection and empowerment of the vulnerable and excluded within a decentralized and democratic environment. Agriculture is a major component of this strategy, and Ghana's Agricultural Development Strategy is rolled out in the Food and Agriculture Sector Development Policy (FASDEP II). The main objective of the policy is to modernize agriculture, culminating in a structurally transformed economy. This transformation is aimed at enhancing food security, among others, in line with the goal set for the sector in the GPRS II paper.

The policy changes in the agricultural sector were prompted by the fact that 80% of Ghana's total agricultural output is predominantly rainfall-dependent and practised on smallholder, family-operated farms using rudimentary technology (MoFA, 2007a). According to the 2000 census, 51% of the labour force is directly engaged in agriculture. The slow growth of agriculture is due to a combination of factors that reduce farmers' incentives to invest. These include lack of technologi-

[1] 'Hard' infrastructure refers to physical facilities such as roads, while 'soft' infrastructure includes institutions and systems that facilitate market transactions.

cal change and poor basic infrastructure. Dissemination of new and improved technologies through extension services is weak, with a high extension worker to farmer ratio (1:1500), which is highly unbalanced between female and male farmers, with as little as 20% of services reaching women.

Annual rainfall varies between 800 and 2400 mm, generally decreasing from south to north. A significant proportion of arable land has soils with poor physical properties and low content of organic matter. As a result, agricultural productivity is low and erratic and vacillates between scarcity, sufficiency and glut. The FASDEP II therefore seeks to address these constraints by the promotion of selected products through improved access to markets, the development of and improved access to technology for sustainable natural resource management, improved access to agricultural financial services, improved rural infrastructure and enhanced human resource and institutional capacity.

The policy targets commodities that are food security-enhancing and facilitate agricultural income diversification, as well as the enhancement of productivity of the commodity value chain, through the application of science and technology. In general, agricultural production outcomes are mixed regarding the achievement of set policy objectives. The structure of the economy remains largely agrarian and agriculture contributes the largest share of gross domestic product (GDP), even though agriculture's share has been declining somewhat (Fig. 8.1). There is an estimated self-sufficiency ratio of 100% for roots and tubers, and 90% for cereals (excluding rice). However, seasonal food insecurity is widespread, due to the almost total dependence on rain-fed agriculture and weak postharvest capacities, which limit the shelf life of many commodities. Estimated self-sufficiency ratios for rice (50%), fish (60%) and meat (30%) are much lower (MoFA, 2007a). Despite the high self-sufficiency ratios for most food crops, the food balance, derived from available supply and demand statistics for key food commodities, shows a

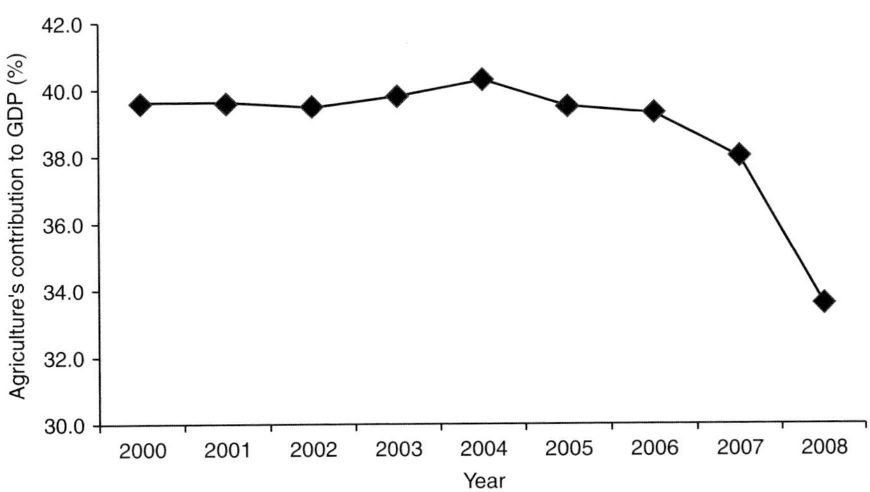

Fig. 8.1. Trends in agriculture's contribution to GDP. (From: Ghana Statistical Service.)

deficit for the major food items, with the exception of cassava, millet, sorghum, plantain, cocoyam and yam. For instance, from 1995 to 2006, Ghana imported, on average, 100% of its wheat and sugar, two-thirds of its rice, half of its beef and one-third of its chicken.

Crop Diversification, Food Self-sufficiency and Food Security

Subsistence production by households could be viewed as a food self-sufficiency strategy (von Braun, 1995; Govereh and Jayne, 2003). Even though economic theory may suggest that this strategy is inferior, the output and marketing constrains faced by farm households in SSA makes it probably the most viable option (von Braun, 1994; Thomson and Metz, 1998; Govereh and Jayne, 2003). Agricultural economists and other researchers have attempted to answer the question as to why rural households in Africa choose the self-sufficiency strategy, even for crops for which they are comparatively disadvantaged. When faced with output and price risks, the profit maximization motive alone cannot explain rural households' crop production choice (Guvele, 2001; Windle and Rolfe, 2005).

Central to the choice of the food self-sufficiency strategy is the attainment of the household food security objective. Thus, given their peculiar economic, agro-ecological and infrastructural circumstances, households would choose a strategy that is most likely to guarantee their food security. The suggestion that subsistence production is an inferior strategy for the attainment of food security is based on the assumption that markets are not missing and that the utility derived by households from participation in markets for the goods and services they produce and those they require for achieving food security exceed the disutility from participating.

The relationship between food self-sufficiency, food security and agricultural diversification would depend on how diversification is defined. If diversification implies increased cultivation or the adoption of cash crops, then risk-averse households may diversify only if they perceive that their food security is not threatened. Fafchamps (1992) presents a model that suggests that, in developing countries, households' cultivation of cash crops is conditional on the attainment of food security, which he suggests could best be realized through food self-sufficiency. The marginal benefit of diversification must exceed the marginal cost if risk-averse households were to diversify (Featherstone and Moss, 1990). Empirical findings by von Braun (1994, 1995), Jayne (1994), Govereh and Jayne (2003) and Joshi *et al.* (2003) have demonstrated that, indeed, households in developing countries strive to achieve food security by maintaining significant levels of subsistence, even when they participate in cash crop production. These empirical investigations, particularly by Bouis (1994), von Braun (1994, 1995) and Govereh and Jayne (2003), also show that diversification into cash crop production has no significant negative effects on household food security. But the cross-sectional data used in most of these studies may not capture the dynamics implied by household food security.

Diversification into cash crop production has been criticized on the basis that households may have to rely more on food purchases as a result, which may lead to deterioration in their food security situation, given the high cost of calories and price instability (Thomson and Metz, 1998). This is not withstanding the fact that this hypothesis was tested by Bouis (1994) in six countries, including Kenya, Rwanda, Malawi, Sierra Leone and Gambia, and no significant change in sources of food consumption was found. There was no significant change in staple crop production for households that diversified into cash crop production. In fact, in most of the African countries, households still grew more than half of their food, even after participation in cash-cropping schemes. On the other hand, if agricultural diversification is defined as multiple agricultural output produced by a household, then it could be interpreted as a food security insurance strategy. The uncertain nature of rainfall patterns explains the observation by Joshi *et al.* (2003) in the Asia region that this type of agricultural diversification is more prominent in rain-fed areas. Thus agro-ecology and water supply are important determinants of multiple crop production (Rahman, 2009). Both output and price risk may be reduced by growing crops that differ in their production and marketing characteristics. Since diversification in this sense is an adaptation strategy to climatic variability (Bradshaw *et al.*, 2005), the more diversified a household is, *ceteris paribus*, the better insurance it has against shocks, particularly where livestock is part of the portfolio mix. But Quiroz and Valdes (1995) noted that, in general, agricultural output prices are positively correlated as a result of substitution possibilities in consumption and production as well as common reaction patterns of macroeconomic and global shocks.

Household wealth or asset endowments also influence the relationship between food self-sufficiency, security and agricultural diversification or specialization. Poorer households are less likely to diversify into cash crop production since they may be unable to cope with the transactions cost of ensuring food security through food purchases (Delgado and Siamwalla, 1997). Diversification into cash crop production may also increase labour productivity and employment, as well as increase hired labour engagements at the village and household level (Joshi *et al.*, 2003). To the extent that this increases household income, food and nutrition security could be enhanced (von Braun, 1995). Increased income could also lead to changes in food consumption patterns. If this happens to a significant extent, then the food security effects of the movement of household resources from the production of staples would be somewhat dampened (Joshi *et al.*, 2003; Windle and Rolfe, 2005; Minot *et al.*, 2006).

The Role of Infrastructure and Institutions

The dominance and persistence of subsistent or semi-subsistent agriculture has been attributed to high transactions cost, among other factors. This situation is mostly the result of poor infrastructure, particularly roads. Improved infrastructure reduces marketing risk, improves marketing efficiency and thus reduces preference for a high degree of self-sufficient levels of production (von Braun *et al.*, 1994; von Braun and Immink, 1994; Quiroz and Valdes, 1995; von Braun, 1995; Joshi *et al.*, 2003; Windle and Rolfe, 2005; Weinberger and Lumpkin, 2007; Rahman, 2009). A successful agricultural diversification that leads to increased and sustainable food security would no doubt require adequate infrastructural development. The development of irrigation facilities would also reduce the reliance on rain-fed agriculture and so lead to a reduction in output risk, which would eventually decrease the need for subsistence and promote diversification into non-staple crops.

Aside from 'hard infrastructure', 'soft infrastructure' in the form of institutions is critical for diversification and improved food security (Goletti, 1999). For example, the development of rural financial institutions which are accessible to smallholders could enhance complementarities between staple crop production and diversification into self-sufficiency. The development of legal and contractual environments, farmers' capacity building, research and extension all reduce the incentive for subsistence and promote diversification.

The state has an important role to play in determining the type and extent of agricultural diversification since the development of 'hard' infrastructure is largely the role of the state. But the development of infrastructure is not enough; promoting technological change in staple food production at the farm level that increases productivity of land and labour plays a parallel role in diversification into cash crops (von Braun, 1994; Joshi *et al.*, 2003). A comprehensive study by the International Food Policy Research Institute, employing case studies from several developing countries in Africa and Asia, concluded that a smooth transition from subsistence-oriented smallholder production systems to diversification into cash crop production requires macro-policy reforms, infrastructure policy, agricultural technology development and dissemination, land tenure reform and rural financial policies, among others (von Braun and Kennedy, 1994a; Pingali and Rosegrant, 1995). Policies that enhance input supply and output marketing would eventually benefit both staple and cash crop production, thereby reducing the insurance price paid by households to maintain food security through their own food supply.

In this regard, the Millennium Development Authority Programme, which is currently being implemented in Ghana with the aim of improving both 'hard' and 'soft' infrastructure, among other things, is most welcome. It has been argued that diversification should be demand-driven rather than policy-induced through the picking and choosing of commodities (Delgado, 1995), but, depending on the extent of market development, the degree of agricultural transformation and the relative importance of agriculture in the economy, it may be necessary for government to promote diversification as a policy objective (Delgado and Siamwalla, 1997).

Analytical Framework

The basic agricultural household model assumes separability of consumption and production decisions (Singh *et al.*, 1986). This assumption may not be plausible under prohibitive transactions cost. Linked to this is household behaviour under output and price risk, which makes consumption and production decisions inseparable. Since an outcome of a production decision made *ex ante* is unknown with certainty, profit maximization alone is an inappropriate behavioural assumption (Guvele, 2001). Rural households would be concerned about meeting food needs through their own production.

Since subsistence food production can be considered 'an insurance and credit market substitute' (von Braun and Kennedy, 1994b:20), hypotheses concerning the allocation of resources for subsistence (self-sufficient production) versus diversification into cash crop production would be made by considering the marginal utility per unit of additional cash crop production and the marginal disutility that could occur if households were to depend on purchased food for meeting their food needs. Jayne (1994) postulates that this decision would be based on the following decision criterion:

$$\textit{cultivate cash crop if } E(\pi) > D[SPP * Q_s - VC_s] + (1 - D)[Q_s * SC * x * PS - VC_s] \qquad (8.1)$$

where E is the expectations operator; π is gross margin from cash crop production; D represents a dummy variable which takes on a value of unity if the household expects to be food self-sufficient and zero otherwise;[2] SPP is staple food crop producer price; Q_s is per hectare expected staple food crop production quantity measured in grain equivalent; VC is the per hectare variable cost of staple food crop production; SC is the proportion of staple food crops consumed over a period of 1 year; x is the extraction rate from grain to meal (per cent); and PS is the acquisition price of staple food meal. The opportunity cost of diversification into non-staple production is given by $Q_s * SC * x * PS - VC_s$. All else held constant, the higher the opportunity cost the less likely it would be for households to diversify. Thus, if the household expects to be food self-sufficient; the decision to diversify (or to allocate more resources to crops other than staples) becomes a comparison of expected gross margins of staples and non-staples. Note that in their choice modelling experiment, Windle and Rolfe (2005) observed risk perception and gross margins as the most important determinants of crop cultivation choice.

Let R_{ijt} be the revenue obtained by household i from choosing to allocate resources to the cultivation of cash crop j in year t, then by ignoring the inequality sign in Eqn 8.1, we can write:

$$R_{ijt} = E(\pi) - D[SPP * Q_s - VC] - (1 - D)[Q_s * SC * x * PS - VC] \qquad (8.2)$$

To test whether household food self-sufficiency plays a significant role in household diversification decisions, it is important to note that decisions affecting food self-sufficiency and resource allocation to non-staples are made simultaneously. Let y be the value of staple crop production, $\hat{y}$ is the predicted value of staple crop production, $\widehat{H_{fss}}$ is estimated household food self-sufficiency, x is a vector of exogenous variables that affect cash crop production decisions, FS is staple food stock at the beginning of the harvest period, CR is household staple food consumption requirements, y' is the share of land planted to non-staple

[2] Jayne (1994) notes that, since the relationship between the factors of production and yields are stochastic, the level of self-sufficiency must be assessed *ex ante*.

crops for sale. Then the following equations could be used to estimate the relationship between food self-sufficiency and diversification into non-staples:

$$y = y(x) + \varepsilon_1 \tag{8.3}$$

$$\widehat{H_{fss}} = \hat{y} + FS - CR \tag{8.4}$$

$$y' = y'(x, \widehat{H_{fss}}) + \varepsilon_2 \tag{8.5}$$

In the above setting, household food stock and consumption requirements are assumed to be known *ex ante*. Jayne (1994) estimated the relationship between oilseed cultivation and grain self-sufficiency by allowing the slope and intercept linking cash crop area to the degree of household self-sufficiency in grain to change at the point at which grain self-sufficiency is reached. Under the null hypothesis that if household food self-sufficiency does not exert a significant effect on diversification into cash crop production, these terms will be significantly different from zero. If market constraints are not binding, a household can diversify into non-staples to increase income without negative impact on household food security. These concepts are investigated empirically using panel data instead of the cross-sectional data applied by Jayne (1994).

The Econometric Models

Two main equations are estimated: the first specifies the relationship between staple crop self-sufficiency and diversification into non-staple crop production while the second quantifies the determinants of household food security. We adopt from Jayne (1994), *mutatis mutandis*, a model derived from Eqn 8.3 to estimate the relationship between food self-sufficiency and diversification into non-staple crop production. The observed dependent variable is left-censored at zero (i.e. it takes on positive values for households that cultivate non-staples over the two periods but zero otherwise).[3] In the second equation the observed dependent variable is binary, taking the value of one if the household is food secure and zero otherwise. Given the panel data structure, it is possible to control for unobserved factors that may influence a household's preference for the self-sufficiency strategy as well as the probability of being food secure. These factors are referred to as unobserved household heterogeneity. If we assume that all household heterogeneity can be captured by the observed explanatory variables, then we can specify the models for household i in period t as

$$y_{it}^{*} = x_{it}\beta + v_{it} \tag{8.6}$$

where $y_{it} = y_{it}^{*}$ when $y_{it}^{*} > 0$ but $y_{it} = 0$ when $y_{it}^{*} \le 0$ in the diversification equation, while $y_{it} = 1$ if $y_{it}^{*} > 0$ and $= 0$ otherwise in the food security equation,

[3] 15% of households did not cultivate any non-staples.

y^* denotes the latent diversification and food security variables respectively, x is a vector of time-varying and time-invariant explanatory variables (including staple food self-sufficiency, infrastructure and institutional variables in the diversification equation), β is the vector of coefficients associated with the vector x, and υ is the composite error term. The above specification leads to the estimation of pooled tobit and probit models for the diversification and food security equations respectively. That is, the data is essentially treated as a cross section: the 2002 and 2008 data are pooled, and tobit and probit models are estimated in each case. These coefficient estimates would be biased if there are significant household unobserved effects. Suppose that the household-specific heterogeneity c is time constant across households, the unobserved effects model can be written as

$$y_{it}^* = x_{it}\beta + c_i\, u_{it} \tag{8.7}$$

where u_{it} is the idiosyncratic error. We treat c_i, x_{it} and y_{it} as random draws from the population of interest but assume that c_i is uncorrelated with the x_{it} (Wooldridge, 2002). The assumption that Cov $(x_{it},c_i) = 0$, $t = 2002,2008$ leads to the estimation of random effects tobit and probit models. This is mainly to allow for the estimation of the coefficient of the infrastructure variable, which is important to our hypotheses but is invariant across observations. The unobserved effects tobit model also assumes that $u_{it} | x_{it},c_i \sim Normal(0, \sigma_u^2)$. The random effects probit model is also estimated under the assumption that $u_{it}|x_{it} \sim IN(0,\sigma_u^2)$ and $c_i|x_{it} \sim IN(0,\sigma_c^2)$ (*IN* refers to independent normal distribution).

A detailed description of the diversification and food security variables is given in the Measurement of Key Variables section. Unlike Jayne (1994), who measured diversification into cash crop production as area under oilseed, this study uses the share of cultivated land allocated by household i to non-staples in period t as the dependent variable in the diversification equation. In the food security equation, the possible exogeneity of the diversification variable is tested. The intuition is that food secure households may be less concerned about meeting consumption needs through own staple production, in which case non-staple crop production would depend on households' expected food security status.

Study Villages

This study is based on household-level panel data collected in 2002 and 2008 from eight villages in Ghana. The crop year, however, covered the years 2001 and 2007. The eight villages are located in two administrative regions (the Eastern and Upper-East regions) and in two distinct agro-ecological zones. The population of the villages, based on estimates by the village key informants, ranges from about 371 in Gyedi to 3800 in Zanlerigu. The research was designed to study four major staple crops: two in each region – cassava and maize in the Eastern region; rice and sorghum in the Upper-East region.

A multi-stage random sampling technique was employed to first select the regions, districts, villages and then, finally, the sample households. The regions were selected based on the major staples cultivated in the villages. The districts in each region were selected based on their agricultural potential as per the researchers' assessment based on information from the Ministry of Food and Agriculture (MoFA). From the focus group discussions, participants described households in all the villages except Akatawia, Asitey and Doba as agricultural households. That is, the primary activity of households in the five villages was crop and livestock production, with crop sales generating the largest share of income. In Akatawia, Asitey and Doba 5%, 10% and 5% of households, respectively, were described as non-agricultural households.

During the 2002 survey, 416 households were surveyed from the Manya Krobo, Fanteakwa, Talensi Nabdam and Kassena-Nankana districts. In 2008, 358 (or 86% of households interviewed in 2002) were successfully contacted. These were made up of 328 'original' 2002 households and 30 descendants of the 2002 households.[4]

All the villages were accessible by public transport, with either tarred roads (in the case of Akatawia, Asitey, Gyedi, Doba) or untarred all-weather roads (in the case of Apaa, Gaane, Zanlerigu and Shia). The villages are varying distances away from the district and regional capitals. Gyedi is a suburb of a district capital, while Asitey is located at the outskirts of another district capital. Apaa, Gaane, Zanlerigu and Shia are relatively remote and served by public transport less frequently.

Measurement of Key Variables

The main variables of interest are food self-sufficiency, food security, agricultural diversification and infrastructure. von Braun (1994) used a rule of thumb figure of 170 kg of cereal equivalent per capita per annum to estimate household food self-sufficiency, while Jolly and Gadbois (1996) applied the FAO's 200 kg of refined cereal equivalent. In this study, food self-sufficiency is estimated using 170 kg.[5] We follow Jolly and Gadbois (1996) and convert all grains and roots to maize equivalent. The maize equivalent of rice and sorghum are calculated to estimate total grain produced and available to the household, i.e. after accounting for postharvest losses. The milled ratios used are adopted from Jolly and Gadbois (1996) and is shown in Table 8.1.

The ratios applied in the conversion to maize equivalent is based on energy (calories) derived from the produce, as reported in Okigbo (1991). We also use ratios from Jolly and Gadbois (1996). Equation 8.4 is combined with information from Table 8.1 to model predicted levels of food self-sufficiency.

[4] Most of these households are made up of adult children of the 2002 households, mostly living in the same dwelling but whose parents (head of households in 2002) had passed away.
[5] We also experimented with 200 kg of cereal equivalent and found most households to be food deficient.

Table 8.1. Maize equivalent ratios for estimating food self-sufficiency. (From: authors' calculations based on sources cited.)

			Maize equivalent ratio[a]	
	Calories/kg	Milling ratio	A	B
Maize	3630	0.85	1.00	1.00
Rice	3340	0.65	0.92	0.99
Sorghum	3350	0.90	0.92	0.97
Cassava	1460		0.40	0.43

[a]The conversion ratio in A is estimated based on energy derived (calories) from the product, while we use ratios in Jolly and Gadbois (1996), after adjusting for cassava, in estimating B.

Due to the lack of actual consumption quantities, we are restricted in our measurement of food security. We construct a food security measure based on the definition in Thomson and Metz (1998), which classified households as food secure and food insecure based on whether or not a household's food entitlement is greater than its needs. First, household food needs are estimated using 170 kg of cereal equivalent per person per year multiplied by household size.[6] To estimate entitlements, own-produced staples are converted into maize equivalent, as in the calculation of self-sufficiency. Let *FE* be household food entitlements and *FN* household food needs, then a household is defined as food secure if $FE > FN$. Let $FN_i = 170\,\text{kg} * HHS$, where *HHS* is household size, and let *HP* be household food needs met by the amount of own-produced food consumed, then $GAP = FN_i - HP_i$ represents the gap that has to be met from elsewhere. Based on the GLSS 5 (see Ghana Statistical Service, 2008)[7] results on the actual share of household expenditure on food, we estimate how much of this gap can be met through the food expenditure share of household total income. If that share of income meets this gap then the household is food secure; at least $FE - FN$ should be positive.

Two measures of agricultural diversification are considered in this study. First, diversification into non-staple crop production is measured as the per cent share of land planted to crops other than staples (maize, cassava, sorghum, rice), which are mainly for sale. These were mainly vegetables, beans, groundnuts, cocoa and oil palm. Given observations for a cross section of households over two periods, we can estimate the relative share of area planted to non-staples in total cultivated area over time. Second, diversification as multiple agricultural output production is measured using the Simpson Index:

$$SID = 1 - \sum_{i=1}^{n} P_i^2$$

[6] The data do not allow the use of an adult equivalent measure.

[7] 26.7% of income is applied to all four villages in the Upper-East region; in Gyedi and Apaa 41.2% is used and in Asitey and Akatawia 48.5% is applied.

where P_i is the proportionate area of the ith crop in gross area cultivated. This measure of diversification has been used severally in the literature (see, for example, Joshi *et al*., 2003 and Minot *et al*., 2006).

Infrastructure is measured by distance to market outside the village (km) and marketing cost (US$/kg/km), while the effect of institutions is captured by agricultural extension contact, active membership of farmer-based organizations (FBOs) and a dummy measuring land tenure security. It takes on a value of unity if the household believes they have full control over land (i.e. they do not need to consult anyone for permission before cultivation) and zero otherwise. All other variables used in the empirical models are described in Table 8.2.

Results

Before presenting the econometric results, some important descriptive statistics are presented. The dynamics of food self-sufficiency and food security are shown in Table 8.3. With regard to food self-sufficiency, the majority of households (68.3%) are in chronic food deficit, with only 8.2% of households being self-sufficient over the period of observation. Less than one-quarter (23%) of the surveyed households were food secure in both 2002 and 2008. A larger proportion (38.5%) of households were transitory food insecure, with about the same percentage being chronic food insecure.

There are also observed changes in agricultural extension contact and FBO membership. About 44.5% of households had frequent extension contacts in both periods, while 12.5% did not have frequent contacts in both periods. It is important to note that extension contact over the two periods is significantly greater among males than females at the 5% level. For example, while 25% of female farmers never had agricultural extension contact over the two periods, only 10% of male farmers had no contact. Gender differences in extension have been observed at the national level (MoFA, 2007b). Though there are similar differences with regard to FBO membership, the differences are not statistically significant, even at the 10% level. Next the econometric results are presented.

Is staple food self-sufficiency necessary for diversification into non-staple crops?

Two main hypotheses are tested: (i) given non-separation of production and consumption decisions, a staple food self-sufficient rural household would allocate a greater share of its resource (land) to the production of non-staples than a food-deficit household; and (ii) improved infrastructure and rural institutions would both improve staple crop productivity and enhance diversification into non-staples. Prior to testing these hypotheses using multivariate econometric models, a two sample *t*-test is performed. Assuming equal sample variance, staple food self-sufficient households allocate statistically significant larger shares of land to non-staples than staple food-deficient households (Table 8.4).

To control for other factors, we estimate a pooled tobit (see Appendix, Table 8A.1) and a household random effects tobit (Table 8.5) model for each

Table 8.2. Description of variables.

	2002		2008	
Variable	Mean	Std dev.	Mean	Std dev.
Share of non-staples in total cropped area (%)	28.19	21.09	26.14	20.27
Simpson Index of Diversification	0.57	0.11	0.53	0.16
Food self-sufficiency dummy (1 = food self-sufficient)	0.24	0.43	0.16	0.37
Food self-sufficiency level (kg)	140.25	199.35	119.05	288.98
Food security dummy (1 = food secure)	0.33	0.47	0.52	0.50
Food security level (kg)	–464.61	138.92	1748.20	633.31
Real marketing cost (US$/kg/km)	0.06	0.04	0.09	0.06
Distance to an all-weather road (km)	0.97	1.44	0.97	0.44
Distance to main market outside village	5.32	3.06	5.32	3.06
Extension contact (1 = received extension advice regularly)	0.73	0.44	0.59	0.49
Active membership of FBO (1 = member)	0.37	0.48	0.26	0.44
Land rights measure (1 = complete control)	0.80	0.40	0.75	0.43
Credit access (1 = has access)	0.40	0.49	0.36	0.48
Sex of farm manager (1 = female)	0.14	0.35	0.16	0.37
Household size	9.04	6.21	7.64	4.63
Proportion of household members below 16 years (%)	37.84	20.19	36.15	21.33
Proportion of household members above 60 years (%)	5.91	12.55	9.91	16.95
Dependency ratio	0.99	0.84	1.15	1.36
Age of farm manager	45.06	14.56	54.05	18.98
Education level of farm manager (years)	4.68	5.11	5.28	5.46
Total area under cultivation (farm size in ha)	2.56	2.16	2.09	1.67
Staple crop farm size (ha)	1.78	1.78	1.35	1.08
Own-produced staples in maize equivalent (kg)	881.29	967.76	713.34	1133.49
Remittance income (US$)	98.98	146.81	116.45	172.72
Other non-farm income (US$)	374.46	793.09	594.38	1258.88
Asset index	0.22	0.13	0.28	0.17
Number of cows (Upper-East region only)	1.79	3.14	2.37	3.97
Number of sheep/goats	5.47	7.89	6.53	9.44
Number of poultry	13.12	18.34	17.80	21.12
Livestock index (poultry equivalent)	83.20	121.78	89.58	118.38

Table 8.3. Food self-sufficiency, food security and institutional dynamics. (From: computed from survey data, 2002 and 2008.)

	Freq.	Per cent	
Dynamics of food self-sufficiency			
Became food deficient	51	15.6	Transitory food deficient (23.5%)
Became self-sufficient	26	7.9	
Chronic food deficient	224	68.3	
Food self-sufficient	27	8.2	
Dynamics of food security			
Became food insecure	31	9.5	Transitory food insecure (38.5%)
Became food secure	95	29.0	
Chronic food insecure	126	38.4	
Food secure	76	23.2	
Dynamics of agricultural extension contact and FBO membership			
Stopped agricultural extension contact	95	28.96	
Began agricultural extension contact	46	14.02	
Never had agricultural extension contact	41	12.50	
Always had agricultural extension contact	146	44.51	
Dropped out of farmer-based organization	77	23.48	
Joined a farmer-based organization	41	12.50	
Never been member of farmer-based organization	167	50.91	
Always been member of farmer-based organization	43	13.11	

Table 8.4. Land allocation to non-staples, by staple crop self-sufficiency status.

	Per cent share of land allocated to non-staples[a]		
	Self-sufficient in staples	Staple crop-deficient	*t*-statistic
Upper-East region	26.3 (25.4)	19.0 (19.1)	1.82
Eastern region	22.3 (28.8)	14.1 (16.6)	3.01***
Combined	23.0 (28.1)	17.0 (18.4)	2.82***

[a]Standard deviation in parentheses.
***Significant at 1% level.

region separately as well as for the entire Ghana sample. These models predict the share of land cultivated to non-staple crops and are statistically significant (as shown by the respective wald chi-squared values). To test hypothesis (i), a joint test is performed on the coefficients of the predicted self-sufficiency dummy and that of its interaction with the predicted level of household food

Table 8.5. Determinants of diversification into non-staple crops (random effects tobit).[a]

	Upper-East region	Eastern region	Combined sample
Expected staple food self-sufficiency level (SFSSL)	−0.0339***	0.0050	0.0059
	(5.66)	(1.35)	(1.95)
Expected staple food self-sufficiency dummy (SFSSD)	−4.3967	−7.2438	−4.7079
	(0.86)	(1.35)	(1.28)
Self-sufficiency dummy × self-sufficiency level (SFSSLD)	0.0014	0.0461***	0.0302***
	(0.08)	(3.63)	(2.98)
Distance to main market outside village (MD)	−0.5864	−19.6320**	−0.9225**
	(1.00)	(2.22)	(2.05)
Year × distance to market	1.2348	11.5624	−0.3087
	(1.19)	(1.00)	(0.56)
FBO membership (FBO)	−1.3486	8.9503**	2.6470
	(0.55)	(2.53)	(1.30)
Agricultural extension contact (AE)	11.9495***	5.5867	6.2060***
	(4.03)	(1.58)	(2.79)
Extension contact × sex of household head (AE_Sex)	−0.6797	−4.7856	−2.7811
	(0.10)	(0.63)	(0.55)
Credit access (Cr)	−0.5530	−0.4857	−0.2637
	(0.23)	(0.17)	(0.14)
Complete control over cultivated land (LR)	−5.5178	4.8885	0.3596
	(1.60)	(1.57)	(0.16)
Female-headed household	−18.7997***	−1.6176	−1.3626
	(3.24)	(0.29)	(0.35)
Adult-equivalent labour unit (L)	−5.4057***	2.0981**	1.2021**
	(5.91)	(2.29)	(2.56)
Dependency ratio (DR)	−2.9292**	−2.2588	−0.5248
	(2.43)	(1.43)	(0.63)
Education level of household head (EDUC)	−0.6461**	−0.1392	−0.0800
	(2.38)	(0.42)	(0.38)
Age of household head (Age)	−0.1393	0.1372	−0.1284
	(0.41)	(0.31)	(0.47)
Square of age of household head (Agesq)	0.0027	−0.0021	0.0011
	(0.92)	(0.55)	(0.45)
Remittances (RI)	−0.0213	−0.0084	−3.9896
	(1.46)	(0.40)	(1.86)
Other non-farm income (Masakure *et al.*, 2008)	−0.0036	0.0043**	0.0014
	(1.32)	(2.21)	(0.92)
Physical asset index (AI)	−1.1792	20.3720**	14.1480**
	(0.14)	(2.04)	(2.18)
Year dummy (2008 = 1)	−11.8409***	−0.7923	0.2959
	(4.70)	(0.12)	(0.08)
Constant	67.9623***	3.9483	12.0961
	(6.64)	(0.30)	(1.42)
Number of observations	375	280	613
Log likelihood value	−1476.70	−1108.11	-2417.86

Continued

Table 8.5. Continued.

	Upper-East region	Eastern region	Combined sample
Wald chi-squared	90.62	77.16	124.74
R^{2b}	0.218	0.270	0.180
Rho[c]	0.080	0.034	0.086
	(0.078)	(0.369)	(0.058)
Ho: SFSSD = SFSSLD = 0	$\chi^2 = 2.10$	$\chi^2 = 14.21$	$\chi^2 = 9.92$
	[Prob = 0.350]	[Prob = 0.001]	[Prob = 0.007]

**Significant at 5% level; ***significant at 1% level.
[a]Absolute value of *t*-statistics in parentheses.
[b]R^2 between the predicted and observed values.
[c]Standard errors in parentheses.

self-sufficiency. In the combined sample, the hypothesis that these coefficients are jointly equal to zero is rejected at the 1% level, suggesting that staple food self-sufficient households have a higher propensity than staple food-deficient households to allocate more resources (land) to non-staples. This result is consistent with Jayne (1994), who applied cross-sectional data from Zimbabwe and used land allocated to oilseed production rather than its share in total land cultivated as the dependent variable.

The regional subsample estimates produce a similar outcome in the Eastern region (F = 6.96, *P* value = 0.001) but not in the Upper-East (F = 1.67, *P* value = 0.189). A priori, it was expected that, given the monomodal rainfall pattern in the Upper-East region, the self-sufficient strategy would be relatively more important than in the Eastern region.

A possible reason for the contrary outcome is the already very low level of per capita staple crop output in the region: the majority of households are staple food deficient even though they consume more than 95% of their staple output. In some of the villages, however, some farmers participate in an irrigation scheme and dry-season vegetable cultivation.

The second hypothesis is confirmed in the combined sample estimates: the further the distance to markets outside the village the less likely it is for households to diversify into non-staples; FBO membership and regular contact with agricultural extension services increases the predicted share of land allocated to non-staples by about 4% and 5% compared to non-FBO members and farmers who rarely have contact with agricultural extension respectively, *ceteris paribus*. In the regional subsample, the distance effect is not significant, even at the 10% level, in the Upper-East villages. FBO membership is relatively more important in the Eastern region, while agricultural extension contact is more important in the Upper-East for diversification into non-staple crop production.

Other significant predictors of non-staple production in the entire sample are number of adult-equivalent labour units and physical assets index (human capital and wealth indicators respectively). This is consistent with the literature, which predicts a positive relationship between household wealth variables and cash crop production (Delgado and Siamwalla, 1997). However, regional differences exist, as shown in the regionally disaggregated estimates. While an adult-equivalent labour unit has a significant positive effect in the Eastern region, the opposite is the case in the Upper-East region. The long lean season in the Upper-East, which results in low labour productivity, may account for this situation. Dependency ratio has a strong negative effect on resource allocation to non-staples in the Upper-East. Indeed, a calculation of the marginal effects on the probability that a household would diversify into self-sufficiency show that a unit decrease in dependency ratio increases this probability by 0.18, *ceteris paribus*. Given that the household allocates some land to non-staples (i.e. if the household is not censored at zero), a unit decrease in dependency ratio increases the per cent share of land allocated to non-staples by 16 units (i.e. 16%). Female-headed households in the Upper-East region are significantly less likely to allocate land to non-staples. Non-farm sources of income exert a significant positive effect on diversification into non-staples in the Eastern region. Thus, there appear to be complementarities between non-farm activity and diversification into 'high value' crops in those villages. The time dummy is negative and statistically significant at the 1% level in the Upper-East region but not in the Eastern region, indicating significant reduction in resources allocation to non-staple production in 2008 compared to 2002 in that region. This is partly attributable to floods that affected some of the villages in the region.

Does diversification into self-sufficiency hurt or enhance household food security?

Next, the determinants of household food security are estimated to test two main hypotheses: (i) if markets are incomplete, the allocation of resources to the production of non-staples would hurt household food security; and (ii) households with multiple crop portfolios are more likely to be food secure. Since we fail to reject the exogeneity of diversification into non-staples in the food security equation, we estimate random effects and pooled probit models using the latent binary measure of food security as the dependent variable. The coefficients of both models were approximately the same,[8] but since we fail to reject the null hypothesis that ρ (rho) = 0 in the random effects probit estimation, even at the 10% level, the pooled probit model is favoured.[9]

The conclusion on our hypotheses depends on the regional location of villages (Table 8.6 and Table 8A.2). The results from the combined sample estimates

[8] Given the correlation structure between two successive error terms $\frac{\sigma_c^2}{\sigma_c^2 + \sigma_u^2}$ where σ_c^2 and σ_u^2 are the variance of the random unobservable component and the idiosyncratic error respectively, then if $\sigma_c^2 = 0$ the pooled probit parameters would be equal to those estimated by the random effects probit.

[9] The fact that we reject the null hypotheses of $\rho = 0$ suggests the absence of significant household unobserved heterogeneity.

Table 8.6. Determinants of rural household food security (random effects probit.)[a]

	Marginal effects		
	Upper-East region	Eastern region	Combined sample
Share of land cultivated to non-staples	−0.0044	0.0120**	0.0045
	−0.89	2.26	(1.28)
Staple crop farm size	0.1151	0.3286***	0.1770***
	1.22	3.46	(2.80)
Simpson's Index of Diversity	0.3133	1.0665	0.2520
	0.50	1.28	(0.52)
Age of household head	−0.0863***	−0.0281	−0.0607***
	−3.42	−0.87	(−3.09)
Age of household head squared	0.0008***	0.0003	0.0005***
	3.40	0.86	(3.10)
Sex of household head	0.2863	−0.3742	−0.3732
	1.06	−1.49	(−1.58)
Upper-East female head of household			0.5281
			(1.50)
Education of household head	0.0479**	0.0328	0.0285**
	2.53	1.60	(1.97)
Dependency ratio	−0.4795***	−0.3861***	−0.3932***
	−3.02	−3.62	(−4.51)
Physical asset index	1.1548	0.8002	1.0723**
	1.76	1.02	(2.24)
Remittance income	0.0056***	0.0044**	0.0044***
	3.08	2.04	(4.27)
Other non-farm income	0.0045***	0.0068***	0.0049***
	3.50	3.1	(3.16)
Number of cows owned	−0.0262		−0.0452
	−0.86		(−1.56)
Number of sheep and goats owned	0.0014	0.0110	0.0054
	0.12	0.42	(0.58)
Number of poultry owned	0.0048	0.0074	0.0051
	1.00	1.04	(1.35)
Social capital	0.1558	−0.0901	0.1532
	0.80	−0.37	(1.01)
Credit access	−0.2133	0.1891	0.0513
	−1.07	0.94	(0.38)
Distance to market	−0.1066**	0.0243	0.0337
	−2.19	0.86	(1.36)
Year dummy (2008 = 1)	−0.2057	0.8325***	0.3090**
	−0.95	3.51	(2.05)
Number of observations	375	277	652
Log likelihood	−130.41	−119.89	−256.2
Per cent correctly predicted	79.5	77.5	76.0
Wald chi-squared	64.59	71.60	202.48
Rho[b]	1.1×10^{-5}	1.8×10^{-6}	2.8×10^{-6}
	(2.5×10^{-4})	(4.8×10^{-4})	(5.3×10^{-6})

[a]Absolute value of robust *z*-statistics in parentheses.
[b]Standard errors in parentheses.
**Significant at 5% level; ***significant at 1% level.

reject the hypothesis that resource allocation to non-staples hurts household food security but is also inconclusive on whether non-staple production enhances food security. A priori, a negative sign was expected on the coefficient of the self-sufficiency crop production variable. Even though we observe a negative sign in the Upper-East region, the effect is not statistically significant, even at the 10% level. In the Eastern region, however, we find a significant positive relationship between non-staple crop production and household food security. The predicted probability of food security for a household that devotes all its land to the production of staples in this region is about 0.51, while for a household that cultivates the sample mean share of land (23.7%) to non-staples, the predicted probability of food security is 0.67, *ceteris paribus*. The different outcomes in the two regions may be attributable to better market conditions (higher prices and higher potential demand due to proximity to larger urban centres) in the Eastern region. The second hypothesis test is carried out on the coefficient of the Simpson's Index of Diversity – a positive sign was expected a priori. This was observed but was not statistically significant at the 5% level. Thus, in general, there is no evidence that multiple agricultural portfolios necessarily enhance food security. It is possible that there are no significant negative or weak positive correlation between agricultural-based portfolios.

Other important predictors of rural household food security include resource allocation to staple crop production, household characteristics (sex, age, household composition and education), physical asset wealth, remittances, other non-farm income and distance to main market outside the village. In general, female-headed households are more likely to be food insecure than their male counterparts, but the difference is statistically significant only in the Eastern region. In this region, the reported marginal effects (of the pooled probit model) show that female-headed households are 9% less likely to be food secure than male-headed households.

Human capital assets are important for household food security. This is measured by dependency ratio and education. The probability of food security decreases with increasing dependency ratio, while education is a significant positive predictor of food security in both regions. Household wealth indicators – physical asset index and small ruminant ownership (in the Upper-East region) – are positively associated with household food security. It appears food security is more responsive to physical asset ownership in the Upper-East than the Eastern region. Even though at the sample mean the predicted probability of food security is 0.65 for the Eastern region sample and only 0.15 for the Upper-East region, a 10% increase in asset index (from the mean), however, increases the probability of food security by only 0.4% in the Eastern but 13% in the Upper-East region. Households who own sheep and goats in the Upper-East region are more likely to be food secure. This is not surprising because, during the lean season, the sale of livestock becomes very important in those villages. Finally, the transactions cost variable shows the expected negative association with food security in the Upper-East region villages, suggesting that the higher the transactions cost, the less likely it is for households to be food secure.

Conclusions and Policy Implications

Using panel data from eight villages in two distinct agro-ecological zones in Ghana this study has examined whether or not rural households seek food security insurance through production of their own staples as a priority before diversifying into the production of non-staples. Since transactions cost has been noted as an important reason for which rural households may choose the self-sufficiency strategy, we have explored the role road infrastructure and institutions play in this relationship. Secondly the study has estimated the determinants of household food security in order to verify if the allocation of resources to the production of non-staples hurts or enhances rural household food security. The results suggest that geographic location is important in the determination of the nature of the relationships. Overall, households in the study villages (particularly in the Eastern region) are more likely to allocate resources to the production of non-staples when household food requirements are met. Even though, by rural African standards, roads linking the villages are fairly good, we find some evidence that transactions cost significantly influences this relationship. Institutions (regular contact with agricultural extension and FBO membership) significantly reduce the need for self-sufficiency in staples and increase the probability of resource allocation to the production of non-staples. This is consistent with the literature, suggesting that the development of both 'hard' and 'soft' infrastructure are necessary for diversification into self-sufficiency (Goletti, 1999; Govereh and Jayne, 2003). This may be because regular access to agricultural extension advice and FBO membership are likely to increase staples crop productivity through technology adoption, which increases the ability of the household to meet its food requirements. Other important determinants of self-sufficiency crop production are adult-equivalent labour resource, dependency ratio and wealth indicators.

Two hypotheses were advanced regarding the second research issue. Overall, we found no evidence that the allocation of resources to non-staple production hurts or enhances household food security. In the Eastern region, however, there are significant synergies; the allocation of resources to non-staples had a positive and significant effect on food security. The second hypothesis, that a more diverse crop portfolio enhanced household food security, was inconclusive. The sign of the coefficient was positive but not statistically significant. Even though all households had diverse crop portfolios, a more diverse crop portfolio is not associated with a higher probability of being food secure. Other important predictors of rural household food security in the entire sample estimates include age, education, household composition, wealth, remittance income and other non-farm sources of income.

In order to speed up poverty reduction in rural Ghana through income growth, there would be the need for farmers to participate more in both staple and non-staple ('high-value crop') markets. But this would not just happen. It would be conditioned on, among other things, increased staple crop productivity. This is because, as productivity of staples increases, households are more likely be food secure, which is important for both staple crop market participation and the allocation of resources to the production of non-staples for the growing urban markets. A policy approach that aims at increasing staple crop productivity is likely to have two effects: first, household food security would be enhanced and, second, households would then allocate more resources towards the production of 'high-value' crops to increase household income and reduce rural poverty.

References

Bouis, H. (1994) Consumption effects of commercialization of agriculture. In: von Braun, J. and Kennedy, E. (eds) *Agricultural Commercialization, Economic Development, and Nutrition*. Johns Hopkins University Press, Baltimore, Maryland, pp 65–78.

Bradshaw, B., Dolan, H. and Smit, B. (2005) Farm-level adaptation to climatic variability and change: crop diversification in the Canadian prairies. *Climatic Change* 67, 119–141.

de Janvry, A., Fafchamps, M. and Sadoulet, E. (1991) Peasant household behaviour with missing markets: some paradoxes explained. *The Economic Journal* 101, 1400–1417.

Delgado, C.L. (1995) Agricultural diversification and export promotion in Sub-Saharan Africa. *Food Policy* 20, 225–243.

Delgado, C.L. and Siamwalla, A. (1997) *Rural Economy and Farm Income Diversification in Developing Countries*. MTID Discussion Papers No. 20. International Food Policy Research Institute, Washington, DC. Available at: http://ideas.repec.org/p/fpr/mtiddp/20.html (accessed 30 July 2009).

Di Falco, S. and Chavas, J.P. (2009) On crop biodiversity, risk exposure, and food security in the highlands of Ethiopia. *American Journal of Agricultural Economics* 91, 599–611.

Dzanku, F.M. (2009) Land Rights, Sustainable Natural Resource Use and Agricultural Productivity in Ghana. Technical Publication No. 85. Institute of Statistical, Social and Economic Research, University of Ghana, Accra, Ghana.

Fafchamps, M. (1992) Cash crop production, food price volatility, and rural market integration in the third-world. *American Journal of Agricultural Economics* 74, 90–99.

Featherstone, A.M. and Moss, C.B. (1990) Quantifying gains to risk diversification using certainty equivalence in a mean variance model: an application to Florida citrus. *Southern Journal of Agricultural Economics* 12, 191–197.

Finkelshtain, I. and Chalfant, J.A. (1991) Marketed surplus under risk: do peasants agree with Sandmo. *American Journal of Agricultural Economics* 73, 557–567.

Ghana Statistical Service (2008) *Ghana Living Standards Survey: Report of the Fifth Round (Glss 5)*. GLSS 5. Accra, Ghana.

Goletti, F. (1999) *Agricultural Diversification and Rural Industrialization as a Strategy for Rural Income Growth and Poverty Reduction in Indochina and Myanmar*. MTID Discussion Paper No. 30. International Food Policy Research Institute, Washington, DC. Available at http://ideas.repec.org/p/fpr/mtiddp/30.html (accessed 30 July 2009).

Govereh, J. and Jayne, T.S. (2003) Cash cropping and food crop productivity: synergies or trade-offs? *Agricultural Economics* 28, 39–50.

Guvele, C.A. (2001) Gains from crop diversification in the Sudan Gezira scheme. *Agricultural Systems* 70(1), 319–333.

Jayne, T.S. (1994) Do high food marketing costs constrain cash crop production: evidence from Zimbabwe. *Economic Development and Cultural Change* 42, 387–402.

Jolly, C.M. and Gadbois, M. (1996) The effect of animal traction on labour productivity and food self-sufficiency: the case of Mali. *Agricultural Systems* 51, 453–467.

Joshi, P.K., Gulati, A., Birthal, P.S. and Twari, L. (2003) *Agricultural Diversification in South Asia: Patterns, Determinants, and Policy Implications*. MSSD Discussion Paper No. 57. International Food Policy Research Institute, Washington, DC.

Masakure, O., Cranfield, J. and Henson, S. (2008) The financial performance of non-farm microenterprises in Ghana. *World Development* 36, 2733–2762.

Minot, N. (1999) Effect of Transaction Costs on Supply Response and Marketed Surplus: Simulations Using Non-separable Household Models. MSSD Discussion Paper No. 36. International Food Policy Research Institute, Washington, DC.

Minot, N., Epprecht, M., Anh, T.T.T. and Trung, L.Q. (2006) *Income Diversification and Poverty in the Northern Uplands of Vietnam*. Research Report 145. International Food Policy Research Institute, Washington, DC.

MoFA (2007a) *Agriculture in Ghana in 2006. Annual Report*. Ministry of Food and Agriculture, Accra, Ghana.

MoFA (2007b) *Food and Agriculture Sector Development Policy (FASDEP II)*. Ministry of Food and Agriculture, Accra, Ghana.

Okigbo, B.N. (1991) *Nutritional Implications of Projects Giving High Priority to the Production of Staples of Low Nutritive Quality: the Case for Cassava (Manihot esculenta, Crantz) in the Humid Tropics of West Africa*. Research Report. International Institute of Tropical Agriculture Ibadan, Nigeria. Available at: http://www.unu.edu/unupress/food/8F024e/8F024E01.htm (accessed 4 August 2009).

Pingali, P.L. and Rosegrant, M.W. (1995) Agricultural commercialization and diversification: processes and policies. *Food Policy* 20, 171–185.

Quiroz, J.A. and Valdes, A. (1995) Agricultural diversification and policy reform. *Food Policy* 20(3), 245–255.

Rahman, S. (2009) Whether crop diversification as a desired strategy for agricultural growth in Bangladesh? *Food Policy* 34(4), 340–349.

Shome, S. (2009) An analysis of crop diversification: experience in the Asia-Pacific region. *ICFAI Journal of Agricultural Economics* 6(1), 7–30.

Singh, I., Squire, L. and Strauss, J. (eds) (1986) *Agricultural Household Models: Extensions, Applications and Policy*. Johns Hopkins University Press, Baltimore, Maryland.

Thomson, A. and Metz, M. (1998) Implications of economic policy for food security: a training manual. *Training Materials for Agricultural Planning 40*. Food and Agricultural Organization, Rome, Available at: www.fao.org/DOCREP/004/X3936E/X3936E00.HTM (accessed 24 July 2009).

von Braun, J. (1994) Production, employment, and income effects of commercialization of agriculture. In: von Braun, J. and Kennedy, E. (eds) *Agricultural Commercialization, Economic Development, and Nutrition*, Johns Hopkins University Press, Baltimore, Maryland, pp. 37–64.

von Braun, J. (1995) Agricultural commercialization: impacts on income and nutrition and implications for policy. *Food Policy* 20, 187–202.

von Braun, J. and Immink, M.D.C. (1994) Nontraditional vegetable crops and food security among smallholder farmers in Guatemala. In: von Braun, J. and Kennedy, E. (eds) *Agricultural Commercialization, Economic Development, and Nutrition*. John Hopkins University Press, Baltimore, Maryland, pp. 189–203.

von Braun, J. and Kennedy, E. (1994a) Conclusions for agricultural commercialization policy. In: von Braun, J. and Kennedy, E. (eds) *Agricultural Commercialization, Economic Development, and Nutrition*. Johns Hopkins University Press, Baltimore, Maryland, pp. 365–376.

von Braun, J. and Kennedy, E. (eds) (1994b) *Agricultural Commercialization, Economic Development, and Nutrition*. Johns Hopkins University Press, Baltimore, Maryland.

von Braun, J., Bouis, H. and Kennedy, E. (1994) Conceptual framework. In: von Braun, J. and Kennedy, E. (eds) *Agricultural Commercialization, Economic Development, and Nutrition*. Johns Hopkins University Press, Baltimore, Maryland, pp. 11–33.

Weinberger, K. and Lumpkin, T.A. (2007) Diversification into horticulture and poverty reduction: a research agenda. *World Development* 35, 1464–1480.
Windle, J. and Rolfe, J. (2005) Diversification choices in agriculture: a choice modelling case study of sugarcane growers. *Australian Journal of Agricultural and Resource Economics* 49, 63–74.
Wooldridge, J.M. (2002) *Econometric Analysis of Cross Section and Panel Data*. MIT Press, Cambridge, Massachusetts.

Appendix

Table 8A.1. Determinants of non-staple crop production (pooled tobit).[a]

	Upper-East region[b]	Eastern region[b]	Combined sample[b]
Expected staple food self-sufficiency level (SFSSL)	−0.0315***	0.0044	0.0066**
	(5.22)	(1.25)	(2.21)
Expected staple food self-sufficiency dummy (SFSSD)	−6.0289	−7.9937	−4.9951
	(1.15)	(1.49)	(1.36)
Self-sufficiency dummy × self-sufficiency level (SFSSLD)	0.0044	0.0462***	0.0307***
	(0.24)	(3.66)	(3.06)
Distance to main market outside village (MD)	−0.8180	−16.4420***	−1.0845***
	(1.41)	(3.06)	(3.19)
FBO membership (FBO)	−0.9805	8.1204**	2.9086
	(0.39)	(2.44)	(1.44)
Agricultural extension contact (AE)	14.7088***	6.4812	6.2322***
	(4.93)	(1.86)	(2.80)
Extension contact × sex of household head (AE_Sex)	0.5882	−5.5693	−2.7033
	(0.08)	(0.73)	(0.53)
Credit access (Cr)	−1.1612	−0.1548	−0.3802
	(0.47)	(0.05)	(0.20)
Complete control over cultivated land (LR)	−5.7454	4.5574	0.7743
	(1.61)	(1.53)	(0.35)
Female-headed household	−19.3641***	−1.8148	−1.2506
	(3.26)	(0.33)	(0.33)
Adult-equivalent labour unit (L)	−4.8262***	1.8771**	1.3455***
	(5.28)	(2.15)	(2.97)
Dependency ratio (DR)	−3.2320***	−2.1366	−0.6284
	(2.62)	(1.39)	(0.76)
Education level of household head (EDUC)	−0.6919**	−0.1216	−0.0723
	(2.54)	(0.37)	(0.35)
Age of household head (Age)	−0.2981	0.1437	−0.1555
	(0.87)	(0.32)	(0.58)
Square of age of household head (Agesq)	0.0032	−0.0019	0.0012
	(1.06)	(0.48)	(0.49)
Remittances (RI)	−0.0313**	−0.0072	−4.1845
	(2.11)	(0.35)	(1.96)
Other non-farm income (Masakure *et al.*, 2008)	−0.0050	0.0043**	0.0013
	(1.81)	(2.23)	(0.86)
Physical asset index (AI)	−10.9081	20.9170**	13.6514**
	(1.29)	(2.18)	(2.18)
Constant	67.9526***	3.0021	12.3093
	(6.56)	(0.24)	(1.51)

Continued

Table 8A.1. Continued.

	Upper-East region[b]	Eastern region[b]	Combined sample[b]
Number of observations	375	280	613
Log likelihood value	–1486.39	–1109.52	–2419.51
Wald chi-squared	66.02	74.73	117.73
R^{2c}	0.205	0.259	0.18
Ho: SFSSD = SFSSLD = 0	F = 1.67	F = 6.96	F = 5.15
	[Prob = 0.189]	[Prob = 0.001]	[Prob = 0.006]

*Significant at 10% level; **significant at 5% level; ***significant at 1% level.
[a]District dummies included but not reported.
[b]Absolute value of *t*-statistics in parentheses.
[c]R^2 between the predicted and observed values.

Table 8A.2. Determinants of rural household food security (pooled probit).[a]

	Marginal effects		
	Upper-East region	Eastern region	Combined sample[b]
Share of land cultivated to self-sufficiency	–0.0011	0.0008***	0.0017
	(0.81)	(2.81)	(1.30)
Staple crop farm size	0.0374	0.0139***	0.0610***
	(1.57)	(2.75)	(2.65)
Simpson's Index of Diversity	0.1135	0.0794	0.0940
	(0.60)	(1.71)	(0.43)
Age of household head	–0.0250***	–0.0018	–0.0245***
	(3.48)	(1.04)	(3.11)
Age of household head squared	0.0002***	0.0000	0.0002***
	(3.47)	(1.25)	(3.26)
Sex of household head	0.0836	–0.0392**	–0.1615
	(1.06)	(2.07)	(1.89)
Upper-East female head of household			0.2176
			(1.78)
Education of household head	0.0143**	0.0021	0.0112**
	(2.54)	(1.90)	(2.03)
Dependency ratio	–0.1425***	–0.0208***	–0.1533***
	(3.49)	(3.62)	(4.85)
Physical asset index	0.3065	0.0573	0.4749**
	(1.52)	(1.37)	(2.37)
Remittance income	0.0015***	0.0004**	0.0022***
	(4.29)	(2.52)	(4.93)
Other non-farm income	0.0012***	0.0006***	0.0019***
	(4.67)	(2.80)	(5.95)
Number of cows owned	–0.0071		–0.0187
	(0.86)		(1.62)
Number of sheep and goats owned	0.0005	0.0001	0.0020
	(0.20)	(0.06)	(0.60)
Number of poultry owned	0.0012	0.0007	0.0024
	(0.92)	(1.69)	(1.57)

Table 8A.2.

	Marginal effects		
	Upper-East region	Eastern region	Combined sample[b]
Social capital	0.0555	0.0225	0.0411
	(0.98)	(1.39)	(0.68)
Credit access	−0.0672	0.0123	0.0686
	(1.20)	(1.19)	(0.52)
Distance to market	−0.0315**	0.0016	0.0143
	(2.15)	(1.01)	(1.52)
Number of observations	375	277	652
Log likelihood	−130.89	−127.00	−258.3
Per cent correctly predicted	82.1	80.1	78.0
Wald chi-squared	77.74	52.09	232.47
Pseudo R^2	0.348	0.256	0.419

[a]Absolute value of z-statistics in parentheses.
[b]District dummies included but not reported.
*Significant at 10% level; **significant at 5% level; ***significant at 1% level.

9 Conditions for Achieving Sustained Agricultural Intensification in Africa: Evidence from Kenya

Stephen K. Wambugu,[1] Joseph T. Karugia[2] and Willis Oluoch-Kosura[2]

[1]*Department of Agribusiness Management and Trade, Kenyatta University, Nairobi, Kenya;* [2]*Department of Agricultural Economics, University of Nairobi, Nairobi, Kenya*

In sub-Saharan Africa (SSA), agriculture is seen as a strong option for spurring economic growth, overcoming poverty and enhancing food security. Growth in agricultural productivity is vital for stimulating growth in other sectors of the economy. Agriculture alone cannot achieve the magic of massively reducing poverty, but it has proven to be powerful for that task. While the worlds of agriculture are vast, varied and rapidly changing, with the right policies and supportive instruments at local, national and global levels, today's agriculture offers new opportunities to hundreds of millions of the rural poor to move out of the poverty trap. Pathways out of the poverty trap offered by agriculture include smallholder farming, animal husbandry, employment in the 'new agriculture' of high-value products, and entrepreneurship and jobs in the emerging rural, non-farm economy (World Bank, 2008).

However, despite the importance of agriculture, its growth in SSA is constrained by a number of factors. According to Kherallah *et al.* (2000) agricultural growth in SSA has generally been constrained by five sets of factors: (i) drought, diseases, war and other exogenous shocks; (ii) structural factors – such as research, extension, transport and communications – that were neglected during the reforms; (iii) inadequate legal and other regulatory systems relating to standards, contracts and property rights, as well as lack of good governance; (iv) partial implementation of reforms; and (v) the tendency to think that reforms amounted to one-shot events rather than long-term processes of learning by doing.

Kenya, like other SSA countries, displays the hallmarks of a developing economy. Agriculture dominates the national economy, employing, directly or indirectly, over 80% of the active population; accounts for over 60% of the gross domestic product (GDP); and provides nearly all the food requirements for the country and the bulk of raw materials for the industrial sector. Agriculture is among the main productive sectors of Kenya's economy. About 80% of the total agricultural output in Kenya comes from small-scale producers (Kenya, 2008). Agricultural intensification is seen as a prerequisite for achieving productivity increases, more so in areas where the farm sizes are dwindling as population increases. Agricultural intensification implies an agricultural production system characterized by high inputs of capital and labour and/or heavy usage of technologies such as pesticides and chemical fertilizers relative to land area. It is usually seen as one of the best ways of achieving productivity increases, especially in situations where farm sizes are declining as rural populations increase, as in the rest of Africa.

In the early days of independence, Kenya's agriculture performed well and was the most important driver of economic growth. The country's GDP grew at an annual average of 6.6% between 1963 and 1973, while agricultural production grew by 4.7% annually. Thereafter, a persistent downward trend in per capita growth rate occurred, with the rate turning negative over the 1990s. Specifically, from 1991 to 1993, the economic performance hit the lowest since independence, with growth in GDP stagnating and agricultural production declining at an annual rate of 3.9% (Kenya, 2008). After the inauguration of a new government in 2003, the economy started recovering and the GDP growth rate improved from −0.2% in year 2000 to 3% in 2003, 6.1% in 2006 and 7.1% in 2008 (see Fig. 9.1).

This paper examines the conditions for achieving sustained agricultural intensification using evidence from micro- and macro-data from Kenya. The analytical framework adopts the 'six Is' framework, which represent significant proximate variables influencing agricultural performance, namely innovations (e.g. agricultural research and extension, constituency development funds (CDF), private–public partnerships), inputs (e.g. fertilizer, certified seeds), infrastructure (e.g. roads, irrigation), institutions (e.g. rules of the game, governance), information (e.g. information on markets and agricultural technologies) and incentives (e.g. input and output prices and conducive policies for growth). This framework has been explicitly and implicitly adopted by a number of

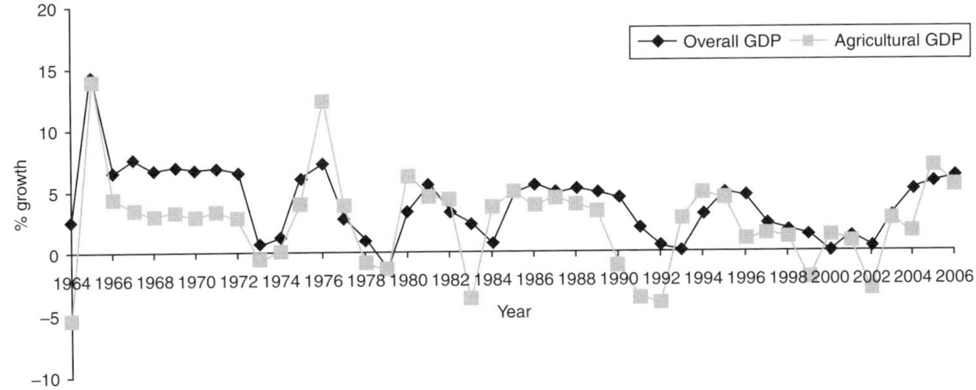

Fig. 9.1. Economic and agricultural growth rates. (Adapted from: Kenya, 1964–2008.)

studies dealing with agricultural development (e.g. Johnston, 1989; Kibaara *et al.*, 2008). The paper further shows how a change in these 'I's affects agricultural productivity and competitiveness. The paper considers the relationship between a number of public interventions and agricultural intensification, and their implications on the realization of the Millennium Development Goals (MDGs) of halving, by 2015, the share of people suffering from extreme poverty and hunger. Emphasis is, however, laid on maize production because it is the most important food staple in the country.

The paper elucidates on the socio-political and economic shocks that have affected the agricultural sector, showing how Kenya recorded an upsurge and impressive growth in GDP and in agriculture sector performance from 2002 up to 2007, after which a sharp decline was observed in 2008. In the absence of mitigating interventions, this decline may signal a downturn, which could have far-reaching negative implications on agricultural intensification and achievement of the MDGs. The exposition is in three main sections. It starts by providing an overview of how the six 'I's facilitated rapid increases in maize productivity from 1963 to 1985. This is followed by another brief overview of how a change in the six 'I's resulted in the above gains not being sustained in the period 1986 up to 2002. The third part of the paper examines the conditions that led to a revitalization of increased agricultural productivity in the period 2003–2007, after an enabling policy environment that favoured the six 'I's was put in place. In doing, this the paper relies on *Afrint* macro- and micro-data collected in two surveys in 2002 and 2008 in Kakamega and Nyeri Districts of Kenya. The paper also presents scenarios likely to emerge after the skirmishes that rocked the country soon after the December 2007 general elections. The paper concludes by offering lessons for sustainable agricultural intensification.

Agriculture and Economic Performance in Kenya

Agriculture was the main economic activity for many years after independence, a situation that led to Kenya's good economic performance, reflected in the country's growth in GDP being closely associated with the performance of the agricultural subsector. Whenever there was an improvement in agricultural performance, a resultant improvement in economic conditions was experienced. A case in point was during the 1986 coffee boom, which turned round the declining economic performance during the early 1980s. This illustrates the interrelationship between agricultural productivity and economic performance in Kenya (see Fig. 9.1).

Brief Overview of the Poverty Situation in Kenya

Deteriorating economic performance, especially during the 1980s and 1990s, had significant implications on the economic well-being of the Kenyan population, which resulted in increased poverty. This is comparable to many SSA countries, where the number of poor people increased from 168 million in 1981 to 298 million in 2004. This is unlike the global situation, where indicators of poverty and food security improved and the number of poor people declined from 1470 million to 969 million over the same period (Kates and Dasgupta, 2007).

In Kenya, poverty levels have been increasing. In 1990 it was at 48.8% but rose to 56% by 2003, a situation that placed more than 14.3 million people below the poverty line (Kenya, 2000, 2004). The condition improved slightly by 2005, with about 52.9% of the rural population and 49.2% of those in urban areas being considered as poor. Statistics further show that about 34.8% of the rural population and 7.6% of the urban population live in extreme poverty, such that they cannot meet their food needs even with their entire resources devoted to food (Manda *et al.*, 2000). Though economic stagnation is a major contributor to high poverty levels, low agricultural productivity, poor marketing of agricultural products, unemployment and low wages, inaccessibility of productive assets (particularly land), poor infrastructure, gender imbalance, high costs of social services, bad governance and diseases such as HIV/AIDS are the major contributing factors (IPAR, 2005).

The government has set the objective of reducing poverty by half by 2015, in accordance with the MDGs. The strategy to achieve it is through the provision of basic needs using targeted programmes, which will be supported by policies that focus on education, health and agricultural production.

Maize Production Trends

The Kenyan government has its policy objectives geared towards making available adequate, nutritionally balanced food in all parts of the country, which is to be achieved by increasing food production through land-use intensification, increased use of high-yielding seed varieties and other inputs and increasing processing capabilities, as well as through the promotion of inter-district trade. At household level this is to be achieved through increasing opportunities to generate cash income and providing incentives to farmers to improve agricultural productivity (Kenya, 2008).

The rationale behind the policy objectives is that food insufficiency has made Kenya a net importer of cereals (maize, wheat, rice, sorghum), especially during drought years, when production does not meet consumption requirements (Haan *et al.*, 2001). The government of Kenya declares the lack of national food self-sufficiency as an indicator of food insecurity, which is estimated to affect about 50% and 38% of the rural and urban population respectively. The situation becomes critical if the deficit relates to maize production, which is the main staple crop and source of sustenance for the majority of households and accounts for nearly half of the calories consumed (Kibaara, 2005). This has led to a deficiency of maize being highly associated with food insecurity even if other food grains may be available.

Maize is the dominant staple food for over 95% of people in Kenya (Wambugu, 2005). It also doubles as a main source of income for the producers in the maize surplus regions. As a food commodity, maize provides 40% of daily caloric needs to the majority of consumers in urban and rural areas

(Okuro *et al.*, 2000). It accounts for 75% of the total cereal area and over 60% in value terms of total marketed cereals. As a source of income, it constitutes 3% of the country's GDP, 12% of the agricultural GDP and 21% of the total value of agricultural commodities (Wangia *et al.*, 2001). Maize is grown in almost all agro-ecological zones and in almost all arable areas, under either a mono-crop or intercrop system.

The trend in maize production has been fluctuating over time, with an average annual production of about 2.7 million t, which is slightly lower than consumption needs. This is attributed to the disparity between actual farm production and the productivity potential recommended through research (Mwangi *et al.*, 2001). The trend in maize production from 1963 to 2007, as shown in Fig. 9.2, indicates that increasing quantities of maize were produced in the 1960s and 1970s. Thereafter, a period of maximum production was realized in the 1980s, except during 1984, when significant decline occurred as a result of the drought condition that characterized that year. However, production gained momentum and reached almost 3 million t in the period 1986–1989. In the 1990s maize production almost stagnated, except in 1994, when a record high of 3.06 million t was attained. The trend, however, appears to improve after the year 2000, owing to government interventions in promoting maize production by enhancing maize marketing, limiting importation and improving prices to farmers.

Figure 9.2 further shows the estimated area under maize cultivation over the same period. Like production, the area increased in the 1960s and 1970s, hitting a record high (>1.6 m ha) around 1976–1977. However, between 1985 and 2002 the area under cultivation appeared to stagnate at between 1.4 and 1.6 million ha. Karanja and Oketch (1992) argue that probably, by then, almost all the suitable land had been put under cultivation. This is, however, disproved by the increase recorded in 2005 and 2006 to over 1.7 million ha, which could have resulted from farmers putting additional emphasis on maize production

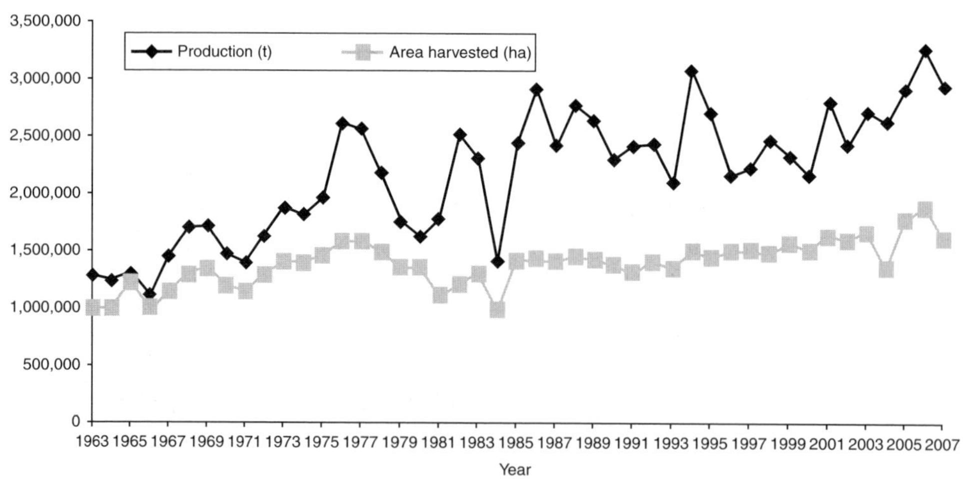

Fig. 9.2. Maize production and area cultivated since 1963, based on FAOSTAT data, 2009.

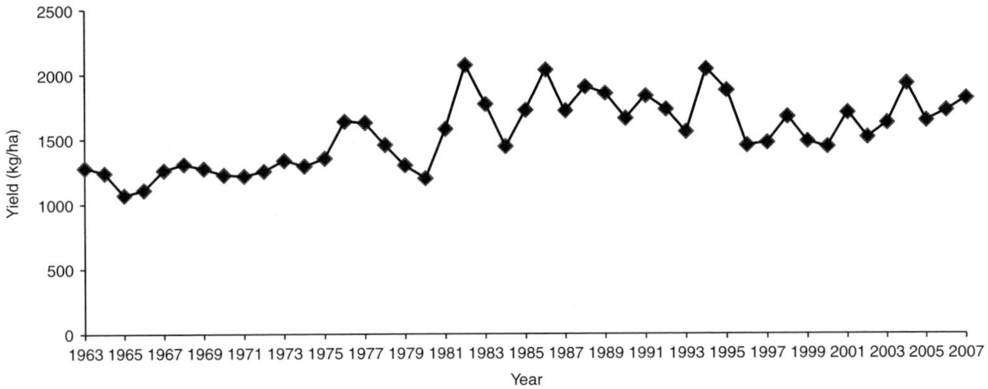

Fig. 9.3. Maize yield per hectare since 1963, based on FAOSTAT data, 2009.

and therefore converting land previously under other crops to maize in order to benefit from the market reforms and better prices.

Maize yield per unit area has also been fluctuating over the years, as shown in Fig. 9.3. The figure indicates the highest yield having been achieved in the 1980s and declining later on. This productivity is illustrated by period, as presented in Fig. 9.4, which shows the period 2003–2007 having recorded the highest average yield: 1.75t/ha, compared to 1.39t/ha in the period 1963–1985. The scenario has many underlying causes, including policies, most of which are outside the scope of the farmer's decision making.

Globally, the challenge of food insufficiency and poverty has been tackled through various strategies, with varying degrees of success. This has included

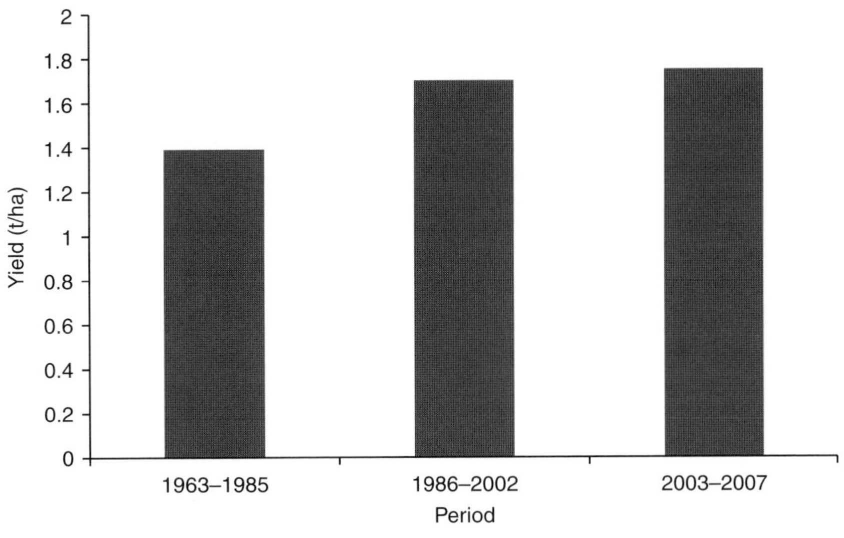

Fig. 9.4. Maize yields over three consecutive periods. (Adapted from: FAOSTAT data, 2009.)

investing in urban areas to raise incomes, with the assumption that wealth will trickle down to the poor people in rural areas, although no sufficient evidence abounds to support it. The alternative has been promotion of the agricultural sector, and varying degrees of success has been reported. For example, Ravallion and Datt (1996), focusing on India, showed that investment in the rural sector reduced poverty in both rural and urban areas. Improving agricultural production also has the benefit of increasing non-agricultural activities, such as processing and small-scale industries that insulate households from poverty (Sarris, 2001). This could explain the commitment by the Kenyan government to revitalize the agricultural sector through planned investment, as documented in the poverty reduction strategy paper (IMF, 2005) and the Vision 2030 (NESC, 2007).

Agriculture Sector Performance During the 1963–1985 Period: a Brief Overview of the Conditions that Enabled Rapid Increases in Agricultural Productivity

This section examines agricultural development in Kenya, where success was more conspicuous than failure compared to many SSA countries. To a striking degree, Kenya's better performance was related to the influence of certain 'strategic notions' that shaped her development strategies. These strategic notions influenced both agricultural sector policies and macroeconomic policies, which in turn had interacting impacts on agricultural development in Kenya. This paper identifies six 'I's that represent significant proximate variables influencing agricultural performance, namely investments, inputs, infrastructure, institutions, information and innovations.

Through the Land Development and Settlement Board, a plan was laid out to facilitate African small- and large-scale commercial farmers' entry to the former white highlands (dubbed the Million Acre Programme). The result was a monetized smallholder sector that contributed significantly to the total agricultural production and marketed volume, especially in cash crops. The number of farmers engaged in commercial agriculture increased substantially. The government also encouraged the development of the smallholder cooperative sector to facilitate access to credit, inputs and marketing services for farm produce. To meet the challenges of an increased clientele with diverse interests and complex farming systems, the research system was expanded. A network of research stations covering all important commodities and most of the agro-ecological zones was established. During the same period, the government, through the Ministry of Agriculture, devoted about 10% of its annual budget to agricultural research (Nyangito and Okello, 1998). As a result, there were major breakthroughs in the release of high-yielding varieties of maize and wheat. Cash crops (coffee, tea, sugarcane and cotton) enjoyed special research programmes funded through their respective parastatals. Similar investments were made in the development of extension services, including training and hiring of a large cadre of staff at certificate, diploma, degree and postgraduate levels, and their deployment down to sub-location level. The government also made substantial investments in support of institutions such as the National Cereals and Produce Board (NCPB), Agricultural Development Corporation (ADC), Agricultural Finance Corporation (AFC), Kenya Tea Development Authority and other commodity parastatals. In 1983, the government investment in the agricultural sector amounted to 13% of the national budget. As a result of all the above actions, the agricultural sector recorded high annual agricultural GDP (agGDP) growth rates, averaging approximately 4% between 1964 and 1986.

Rapid growth of agricultural exports was a particularly dynamic component of the rise in GDP and of the growth of farm cash income among Kenya's small-scale farmers. In 1954 the impressive expansion of export crops by Kenya's smallholders began with the launching of 'A Plan to Intensify the Development of African Agriculture in Kenya', commonly referred to as the Swynnerton plan. The plan was aimed at giving farmers security of tenure and incentives to improve their farm holding or layouts that would maintain soil fertility, avoid soil erosion and achieve a dramatic increase in farm incomes. This worked very well in favour of agricultural growth.

The other sources of agricultural growth during this period included area expansion, expansion of cash crops and dairy, adoption of high-yielding crop varieties and livestock breeds, availability of affordable credit and inputs, effective state and commodity extension, and favourable commodity prices, both internal and external. Generally, the considerable dynamism in Kenya's agricultural sector during the period 1963–1985 can largely be attributed to a relatively favourable policy environment. That environment and the associated continuity of policy and institutions from the colonial regime had positive effects on all of the six 'I's. The principal shortcoming was the contrast between the impressive growth of output and farm cash incomes in the high-potential areas and much more limited progress in the areas of medium and low potential.

The impressive performance of Kenya's agriculture was sustained up to the mid-1980s, when the structural adjustment programmes (SAPs) were implemented, after which a downturn was experienced. The next section gives an overview on how the donor-instigated SAP had a negative impact on the six 'I's and on agricultural intensification in Kenya.

Agriculture Sector Performance during the 1986–2002 Period: a Brief Overview of the Conditions that Hindered Sustained Agricultural Productivity

The gains in agricultural performance achieved in the period 1963–1985 were lost with the implementation of the SAPs. The reform programmes in the agricultural sector were part of the wider structural adjustment programmes. The impact of the SAPs on input use and productivity growth in Kenya was negative. Fertilizer prices rose in response to subsidy removal and depreciation of the Kenyan shilling. Meanwhile, fertilizer crop price ratios increased, particularly for non-tradable crops such as maize. As a result fertilizer use declined, especially for maize. Access to credit for input use declined because state-sponsored credit systems through the AFC collapsed. The private sector was not able to provide input credit to farmers due to its inability to enforce loan repayments. Access to extension services substantially declined because the government cut public expenditures in the agricultural sector from 13% in 1983 to 3% in 2000 (Kenya, 2003).

A study by Oluoch-Kosura and Karugia (2005) shows that the initial promise in maize yield growth was not sustained and from the mid-1980s yields declined. While climatic factors, such as incidences of drought, may have contributed to yield decline, there is overwhelming evidence that policy- and institutional-related factors stand out as the major reasons for not sustaining the increases witnessed in the 1960s and 1970s.

Weak institutional support for agriculture, policy failures, low levels of adoption of improved technologies and poor infrastructure were identified as the major constraints to agricultural intensification in the SAP and post-SAP period (Oluoch-Kosura and Karugia, 2005). The factors related to weak institutional support for agriculture included small allocation and declining government expenditure in the sector. It was observed that only about 40% of the government's expenditure on the agricultural sector was spent on agricultural research and market information, animal health services, crop protection, seed inspection, mechanization services and farm planting services, while about 60% was spent on recurrent expenditure. Other weak institutional support for agriculture documented by Oluoch-Kosura and Karugia (2005) since the introduction of SAPs includes weaknesses in research and extension, weak agricultural credit schemes and liquidity constraints which limited demand for key productivity-enhancing inputs.

A number of policy failures, especially in the maize subsector, contributed to a decline in agricultural productivity. Policies on maize production, pricing and marketing have been major concerns for the government of Kenya. These policies ranged from government controls on maize production, pricing and marketing up to 1994, when the current policy of liberalized markets was enacted. The liberalization policies were not properly sequenced and coordinated, and as a result it had adverse effects on the subsector.

Low levels of adoption of improved technologies have also been cited as a contributory factor to declining agricultural productivity, especially during the SAP and post-SAP periods (Karugia, 2003). Farmers adopted parts of the technology packages introduced in Kenya in the late 1960s and early 1970s but missed out on the synergies to be derived from the use of these technology packages. During the era of SAPs, input use among farmers, particularly smallholders, was low and declining due to withdrawal of subsidies and high prices occasioned by the depreciation of the Kenyan shilling, among other factors (Oluoch-Kosura and Karugia, 2005).

During the era of SAPs many poor smallholders could not access markets, due to poor infrastructure, among other factors. Roads deteriorated to the extent that it became a hindrance to growth. The infrastructure was characterized by the poor state of the road network, unreliable and costly electricity, inadequate housing and poor quality of water supply, poor telecommunications and an inadequate information and communication technology (ICT) infrastructure. Karugia *et al.* (2003) noted that infrastructural constraints (including storage facilities, market centres, financial institutions, market information and transport infrastructure) have impeded efficient marketing of maize in Kenya.

The period 1986–2002 saw the reversal of all the favourable attributes for agricultural development generally and for the six 'I's in particular, leading to a dismal performance of the sector. During this period politics took centre stage and resources were diverted from the key sectors of the economy for political survival, to the detriment of sound development policies. Other factors that impacted negatively on agricultural growth and intensification included: mismanagement of farmer support institutions, e.g. Kenya Farmers Association, Kenya Cooperative Creameries, ADC, AFC; dumping of agricultural commodities in the local markets, which acted as a disincentive for farmers to produce more; suspension of the international coffee agreement; depreciation of the Kenyan shilling, which resulted in large increases in the cost of imported inputs; withholding of donor funds over disagreements on democracy, governance and accountability; implementation of SAPs without proper planning; and a decline in budgetary allocation to the agricultural sector (Kenya, 2003). However, as demonstrated in the next section, the government implemented a number of initiatives which led to the revitalization of agriculture.

Agriculture Sector Performance during the 2003–2007 Period: Interventions Leading to Revitalization of Agriculture

The implementation of SAPs in the 1980s and 1990s had negative impacts on markets and prices, which led to declining production of major food crops as well as some cash crops. As a result, food security and household incomes were declining. Significant progress in reversing the trend was made by the government between 2003 and 2007 through agricultural revitalization. Increased maize and rice production was achieved; national GDP and agricultural GDP grew during the period; and poverty declined from 56% in 2003 to 46% in 2006. During this period, Kenya's agricultural productivity, as compared with a number of African countries, performed better, as shown in Table 9.1.

The share of resource allocation to the agricultural sector, which had declined over the years, especially during the SAP era, was improved and the trend was reversed in response to the government's renewed realization regarding the importance of agriculture for economic growth and the need to adhere to the African heads of state and governments, Maputo declaration of increasing the budget allocation to agriculture to at least 10% of the total government budget by 2010. Although 10% allocation has not yet been achieved, there is an increase in total budget allocation compared to the 1990s. The proportion of government expenditure in the agriculture sector increased from about 4% in the 1990s to more than 5.6% in the year 2003. Similarly, there has been a shift in resource allocation at the Ministry of Agriculture from the previously

Table 9.1. Comparison of Kenya's agricultural productivity with other countries (1997–2007). (Adapted from: Kibaara *et al.*, 2008.)

Commodity	Productivity		
	Kenya	Other countries	
Maize yield (bags/acre)	9	Uganda	7
		Tanzania	4
		South Africa	13
		Malawi	7
		Argentina	31
Coffee yield (kg/acre) of green coffee	214	Brazil	345
		Columbia	436
		Uganda	213
Sugarcane yield (t/acre)	25	Egypt	40
		Malawi	43
		Sudan	42
Tea yield (kg/acre) of green tea	4507	Malawi	3523
		India	2774
		Uganda	2601
		Tanzania	2348
		China	1369
Milk yield (kg/cow) per year	1371	Argentina	4773
		South Africa	3093
		Malawi	461
		Uganda	331
		Lesotho	245
		Tanzania	173

huge recurrent expenditure and less for development expenditure. Although a substantial amount still went to recurrent expenditure, the trend changed, as shown in Table 9.2. This has allowed the government to undertake and provide agricultural research and extension, animal health services, crop protection, seed inspection, mechanization services and farm planning services.

From 2002, the Ministry of Agriculture has also focused much of its budget allocation towards priority programmes that will ensure higher returns to investments. For example, during the 2008/09 financial year about 50% of the budgetary allocation made to the agricultural sector ministries was allocated to development activities, targeting projects and programmes in research, extension, training, value addition and development of market infrastructure, among other essential services. The importance of these services is that they help farmers enhance their agricultural productivity by providing them with important information, such as patterns in crop prices, new seed varieties, crop management and marketing, and therefore increasing farmers' ability to optimize the use of their resources. Extension services also create awareness of existing technologies, which generates effective demand by providing a critical signal to input distribution systems (Kenya, 2005).

Table 9.2. Expenditure by three agricultural sector ministries in Kenya (Kshs million). (Adapted from: Kenya 2004, 2006, 2008.)

	Actual					Projected			
	2000/01	2001/02	2002/03	2003/04	2004/05	2005/06	2006/07	2007/08	2008/09
Recurrent	5,438	5,485	5,869	6,404	6,236	8,304	10,497	11,096	11,997
Development	1,652	1,052	1,202	2,858	2,721	4,555	6,522	9,712	11,655
Total	7,090	6,537	7,071	9,262	8,957	12,859	17,019	20,808	23,652
Recurrent as % total	76.7	83.9	83.0	69.1	69.9	64.6	61.7	53.3	50.7
Agriculture as % total GoK expenditure	4.2	3.8	3.8	3.6	2.9	3.7	4.4	4.8	5.2
Agriculture as % total GDP	0.8	0.7	0.7	0.7	0.6	0.8	1.0	1.1	1.2

The other set of policies that has brought changes to extension service delivery comprises the National Agricultural Extension Policy, the National Agricultural and Livestock Extension Programme and the National Agricultural Sector Extension Policy, which emphasize commercialization and privatization of services. These policy changes have altered the previous role played by the government by introducing a multitude of actors, among them the private sector, non-governmental organizations (NGOs), community-based organizations, faith-based organizations and civil society players (Kenya, 2005). They offer working examples of public–private–community partnership arrangements, which should be encouraged countrywide. The changes have several implications on how extension is managed, the approaches and methods used and coordination and linkage of key stakeholders, as well as financing of extension service in the country. In addition, since such players have their geographic preferences, there are incidences where some areas are preferred or draw more attention than others, bringing in disparities in geographical representation. Moreover, use of different approaches to extension management by some actors may sometimes result in contradictory messages to the clientele and in others duplication of efforts and wastage of resources. Nevertheless, if well organized, the entry of multiple extension service providers as a result of policy reforms has the potential of creating complementary synergies among collaborators and thereby leading to better services to the clientele.

Karugia (2003), in a study in Nyeri and Kakamega districts, found that extension services were absent in most villages, and where they were available, they were often provided irregularly. Another study by Karugia and Wambugu (2008) in the same districts found that, in all the surveyed villages, extension services are currently available to the farmers, provided by government agencies (Ministry of Agriculture). Although extension services are available in all the sampled villages, the service does not cover all the farmers. In one of the villages, the key informants reported that extension services target certain categories of villagers, mainly the progressive farmers. This may explain why only 43% and 34% of the sampled farmers had received extension services from the government and NGOs respectively. Information from the Ministry of Agriculture, Department of Extension service revealed that although there is emphasis on commercializing and privatizing extension services in Kenya, the government will continue offering free extension service for food crop production. However, cost recovery strategies are exercised for cash crops and large-scale farming. It also became evident that since most large-scale farmers have close association with certain companies where they sell their produce, they are able to acquire extension service through such companies. This implies that most of the extension service provided by the government goes to small-scale farmers who produce for subsistence.

The fiscal policy reforms had serious impacts on budgetary allocation for rural infrastructure development. Spencer and Badiane (1994) noted that rural infrastructure, comprising rural roads, markets, irrigation systems, water supply, and health and educational facilities, is basic to quality of life in rural areas, in addition to being an important facilitator of economic development. It is also central to agricultural intensification. The deplorable condition of the roads was acknowledged by the government as a major hindrance towards achieving the desired economic progress. This led to concerted efforts to solicit funds from various donor agencies for repairing existing roads and developing new ones. A substantial amount of money was spent in developing new roads as well as maintaining and repairing existing ones. For instance, there was a tremendous increase in budgetary allocation, from Kshs 8.62 billion in 2002 to Kshs 51.18 billion in 2007 (Kenya, 2008). Confirming the situation, Karugia and Wambugu (2008) showed that all the sampled villages had regular public transport, with all of them being serviced more than once a day. Although the villages are fairly well served by road infrastructure, the state of infrastructure in Kenya is still poor and inadequate, leading to increased cost of transport. High transport costs act as a disincentive for small-scale farmers to commercialize. Provision of adequate road infrastructure is essential for integration and agricultural development.

Historically, inadequate rainfall has been one of the main limiting factors in African agriculture. Given that some sampled villages reported below-average rainfall conditions (Karugia and Wambugu, 2008), there is need to devise alternative means of enhancing water availability, such as irrigation and rainwater harvesting. Irrigation and water harvesting hold some promise for enhancing agricultural productivity and intensification in Africa.

Although irrigation investments are a basic component of agricultural intensification, most of the smallholder farmers have not invested in irrigation. Karugia and Wambugu (2008) found that irrigation is practised on maize, where farmers irrigated on at least one-half of the portion planted to maize. Irrigation of maize has enabled 74% of the farmers practising it to harvest more than one maize crop per year, after which the land is used to grow other food crops. Although the number of farmers practising irrigation increased from 5% in 2002 to 14% in 2008 (Karugia and Wambugu, 2008), there is potential for irrigation expansion, which is far from utilized. Investments in irrigation were a basic component of the Asian Green Revolution (Jirström, 2005). It is this underutilized potential that holds some promise for the future in Kenya, given its possibly greater need for irrigation due to serious problems of erratic and inadequate rainfall, high evapo-transpiration and climate change. The slight increase in irrigated land was attributed to associations of small-scale farmers constructing water-control devices and judiciously managing them.

In addition to irrigation, agriculture credit is also central to agricultural intensification. The deteriorating economic situation in the 1990s after the implementation of SAPs hampered the government's role in providing agricultural credit. Public institutions like the AFC were rendered ineffective, a situation that reduced farmers' access to agricultural loans. In other instances, loans for agricultural activities were unavailable and farmers could not purchase the necessary inputs for production. This could partly explain the declining agricultural production recorded in the 1990s. However, there have been efforts to revive such institutions and to provide credit to farmers. Moreover, as stated in the poverty reduction strategy paper (IMF, 2005), the government has shown concern in investigating and selecting options that would enhance the financial credit market for small agricultural borrowers and ensuring that the rural Savings and Credit Cooperatives (SACCOs) play a major role in mobilizing savings for onward lending to their member farmers. This will supplement agricultural credit facilities provided by agricultural companies and other microfinance institutions supporting agriculture.

A study in Nyeri and Kakamega districts by Karugia and Wambugu (2008) found that in all the surveyed villages farmers had opportunities to obtain credit. A number of institutions extend credit to the farmers, mainly the SACCOs, microcredit institutions and rotation savings clubs/self-help groups (Rotating Savings and Credit Associations). About 60% of the respondents reported that they were able to obtain credit, an improvement since 2002, when only 35% of the farmers could access credit facilities. Further, the study revealed that farmers had access to credit for staple food production. Land title deeds, cattle and household assets were reported to be the most important collateral required to obtain credit for staple food production.

Credit has been found to be one of the institutional factors that affect agricultural intensification, since liquidity constraints limit demand for key productivity-enhancing inputs. Availability of credit may explain why there was an increase in the number of farmers using the various technologies in 2008 compared to 2002 (Karugia and Wambugu, 2008). A good example is the proportion of farmers growing hybrid maize varieties, which reportedly increased from 75.3% in 2002 to 86.6% in 2008. Similarly, the increased adoption of other technologies (Table 9.3) is a good indicator that farmers are intensifying their agricultural practices.

Agricultural subsidies in the form of fertilizers, certified seeds, agricultural credit, etc. are important incentives for enhanced agricultural productivity. Although Kenya, like most other countries, had removed agricultural subsidies during the SAP period, the government reinstated fertilizer and agricultural

Table 9.3. Technologies used by the farmers (% of the respondents reporting). (Adapted from: Karugia and Wambugu, 2008.)

	Maize		Cassava		Sorghum	
Technology	2008	2002	2008	2002	2008	2002
Pesticide	27.7	6.0	0.1	0	0.7	0.3
Crop rotation	54.0	48.0	5.8	0.7	6.0	1.3
Intercropping with N-fixing crops	88.0	6.3	2.8		5.7	0.7
Animal manure	91.3	80.7	6.5	0.3	6.7	0.7
Manure/compost/residue incorporation	78.7	50.3	5.8	1.3	4.7	1.0
Agro-forestry	66.7		2.5		3.3	
Traditional varieties	9.6	21.7		2.0	7.6	2.3
Improved varieties	3.3	1.0		1.0	0.3	0
Hybrid varieties	86.6	75.3		0	1.0	0
Irrigation	14.0	5.0	0.0	0	0.0	0

subsidies during the post-2002 period. This policy reversal has played a great role in revitalizing agriculture, since the farmers have easy access to this productivity-enhancing input.

The trend in the proportion of smallholder households using fertilizer on maize has been upward in both Kakamega and Nyeri, as shown in Fig. 9.5. The increasing trend in fertilizer use points to greater access and affordability, which are important aspects in agricultural intensification among smallholder farmers.

An important indicator of intensification is the degree of commercialization, and with the advent of market liberalization, various forms of production and marketing innovations, including contract farming for certain crops, have emerged in Kenya. Through the terms of contract, there is specification on how much produce the contractor will buy and at what price; and normally the contractor provides credit inputs and technical advice to enhance production. While this has been common in Kenya, especially for seeds, sugarcane, tobacco and horticultural crops, it is limited in the case of maize.

Karugia and Wambugu (2008) reported the presence of contract farming and out-grower schemes which only targeted non-food cash crops and horticultural crops. This form of marketing was also available in 2002 and is probably the reason why the proportion of farmers growing cash crops had not changed between 2002 and 2007. The companies also provided a number of services and inputs to the contracted farmers, which included provision of seeds, fertilizers, pesticides, quality control and confirmation of standards, land fumigation and preparation. The proportion of the farmers engaged in contract farming ranged from 3% in Ekero to 98% in Thegenge/Gatondo village. The difference in participation in contract farming can be attributed to distance from the villages to major urban centres, which are major consuming areas, and export routes, e.g. airports.

Although the proportion of farmers selling food staples increased between 2002 and 2007, it has remained very low, with most of the produce being sold

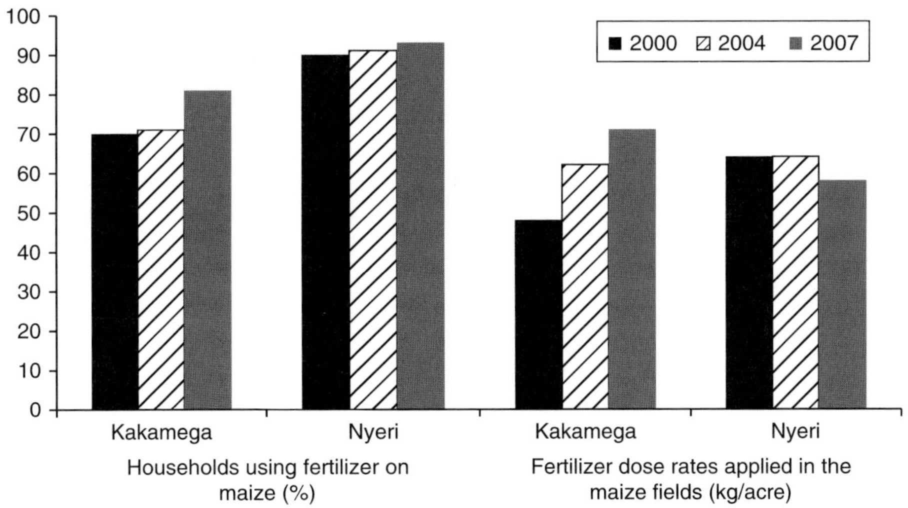

Fig. 9.5. Trends in fertilizer use in Nyeri and Kakamega districts. (Adapted from: Ariga *et al.*, 2008.)

Table 9.4. Main marketing outlets for food staples (% of farmers selling). (Adapted from: Karugia and Wambugu, 2008.)

	Maize	Cassava	Sorghum
Farm gate	14.3	1.3	0
Village market	10.3	0.7	0.3
Market outside the village	4.3	0.3	0
State marketing board	0.3	0	0
Others	1.3	0	0

mainly at the farm gate and village markets (Table 9.4). Selling through brokers and middlemen was also dominant in some villages. Maize is the only crop which is sold to the state marketing board and other markets. The increased commercialization of food staples is an indicator of intensification.

The CDF is another recent innovation, in which grass-roots people (constituents) identify their development priorities, which are then funded. The CDF has been instrumental in rehabilitating and improving the rural agricultural infrastructure (access roads, cattle dips, building of bridges, sinking boreholes, rural electrification, etc.). This has led to increased agricultural productivity and commercialization.

Table 9.5 summarizes the impact of the six 'I's on agricultural development for each of the periods discussed in this paper. The table shows how a change in the 'I's affect agricultural productivity, degree of commercialization, poverty reduction and the contributions of agriculture to the GDP.

Prospects for Sustained Intensification into the Future: Some Predictions/Likely Scenarios

Although progress was made by the government between 2003 and 2007 towards increasing agricultural production, reducing poverty and improving national and agricultural GDP, the December 2007 post-election violence poses a major challenge. GDP plummeted from a high of 7% in 2007 to a low of 1.7% in 2008, while the percentage growth of agriculture GDP dropped substantially, from 2.3% to −5.1%, as shown in Fig. 9.6. This scenario at national level was also mirrored at household level, where agricultural production also declined.

This is because many farming households have been displaced and/or their property and investments destroyed. Food crops (harvested and those in fields) through which they could have generated incomes were destroyed. The most tragic thing is that some of the most affected regions are the prime maize-producing zones, which implies that it will take time before meaningful production can be attained. Since December 2007 economic activities and farm production have declined tremendously, leading to skyrocketing food prices and inflation. More so, incomes and food sufficiency, at both household and country level, have been severely compromised. Equally, education and health services have been grossly interrupted through the destruction of service facilities or challenges

Table 9.5. Summary of the impact of the six 'I's on agricultural intensification.

Variable	Period 1963–1985		Period 1986–2002		Period 2003–2007	
	Status – brief description	Outcome	Status – brief description	Outcome	Status – brief description	Outcome
Innovations	Institutional innovations	Growth of per capita GDP and AgGDP Increased agricultural exports	Contract farming (targeting large farms)	Small-scale farmers neglected – low yields	CDF Public–private partnerships	Increased yields Increased commercialization
Inputs	Input price subsidization Wide distribution of inputs Subsidized credit	Intensification of African agriculture Enhanced food security Increased maize yields	High input costs Low level of input use Removal of input subsidies	Low yields Low commercialization Increased poverty levels	Increased use of agricultural inputs Restoration of agricultural subsidies	Growth in GDP and AgGDP Increased maize yields Increased commercialization Reduction in poverty
Information	Increased budgetary allocation to agriculture Expansion of agricultural extension	Increased maize yields	Low budgetary allocation to agriculture	Reduced yields Decline in agricultural GDP Decline in agricultural exports	Increased budgetary allocation to agriculture Increased extension officers Increased use of mobile phones	Increased yields Increased commercialization
Infrastructure	Provision of marketing infrastructure Establishment of irrigation infrastructure Increased investments to improve roads	Increased commercialization	Dilapidated roads Unreliable and costly electricity Inadequate ICT infrastructure Collapse of irrigation infrastructure	Low yields Low commercialization	Repair of roads and development of new ones Revival of irrigation infrastructure	Increased yields Increased commercialization Enhanced food security

Continued

Table 9.5. Continued.

Variable	Period 1963–1985		Period 1986–2002		Period 2003–2007	
	Status – brief description	Outcome	Status – brief description	Outcome	Status – brief description	Outcome
Institutions	Establishment of agricultural institutions	Increased yields	Collapse of agricultural institutions	Low yields Low commerciali-zation	Revival of credit, marketing and research institutions	Increased yields Enhanced food security Enhanced commercialization
Incentives	Provision of funds for purchase of land from settlers Establishment of settlement schemes Knowledge and research inherited from the colonial government Ready market outlets	Increased exports	Mismanagement of farmers' support institutions Dumping of cheap imported agricultural commodities Implementation of SAPs	Low yields Low commerciali-zation	Provision of extension services Provision of credit, marketing and research facilities	Improved yields Enhanced food security Enhanced commercialization

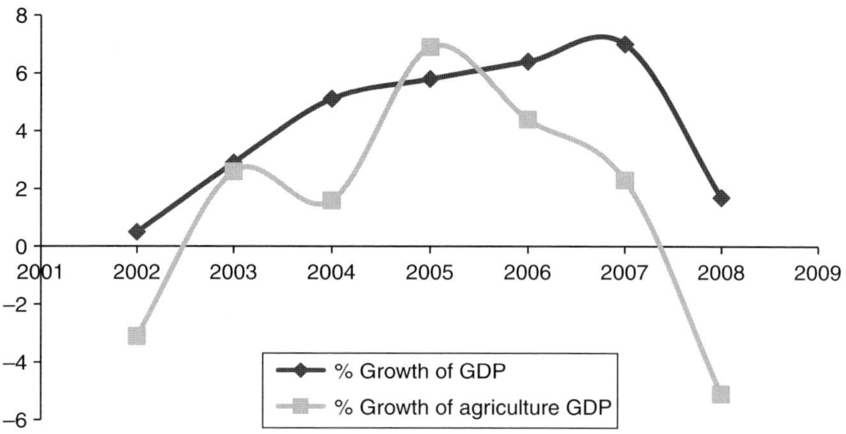

Fig. 9.6. Growth rates for total GDP and agricultural GDP. (Adapted from: Kenya, 2002–2008.)

posed to access to such services. Those living in camps face the challenge of disease incidences due to lack of proper hygiene and nutrition, thereby compromising their safety and welfare. Since it might take time for such households to resettle back on their farms and engage in sufficient production, this has serious implications on the realization of MDGs. There is, therefore, need for vigorous measures to alleviate the situation by supporting the affected households to resettle and engage in economic activities.

The high cost of food is expected to get worse, due to poor weather, political indecision, destruction of the main water towers, high oil and electricity prices and the global downturn. The drought of 2009 has caused crop failures in many parts of the country and water levels in power generating dams have drastically dropped. A power-rationing programme is likely to push more Kenyans out of jobs. The government programme, initiated in 2008, to provide subsidized maize to cushion the poor was a failure after the NCPB ran out of subsidized maize. Rice, which is considered the second staple food after maize, is in short supply owing to an acute water shortage. The effect of bad weather and migration following the post-election chaos are expected to continue to be felt in terms of food shortages and losses in income until the government provides incentives to small-scale farmers.

Conclusion: Lessons for Sustainable Agricultural Intensification in Africa

A number of lessons can be learnt from the agricultural performance in Kenya from independence to the present if Africa, which missed out on the Green Revolution, is to intensify her agricultural practices. From 1963 to the mid-1980s Kenya intensified her agricultural practices, on account of the legacy she inherited from the colonial era. Also intensification was closely associated with the six 'I's, which all worked in favour of Kenya, especially in the high-potential areas. Investment in research, adoption of improved crop varieties, adoption of inorganic fertilizers and good agronomic practices all worked in favour of agricultural intensification. Area expansion and the smooth transfer of land from the Europeans to Africans, coupled with diversification towards high-value crops, such as horticultural crops and dairy farming, also contributed to agricultural intensification.

However, the gains made towards agricultural intensification were not sustained in the period 1986–2002.This is largely attributed to the negative impacts of the donor-instigated SAPs. These negative impacts relate to erosion of functional institutions and incentives, macroeconomic imbalance, declining investment in research and in agriculture generally, and in social capital. A poor infrastructure and information network also worked against agricultural intensification in Kenya during this period. Innovative institutions that had worked very well for the country were eroded through mismanagement and lack of transparency and accountability.

To arrest the declining trend, Kenya revitalized its agriculture after the inauguration of a new government, which lasted between 2003 and 2007. The government realized the failures of the previous regime and adopted measures, especially in extension service provision, credit provision, infrastructural development and budgetary allocation to the agricultural sector. The new government revitalized agricultural institutions that had hitherto become moribund. Irrigation schemes were resuscitated and the government launched the National Economic Stimulus Project on Food Production under Irrigation for Kenya. The project was a new paradigm and impetus toward irrigated agriculture in the country. The specific objectives of this project are to develop irrigation infrastructure, increase area under irrigation, produce more food and create employment for more people. While such efforts yielded progress, there is a need for concerted effort in policy formulation to address gender issues in agricultural production, contract farming and measures to correct market distortions, in order to intensify crop production for poverty reduction and food security.

The outcomes of macroeconomic policies also pose major challenges to the realization of the MDGs. Although the post-2002 government interventions in agriculture and the general economy have shown that progress could be made by correcting some of the policy failures and thereby accelerating the realization of the MDGs, the December 2007 post-election violence poses major challenges. These challenges pertain to implementing measures to support displaced households to resettle and resume agricultural production, and ensuring that existing policy failures are dealt with in order to accelerate agricultural production as a step towards realizing the MDGs.

Acknowledgement

The assistance provided by Lucy Ngare of Kenyatta University in the development of this chapter is gratefully acknowledged.

References

Ariga, J., Jayne, T.S., Kibaara, B. and Nyoro, J.K. (2008) *Trends and Patterns in Fertilizer Use by Smallholder Farmers in Kenya, 1997–2007*. Tegemeo Institute of Agricultural Policy and Development, Nairobi.

FAOSTAT (2009) Production data. FAO Statistics Division. Available at http://faostat.fao.org (accessed 12 August 2009).

Haan, N., Farmer, G. and Wheeler, R. (2001) *Chronic Vulnerability to Food Insecurity in Kenya – 2001: a WFP Pilot Study for Improving Vulnerability Analysis*. World Food Programme, Nairobi.

IMF (2005) Kenya: poverty reduction strategy paper. IMF Country Report No. 05/11, Kenya.

IPAR (2005) Economic growth and poverty in Kenya: a comparative analysis of effects of selected policies. IPAR Policy Brief Vol. 11, Issue 5, Nairobi.

Jirström, M. (2005) The state and green revolution in East Asia. In: Djurfeldt, G., Holmén, H., Jirström, M. and Larsson, R. (eds) *The African Food Crisis: Lessons from the Asian Green Revolution*. CAB International, Wallingford, UK, pp. 25–42.

Johnston, B.F. (1989) *The Political Economy of Agricultural Development in Kenya and Tanzania*. Food Research Institute Studies 21, pp. 205–263.

Karanja, D.D. and Oketch, A.G.O. (1992) The impact of maize research in Kenya. In: Proceedings of a Workshop. Review of the National Maize Research Program. KARI/ISNAR. Management Training Linkage Project, Kenya.

Karugia, J.T. (2003) *A Micro-level Analysis of Agricultural Intensification in Kenya: the Case of Food Staples*. Afrint I report, Lund University, Lund, Sweden.

Karugia, J.T. and Wambugu, S.K. (2008) *The Millennium Development Goals and the African Food Crisis: a Meso and Micro Level Analysis of the Drivers of Agricultural Intensification of Food Staples in Kenya*. Afrint II report, Lund University, Lund, Sweden.

Karugia J.T., Wambugu, S.K. and Oluoch-Kosura, W. (2003) *The Role of Infrastructure and Government Policies in Determining the Efficiency of Kenya's Maize Marketing System in Post-liberalization Era*. A research report submitted to the International Food Policy Research Institute (IFPRI) 2020 Vision Network for Eastern Africa, Kenya.

Kates, R.W. and Dasgupta, P. (2007) Poverty and hunger special feature: African poverty: a grand challenge for sustainability science. *PNAS* 104, 16747–16750.

Kenya, Republic of (1964–2008) *Economic Survey: Various Issues*. Government Printers, Nairobi.

Kenya, Republic of (2003) *A New Strategy for the Agricultural Sector, 2003–2013*. Ministry of Agriculture and Ministry of Livestock and Fisheries Development. Draft strategic plan.

Kenya, Republic of (2004). *Investment Programme for the Economic Recovery Strategy for Wealth and Employment Creation 2003–2007*.

Kenya, Republic of (2005) *National Agricultural Sector Extension Policy (NASEP)*. Ministry of Agriculture; Ministry of Livestock and Fisheries Development and Ministry of Cooperative Development and Marketing.

Kherallah, M., Delgado, C., Gabre-Madhin, E., Minot, N. and Johnson, M. (2000) *Agricultural Market Reforms in Sub-Saharan Africa: a Synthesis of Research Findings*. Markets and Structural Studies Division, IFPRI, Washington, DC.

Kibaara, B.W. (2005) Technical efficiency in Kenyan's maize production: an application of the stochastic frontier approach. MSc thesis, Colorado State University, Colorado.

Kibaara, B., Ariga, J., Olwande, J. and Jayne, T.S. (2008) *Trends in Kenyan Agricultural Productivity: 1997–2007*. Tegemeo Institute of Agricultural Policy and Development, Nairobi.

Manda, K.D., Kimenyi, S.M. and Mwabu, G. (2000) *A Review of Poverty and Antipoverty Initiatives in Kenya*. Background paper prepared for Poverty, Education and Health Project, Social Sector Division, Kenya Institute for Public Policy Research and Analysis, Nairobi.

Mwangi, J.N., Mboya, T.O. and Kihumba, J. (2001) Improved maize production in central Kenya with adoption of soil and water conservation measures. Seventh Eastern and Southern Africa Regional Maize Conference, Kenya, pp. 299–300.

NESC (2007) *Kenya: Vision 2030*. National Economic and Social Council of Kenya (NESC), Nairobi.

Nyangito, H. and Okello, J. (1998) *Kenya's Agricultural Policy and Sector Performance 1964 to 1996*. IPAR Occasional Paper Series 4, pp.1–32, Kenya.

Okuro, J.O., Murithi, F.M., Verkuijl, H., Mwangi, W., De Groote, H. and Gethi, M. (2000) Factors affecting adoption of maize production technologies in Embu district, Kenya. In: Mukisira, E.A., Kinro, F.H., Wamae, J.W., Murithii, F.M. and Wasike, W. (eds) *Collaborative and Participatory Research for Sustainable Improved Livelihoods*. Proceedings of the 7th KARI Biennial Scientific Conference, Nairobi, Kenya, pp. 45–55.

Oluoch-Kosura, W. and Karugia, J.T. (2005) Why the early promise for rapid increases in maize productivity in Kenya was not sustained: lessons for sustainable investment in agriculture. In: Djurfeldt, G., Holmén, H., Jirström, M. and Larsson, R. (eds) *The African Food Crisis: Lessons from the Asian Green Revolution*. CAB International, Wallingford, UK, pp. 181–196.

Ravallion, M. and Datt, G. (1996) How important to India's poor is the sectoral composition of economic growth? *World Bank Economic Review* 10, 1–25.

Sarris, A.H. (2001) The role of agriculture in economic development and poverty reduction: an empirical and conceptual foundation. Rural Strategy Background Paper 2, Rural Development Department, World Bank, Washington, DC.

Spencer, D.S.C. and Badiane, O. (1994) Agriculture and economic recovery in African countries. In: Peters, G.H. and Hedley D.D. (eds) *Proceedings of the Twenty-second International Conference of Agricultural Economics*. Dartmouth, Aldershot, UK.

Wambugu, S.K. (2005) Analysis of the nature and extent of integration of Kenya's maize markets in the post liberalization era. PhD thesis, Kenyatta University, Nairobi, Kenya.

Wangia, C., Wangia, S. and DeGroote, H. (2001) Review of maize marketing in Kenya: implementation and impact of liberalization 1989–1999. Seventh Eastern and Southern Africa Regional Maize Conference, Kenya, pp.12–21.

World Bank (2008) *World Development Report 2008: Agriculture for Development*. The World Bank, Washington, DC.

10 The Fertilizer Support Programme and the Millennium Development Challenge in Zambia: Is Government a Problem Solution?

Hyde Haantuba,[1] Mukata Wamulume[2] and Richard Bwalya[2]

[1]*Agricultural Consultative Forum;* [2]*Institute of Economics and Social Research, University of Zambia, Lusaka, Zambia*

Zambia is committed to contributing towards meeting the Millennium Development Goals (MDGs). According to the 2005 MDG report, halving the proportion of people living in extreme poverty and suffering from hunger is one of the targets perceived as likely to be achieved by 2015. It is in this vein that the government has embraced the *Zambia Human Development Report* (UNDP and GRZ, 2003) and its focus on the reduction of poverty and hunger as the first step towards the fulfilment of the MDGs. At the national level, Zambia has articulated its long-term development objectives in the National Vision 2030. This vision identifies a number of developmental goals, which include reduction of hunger and poverty. Together, these goals call for policies that accelerate and sustain economic growth while enabling the poor to participate in the growth process. Furthermore, Zambia is a signatory of the Maputo Declaration. Based on the view that enhanced agricultural performance has the potential for broad-based poverty reduction, African leaders, through the New Partnership for Africa's Development (NEPAD) initiative, have increasingly underlined the importance of accelerating agricultural growth in Africa (NEPAD, 2005). Furthermore, recognizing the need for public investments to enable agricultural growth, the heads of state agreed to increase their budgetary allocations for agriculture to 10% of total outlays by 2008. In view of the above, the share of the total national budget allocated to the agricultural sector has been on an increase since 2002 (see Fig. 10.1).

However, despite the increase in agricultural budget, the quality of spending matters, as spending in some areas always proves more productive than

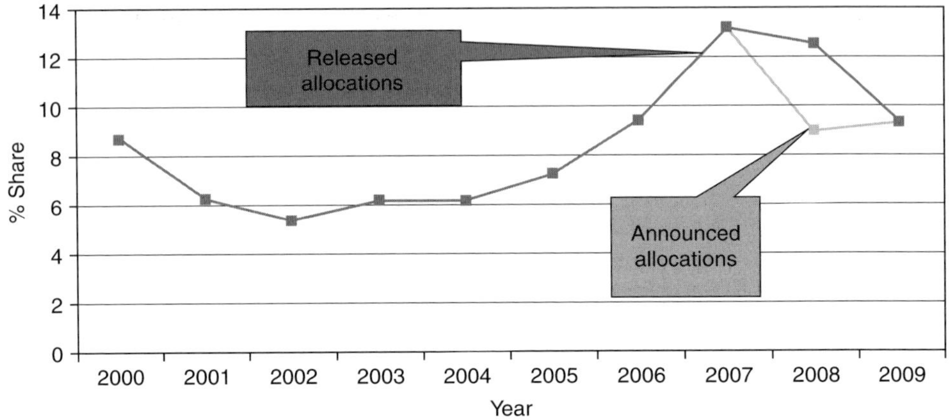

Fig. 10.1. Share of the national budget allocated to agriculture. (Adapted from: ACF/FSRP, 2009.)

others (Haggblade, 2007). As the ACF/FSRP (2009) budget review for 2009 shows, currently the single largest line item in Zambia's agricultural budget is fertilizer subsidies to individual farmers (Table 10.1).

This is despite the fact that the agricultural input and output markets have been liberalized, with the main thrust of policy being economic liberalization and market reforms. This has entailed decontrol of prices and market liberalization for both inputs and outputs. The policy emphasizes government withdrawal from direct involvement in agricultural output marketing and input supply, freeing prices, removing subsidies, privatizing government companies, leasing out public storage facilities to the private sector and overall removal of constraints and distortions to domestic and international trade in farm products. Under this policy framework, it is envisaged that the role of government is confined to policy formulation, legislation and development of support services such as market information, extension and research services, and infrastructural development. While some positive developments, such as increased out-grower schemes and contract farming, crop diversification and changes in land management strategies, have been recorded since liberalization, the private sector has, however, remained constrained in providing input and output marketing services. In response to the above, the government designed the Fertilizer Support Programme (FSP) (GRZ and MACO, 2009).

Under the current agricultural policies, the government's approach has three components: (i) public production of fertilizers; (ii) distribution of free fertilizer through the Food Security Pack Program; and (iii) a 50% (50/50) seed and fertilizer subsidy for hybrid maize production (Jorgensen and Loudjeva, 2005). The government has also been active in output markets through the Food Reserve Agency (FRA). Initially the role of the FRA was specifically to maintain strategic food reserves, but additional roles have since been added, especially those of assisting small-scale farmers to sell their maize, as well as price setting on behalf of government. Analyses of these government programmes

Table 10.1. Resource percentage allocations within agriculture, 2009 budget. (Adapted from: ACF/FSRP, 2009.)

Item	%
Ministry's expenditures including capital expenditure	4
Personal emoluments	10
Recurrent departmental charges	13
Poverty reduction programmes (FSP, FRA and others)[a]	45
Agriculture development programmes[b]	13
Allocation to other ministries	15

[a]Under the Poverty Reduction Program in 2009, 76% was dedicated to the Fertilizer Support Program, 17% to the Food Reserve Agency and only 7% to the remaining programmes, such as livestock development, animal disease control and irrigation development.
[b]The Agricultural Development Programs comprise the Agricultural Support Program and Smallholder Enterprise and Marketing Program.

(Jorgensen and Loudjeva, 2005; Haggblade, 2007; Minde *et al.*, 2008) show that fertilizer programmes have limited the private sector's response to the liberalization reforms, in terms of new entry and investment. The government's distribution of large quantities of poorly targeted fertilizer on loan with recurrently high default rates has undercut private firms' ability to distribute fertilizer commercially. Likewise, government's participation in the output markets has also undermined the private sector's ability to participate.

This paper reviews the operations of the FSP at the macro level to assess its effects on the nation's ability to contribute towards the global goal of attaining MDG 1 as well as the national goal of reducing hunger and poverty, as outlined in the Fifth National Development Plan (FNDP). The paper also uses household-level data from the *Afrint I* and *II* surveys (Wamulume, 2003, 2009), conducted by the Institute of Economic and Social Research (INESOR) to assess micro- and macro-level changes in agricultural productivity, market access, input usage and cropping patterns. The purpose is to analyse both macro-level and micro-level processes unfolding on food and non-food production and productivity initiatives. This will add to the current literature under *Afrint I* studies (Djurfeldt *et. al.*, 2005).

Methodology

Scope

As in *Afrint* (INESOR, 2003 Zambia micro report, unpublished), Mkushi and Mazabuka districts were again (for *Afrint II*) part of the regional sampling frame in the micro-level part of the study. The 2002 survey was treated as a baseline and as much as possible the same households interviewed in 2002 were re-interviewed in 2007. This selection of district (regional) cases was linked to the group of regions located in what we may depict as the 'maize belt' (see Byerlee and Heisey, 1996). The choice of Mkushi in Central Province and Mazabuka in Southern Province was not random but purposive, to ensure sufficient variation in factors assumed to be crucial for agriculture development. Mazabuka and Mkushi districts are sufficiently large so as to contain the prescribed variation of villages along the 'agricultural dynamism' continuum and sufficiently small not to present overwhelming difficulties when it came to survey logistics, costs and time frames.

Methods of data collection

A total of 423 households were interviewed. The macro-level study aimed at clarifying the overall environment in which private entrepreneurs and farmers make their plans and investment decisions and how this environment is shaped by government's action or inaction. The study was therefore based on secondary sources and interviews with key respondents. In addition, country-level analysis of agricultural intensification, i.e. crop yields and crop production and drivers behind contemporary trends analysis, was undertaken by the macro study.

Data analysis

Both qualitative and quantitative analysis methods were used in analysing the primary and secondary data. For quantitative analysis, the Statistical Package for Social Sciences (SPSS) was used. This was used to generate descriptive statistics such as frequencies, as well as *t*-tests for comparison of means between the two time periods in the *Afrint I* and *II* data sets (INESOR, 2003 Zambia micro report, unpublished). Multiple regression analysis was also used to identify the various factors that contributed towards increased yields during these two periods.

Research Findings

Macro-level analysis

At the macro level, the impact of fertilizer subsidies under the FSP in Zambia has also been analysed by various authors (CSPR, 2005; Govereh *et al.*, 2006; Minde *et al.*, 2008). The general conclusion has been that although the programme has increased maize output by up to 12.5%, with smallholder maize yields rising from 2.19 t/ha in 2002/03 to 2.51 t/ha in 2007/08 (see Minde *et al.*, 2008), negative impacts such as crowding-out of private sector participation in the fertilizer markets have been reported during the same period. Govereh *et al.* (2006) report that fertilizer subsidies reduce private sector fertilizer sales by roughly 75% in accessible areas that are well served by private sector fertilizer distributors. Furthermore, to the extent that fertilizer and other farm inputs are private goods, subsidies to individual farmers displace funds these farmers would otherwise spend purchasing inputs. Similarly, despite the fact that one of the objectives of the FSP is to build capacity of the private sector in input marketing, a World Bank study (Jorgensen and Loudjeva, 2005) reports that traders continually complained that uncertainty over the timing, location and volume of fertilizer distributed under the government programmes adds risks and costs to their operations and hence reduces their participation in the market.

Apart from crowding-out private sector participation in fertilizer markets, the FSP fertilizer delivery contributes to late private sector fertilizer delivery and sales in areas where the FSP does not operate. This results from a tendency by the private fertilizer wholesaling firms to stock their fertilizer in Lusaka and wait to see where the government programmes are operating before delivering fertilizer to specific districts (Jorgensen and Loudjeva, 2005). Furthermore, a report by the CSPR (2005) attributed the programme's poor impact to inconsistent supply of inputs. In addition, the inputs are reported to be delivered late, affecting the planting time and consequently yield. The report depicts situations where fertilizers are supplied earlier than seed and cases where top-dressing fertilizer is delivered before basal dressing.

The other issue raised at national level as regards the FSP is of poor beneficiary targeting (GRZ and MACO, 2009). The selection of beneficiaries is done by the District Agricultural Committees. Since most of these are in poor shape or non-existent, the targeting has often been inaccurate. Evidence (CSPR, 2005; Jorgensen and Loudjeva, 2005; Minde *et al.*, 2008) indicates that FSP fertilizer subsidy recipients are typically the better-off smallholder farmers and that their incremental output gain per tonne of fertilizer applied appears to be smaller relative to poor smallholder farmers. Moreover, providing subsidized inputs to relatively well-off farmers may be inconsistent with national policy objectives related to productivity improvement as well as poverty alleviation. For example, the study by MACO, CSO and FSRP (2008), based on Central Statistical Office (CSO) survey data for 2007–2008, indicates that mean maize yield increases per tonne of fertilizer applied are lowest for the largest farm size category (3.32 Mt/ha for farms between 5 and 20 ha). The highest yield increase per tonne of fertilizer was 5.33 Mt/ha for farmers in the 1.7–5 ha category, while farms less than 1 ha averaged 4.55 Mt/ha. Based on this information and the *Afrint II* sample survey regression results alone, one might conclude that improvements in the pass-through of subsidized fertilizer to smallholder farmers and changes in targeting criteria and effectiveness would greatly increase the aggregate benefits of the FSP relative to its cost.

The FSP has also reportedly been biased towards maize and promotes the culture of maize mono-cropping (Saasa, 2003). For example, under the 50/50 scheme, which was introduced in 2002, the government subsidizes 50% of the price of fertilizer and some hybrid seeds. The subsidy is available for maize only and is in the form of a pack comprising of 25 kg maize seed, 4 × 50 kg basal-dressing fertilizer and 4 × 50 kg top-dressing for a 1 ha field. The target is those farmers with capacity to farm between 1 and 5 ha. To receive the subsidy, farmers have to make a 50% down-payment. The idea is to target small farmers with marketing potential.

The volume of subsidies delivered under the FSP has been somewhat larger than during the first four phases of fertilizer subsidies, averaging 66,345 Mt of fertilizer per year compared to 42,505 Mt per year in the previous 8 years (Minde *et al.*, 2008). During the 2009/10 production season, the government targeted to distribute 100,000 Mt of fertilizer to 500,000 small-scale farmers. This represented an increase in both beneficiary farmers (150%) and volume of fertilizer (25%) relative to the previous year. Table 10.2 shows the distribution of fertilizers and maize seed and the number of beneficiaries since 2002.

Two factors have relieved the government's budget constraints and made it easier for them to reinstate and self-finance their fertilizer promotion programmes: firstly, the transition of the World Bank and other donors from conditionality agreements to direct budget support and, secondly, debt forgiveness under the Highly Indebted Poor Countries programme. Both of these recent developments have provided additional discretionary funds to scale-up the farmer fertilizer programmes (Minde *et al.*, 2008).

Policy inconsistencies are another issue raised over the FSP. The programme has been characterized by a number of policy inconsistencies, especially with regards to levels of subsidy and farmer graduation from such programmes. Initially the level of government subsidy per FSP input pack was expected to decrease gradually, from 50% in the first year to 25% in second year, reaching zero in the third year for each beneficiary. Conversely, each FSP beneficiary was expected to contribute 50% of total costs of inputs in the first year, increasing to 75% in the second year, and finally meet the full inputs cost in the third year. For some reason, this has not happened as initially planned. Subsidy levels have instead increased steadily from 50% to 60% in 2007, then to 85% in 2008 and down to 75% in 2009, making it impossible to gradually wean-off beneficiaries from the programme (Jorgensen and Loudjeva, 2005).

Another concern raised is the issue of long-term sustainability and efficiency[1] (Haggblade, 2007). In the absence of a comprehensive analysis of economic efficiency and programme effectiveness, stakeholders are wondering if Zambia is getting the best value for money from FSP interventions at all, especially now that more money is being allocated to FSP every year. Literature (Haggblade,

Table 10.2. Trends in the distribution of inputs under the FSP for 2002–2009. (Adapted from: Ministry of Agriculture and Cooperatives.) NB: In some years, the government increased the amount of distributed fertilizer above these targets.

	Main season input distribution target per agricultural season							
Item	2002/03	2003/04	2004/05	2005/06	2006/07	2007/08	2008/09	2009/10
Number of beneficiaries	305,924	336,000	134,000	186,000	236,292	131,000	200,000	500,000
Maize seed (Mt)	3,333	3,935	2,545	2,938	4,422	2,500	4,000	
Fertilizer (Mt)	66,600	79,445	45,900	55,930	86,792	50,600	80,000	100,000

[1] In 2009, the Zambian government constituted a team to evaluate the Fertilizer Support Program and propose reforms to make the programme more effective. Sustainability was reported to be one of the shortcomings of the programme in its current state.

2007; Minde *et al.*, 2008; FEWSNET, 2009) has shown that these fertilizer programmes have been costly and they reduce financing available for other investments that might have increased more substantially. A policy brief by Haggblade (2007) asserts that although subsidies to individual farmers have produced positive returns similar to the Green Revolution in Asia, subsidies work best where new technologies and good extension support are available. None of these conditions currently hold in Zambia. As such, at macro level, the conclusion is that, although popular, these fertilizer subsidies are typically less effective at stimulating agricultural growth than investment in research, extension, roads and other public goods, because the input subsidies displace private spending that would otherwise occur. Available evidence (Haggblade, 2007) suggests that investment in such public goods constitutes one of the most effective tools available for stimulating economic growth and poverty reduction.

Micro-level analysis

As indicated in the introduction and macro-level analysis, the government is still active in the input markets through the FSP and output markets through the FRA. Using these two programmes, the government has continued to influence the production patterns of smallholder farmers, consequently promoting the maize culture by supporting cultivation and marketing of maize through the entire supply chain: first, by providing subsidized maize input packs to increase maize production and marketed supplies and, secondly, the FRA has revised its mandate from that of managing a strategic reserve to that of marketing surplus maize from small-scale farmers. Consequently, the area dedicated by farmers to maize production has been increasing at the expense of other crops such as sorghum.

At the household level, the FSP has reportedly been having a negative impact on crop diversification. For example, it is interesting to note that there was a significant increase in both the area and overall production of sorghum during the period 1994–1999. This could be attributed to the emphasis on crop diversification during the Agricultural Sector Investment Program (ASIP) period, which coincided with this development (Saasa, 2003). However, this trend seems to have declined recently as a result of the decline in the number of households growing drought-tolerant crops in preference to the subsidized maize. Table 10.3 presents a comparison of cropping patterns among the surveyed households between 2002 and 2007 by gender of household head.

It is clear from Table 10.3 that there was a marked increase in the proportion of farmers growing maize and a corresponding decrease in the percentage of farmers growing other crops such as cassava and sorghum, which have low input requirements and are more resilient to climatic factors such as droughts. This is likely to impact negatively on the government's objectives of attaining food security as households have fewer alternatives in case of failure of the maize crop, which is more sensitive to drought compared to cassava and sorghum.

The FSP has also impacted differently on the mean area, mean yields and mean output of maize and sorghum (see Tables 10.4–10.6).

Table 10.3. Percentage of households growing different crops by gender. (From: authors' computations using survey data.)

	2002		2007	
	Male	Female	Male	Female
Maize	80.4	81.8	96.7	95.6
Cassava	35.2	36.4	21.3	23.3
Sorghum	25.9	20.8	5.7	8.7
Rice	1.7	–	–	1.5
Other food crops	78.1	76.6	86.8	82.2

Table 10.4. Trends in mean area, production and yields for maize (male subsample). (Adapted from: INESOR, 2003, 2007 Zambia micro reports, unpublished.)

	2002	2007	Mean difference	*t*-value	n
Mean area (ha)	1.25	1.50	0.25	3.48**	363
Mean production (kg)	1111.90	2941.66	1829.77	9.21**	364
Mean yields (kg/ha)	889.50	1783.65	894.15	14.19**	364

*Significant at 10% level; **significant at 5% level.

Table 10.5. Trends in mean area, production and yields for maize (female subsample). (Adapted from: INESOR, 2003, 2007 Zambia micro reports, unpublished.)

	2002	2007	Mean difference	*t*-value	n
Mean area (ha)	1.00	0.804	–0.195	–3.090**	102
Mean production (kg)	749.80	1233.33	483.533	3.150**	102
Mean yields (kg/ha)	749.80	1444.20	694.412	7.616**	102

*Significant at 10% level; **significant at 5% level.

Table 10.6. Trends in mean area, production and yields for sorghum (male subsample). (Adapted from: INESOR, 2003, 2007 Zambia micro reports, unpublished.)

	2002	2007	Mean difference	*t*-value	n
Mean area (ha)	0.30	0.60	0.30	3.47**	19
Mean production (kg)	383.8	305.53	–78.27	–1.75*	19
Mean yields (kg/ha)	1279.30	605.00	–674.30	–7.67**	19

*Significant at 10% level; **significant at 5% level.

Table 10.4 shows the trends in mean area, production and yields for maize among the surveyed male-headed households. A comparison of mean area, production and yields of maize between the periods 2002 and 2007 shows a significant increase in these variables. Similarly, for the female-headed households in the sample, a comparison of mean area, production and yields between the years 2002 and 2007 shows a significant increase in mean production and mean yields. However, the mean area under production showed a significant decline (Table 10.5). In addition to the FSP, this, in part, could be attributed to the fact that the Food Security Pack programme (a component of the FSP) targets the vulnerable sectors of society, which include female-headed households.

However, during the same period, although mean area under sorghum increased significantly, mean production and mean yields declined significantly for male-headed households (Table 10.6). The computation and consequently the comparison of mean area, production and yield under sorghum for female-headed households could not be done as only five households had grown the crop during the same period.

The observed increase in the maize yields among the surveyed households corresponds with the national situation at the macro level. Over the past 7 years since the introduction of the FSP programme, smallholder maize yields have shown a marginal rise from 2.19t/ha in 2002/03 to 2.51t/ha in 2007/08 (Minde *et al.*, 2008). One explanation for this is that upon receiving subsidized maize seed and fertilizers under the FSP, most farmers tend to concentrate on maize production at the expense of the other crops. Furthermore, the subsidized maize input packs have resulted in expansion of maize production even into areas that are no longer suitable for the crop due to droughts, such as the Southern province (GRZ, 2007). This, coupled with delays in delivering inputs, explains the marginal rise in productivity despite the massive investments in the programme. Furthermore, efforts by non-governmental organizations (NGOs) and other organizations that are trying to promote diversification into other drought-resistant and low-input crops, such as cassava and sorghum, in low-rainfall areas are being hampered by these maize input subsidies (Minde *et al.*, 2008). It has also been observed that subsidies targeted to particular crops such as maize may reduce output of other crops such as cassava (Zulu *et al.*, 2001). Simatele (2006) shows that such policies provided an incentive to move away from the production of other food crops such as sorghum, millet and cassava during the pre-liberalization period in Zambia. This is despite these alternatives being drought tolerant and more traditional staple crops than maize in certain areas.

Initially the role of the FRA was specifically to maintain strategic food reserves, but additional roles have since been added, especially those of assisting small-scale farmers to sell their maize as well as price setting on behalf of government. The FRA has also been perceived as crowding-out private sector investment in the output market as it has been getting government grants with zero risk. Table 10.7 shows the different channels through which the surveyed farmers sold their maize in 2002 and 2007. Despite the proliferation of farmer cooperatives, most of them are specifically created to meet the government requirement, which states that to benefit from FSP inputs one needs to be a member of a cooperative. Apart from facilitation in obtaining FSP inputs, these cooperatives do not offer any other services, such as marketing. They are only active when inputs are being distributed.

Interesting to note is the changes in the proportion of farmers marketing their output through the state marketing boards and private agents. Whereas in

Table 10.7. Main marketing channels used by households in 2002 and 2007. (Adapted from: INESOR, 2003, 2007 Zambia micro reports, unpublished.)

Channel	2002	2007[a]
Farmer cooperative	4.0	2.8
Private trader	80.0	37.8
State company/board	1.3	53.5
Own piecemeal/local market	6.7	–
Others	8.0	5.9

[a]For 2007, some categories of channels were collapsed to match the 2002 data.

2002 the majority (80%) reported marketing though private traders, this decreased to 37.8% in 2007. On the other hand, even though only 1.3% reported marketing through state agents in 2002, this increased to 53.5% in 2007, showing the crowding-out effects of government participation in the agricultural marketing system.

Usage of improved varieties of seed

The impacts of the one crop message being propagated by the FSP and FRA can also be seen in the adoption patterns of technologies among smallholder households. The survey data showed that usage of hybrid seed is quite high for maize (Table 10.8) and very low for sorghum. The survey finding for maize contradicts the findings of other studies at national level. For example, a postharvest survey (Govereh *et al.*, 2002) revealed that only 20% of the small-scale farmers had access to high-yielding inputs through schemes and programmes like the FSP and the FRA, as well as other donor/non-governmental (NGO)-supported food security packs programmes. For sorghum, the low usage of improved seed also explains the observed declining yields despite the reported increases in area cultivated among the surveyed farmers in 2002 compared to 2007. Impact studies in the region show that adoption rates for sorghum are low. Again, lack of improved seed is cited, together with lack of information, few alternative end uses and poor markets as the main reasons. However, on a comparative basis, South Africa and Tanzania are reported to utilize the crop on a wider scale compared to Zambia (Chisi, 2000). Similar observations have been made by Saasa (2003), who notes that despite the development of high-yielding varieties by the research branch of the Ministry of Agriculture, maize technologies

Table 10.8. Varieties of seed commonly planted in 2007. (Adapted from: INESOR, 2003, 2007 Zambia micro reports, unpublished.)

Variety	Maize (%) (n = 364)	Sorghum (%) (n = 19)
Traditional	15.4	94.7
Improved (OPV)	4.1	5.3
Hybrid	80.5	–

are the only modern crop technologies that have been widely adopted by smallholder farmers.

Extension services

The marginal increases in maize yields and the ever-increasing share of the agricultural budget dedicated to the FSP and FRA have raised concerns on the sustainability and efficiency of these programmes. Particularly, the seemingly limited programme impact on agricultural productivity – and consequently household and national food security – has raised concerns on the efficiency of fertilizer use among the stakeholders. Reports (Saasa, 2003; Haggblade, 2007; GRZ and MACO, 2009) have attributed the poor productivity among smallholder farmers to untimely delivery of inputs and poor farming practices among the farmers. These poor farming practices have further been linked to immobile and demotivated extension staff (Saasa, 2003).

However, literature shows that the payoffs to fertilizer subsidy programmes could be enhanced by improving the aggregate crop yield response rates to fertilizer application. This requires complementary investment in training for farmers on agronomic practices, soil fertility and water management and efficient use of fertilizer, and investing in crop science to generate more fertilizer-responsive seeds. Some studies indicate that, in some areas, improved management practices may have greater impact on yields than fertilizers alone (Haggblade and Tembo, 2003). Indeed, despite having a wide extension network starting from camp to provincial level, factors such as poor resource allocation to extension (estimated at less than 5% of total agricultural budget) results in demoralized extension staff that perform poorly. Low funding levels also limits the Ministry's ability to invest into development of new and innovative extension methods to address new challenges.

Among the sampled farmers for this study, access to extension did not seem to be a serious problem as only 17.1% (Fig. 10.2) reported not receiving any extension advice from government extension workers. Furthermore, another 80.2% reported receiving extension information from private extension workers. The majority of the farmers reported that they did not have to pay for the services of the extension workers. These findings contradict those from other studies at national level (Saasa, 2003; GRZ and MACO, 2009). One plausible explanation is that the interviews for this study (*Afrint I* and *II*) were done by agricultural extension officers. As such, interviewer bias may have influenced the responses, as respondents would not have wanted to be seen as reporting on the extension officers. The findings should therefore be treated with caution.

Access to agricultural credit

Other efficiency-related issues raised concerning the FSP and FRA programmes relate to targeting. Crop forecast survey data from CSO indicates an increase in fertilizer usage by smallholder farmers by 12% since the introduction of the FSP in 2002/03 at national level (Minde *et al.*, 2008). However, at ground level, things look a bit different. In the survey, the respondents were asked whether they had received any form of agricultural

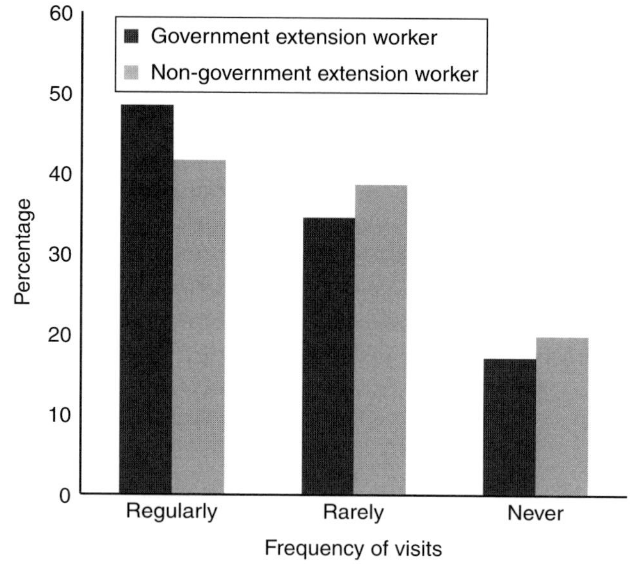

Fig. 10.2. Households' access to extension services. (Adapted from: INESOR, 2003, 2007 Zambia micro reports, unpublished.)

credit. This includes inputs such as seed and fertilizer from sources such as private companies and government programmes such as the FSP. Despite the availability of the FSP programme, only 17.1% (see Fig. 10.3) of the sampled households reported receiving any form of agricultural credit in 2007. This represents a smaller proportion compared to 2002, when 33.6% reported receiving credit. Analysis of the FSP programme by this study team reported poor targeting as one of the weaknesses of the programme, with relatively rich households benefitting the most.

Regression analysis

Whereas the Zambian government has concentrated only on fertilizer as the main constraint to agricultural production, leading to ever-increasing fertilizer subsidies, studies have shown that there are other constraints (other than fertilizer), which would lead to increased food production if addressed. For example, the CSPR (2005) reports that, apart from fertilizers, limited access to improved seed, agricultural credit, farm produce markets and extension services all have contributed to reduced food output among smallholder households. In a study to identify barriers to development among small and medium farms in Zambia, Kimhi and Chiwele (2000) found that maize yields and crop diversification could be promoted by factors such as road construction, developing markets for agricultural products, increasing availability of seeds, credit, draught animals, farm machines, increasing farm work participation by women and increasing the size of land holdings. The same study also found that maize yield was influenced by demographic variables such as age and sex of household head. Earlier studies identified factors such as highly imperfect labour markets (Holden, 1993), credit (Jha and Hojjati, 1993) and support systems

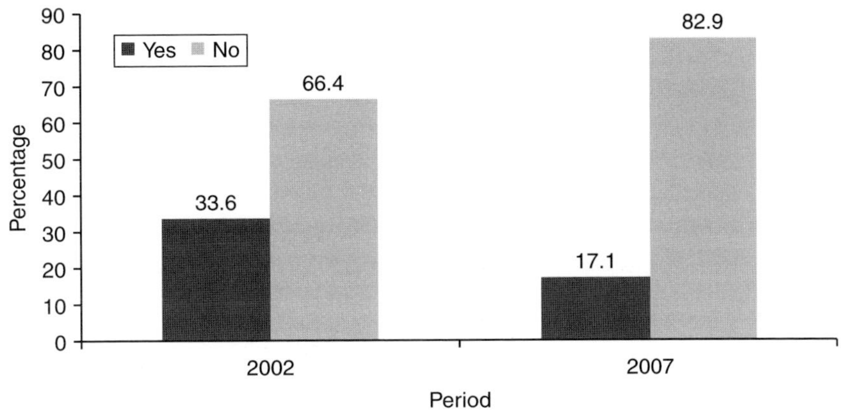

Fig. 10.3. Percentage responses on access to agriculture credit. (Adapted from: INESOR, 2003, 2007 Zambia micro reports, unpublished.)

such as extension, research, infrastructure and markets (Foster and Mwanaumo, 1995). In order to identify the constraints (factors), as well as the effects of the identified constraints on the quantities of maize produced among the sampled households, this study uses linear regression analysis. The findings would provide policy makers with information on additional options that could be used to influence production decisions. The following section shows the regression model, the results and some explanations.

Model specification

Based on the above-mentioned studies, variables falling under categories such as household-specific variables (HHSP), institutional variables (INST) and market access (MKT) variables were used. The HHSP variables included age, gender and education level of household head, active labour force in terms of male and female household members aged between 16 and 65 years, ownership of cattle, means of land cultivation infrastructure and provincial dummy to reflect agro-ecological region, as well as expenditure on artificial fertilizer. The INST variables include membership of farmer organizations, access to extension services, availability of agricultural credit and availability of hybrid seed. MKT-related variables included distance to market centres, market channels used, perceptions on prices and market access compared to baseline period (2002). The dependent variable was the total production of maize grain in kilograms for the most recent agricultural season (PROD) and was assumed to be linearly related to HHSP, INST and MKT variables (Eqn 10.1).

$$\text{PROD} = f(\text{HHSP, INST, MKT}) \tag{10.1}$$

A multiple linear regression equation expressed as Eqn 10.2 was estimated using ordinary least squares.

$$\begin{aligned}\text{PROD} = {} & \alpha_1 + \alpha_2 D_{2i} + \alpha_3 D_{3i} + \alpha_4 D_{4i} + \alpha_5 D_{5i} + \alpha_6 D_{6i} + \alpha_7 D_{7i} \\ & + \alpha_8 D_{8i} + \alpha_9 D_9 + \alpha_{10} D_{10} + \beta_2 \text{MAL} + \beta_3 \text{FEM} + \beta_4 \text{AGE} \\ & + \beta_5 \text{EDU} + \beta_6 \text{EXP} + \beta_7 \text{CAT} + U_1 \end{aligned} \tag{10.2}$$

Where:
D_2 = Perceptions of market access (1 if better than 2002, 0 otherwise)
D_3 = Dummy for province
D_4 = Dummy for market channel (1 if used state agent, 0 otherwise)
D_5 = Dummy for type of seed used (1 if used hybrid seed, 0 otherwise)
D_6 = Dummy for access to input credit (1 if yes, 0 otherwise)
D_7 = Dummy for access to extension in previous year (1 if yes, 0 otherwise)
D_8 = Dummy for household head membership in FO (1 if yes, 0 otherwise)
D_9 = Dummy for gender of household head (1 if male, 0 if female)
D_{10} = Dummy for means of cultivation (1 if hand hoe, 0 otherwise)
PROD = Total production of maize grain (kg) in previous season
AGE = Age of household head
EDU = Education level of household head
MAL = Number of males aged between 16 and 65 in the household
FEM = Number of females aged between 16 and 65 in the household
EXP = Household expenditure on artificial fertilizers (USD)
CAT = Number of cattle owned

Table 10A.1 in the Appendix is a summary of the variables used and the hypothesized relationships with the dependent variable (quantity of maize produced).

Regression results

Table 10.9 shows the results of the regression.

The factors that significantly influenced the quantities of maize produced were household's expenditure on artificial fertilizers, use of oxen and other mechanical tools (such as tractors) for cultivation, market channel used for marketing maize, ownership and number of cattle, and active labour force measured as number of males aged between 16 and 65 years in the household. All the significant variables had the correct hypothesized signs on the coefficients.

The coefficient on the variable for expenditure on artificial fertilizer shows that a dollar increase in fertilizer expenditure resulted in only 7.559 kg increase in maize produced. Although 7.559 kg of maize sells for about US$2.2, this increase may not be sufficient to cover the costs of seed and labour used. However, this observation is in line with earlier observations that returns to fertilizer usage are low among smallholder farmers (Minde *et al.*, 2008; GRZ and MACO, 2009).

The coefficient on the variable for means of cultivation shows that households that used oxen and other mechanical means produced 704.384 kg more maize than those that used hand hoes. This is not surprising as earlier studies (Holden, 1993; Kimhi and Chiwele, 2000) also showed that labour has been one of the major constraints among smallholder households. Ownership of oxen not only allows households to plough more land but also enables them to

Table 10.9. Determinants of household maize production.

Variable	Coefficient	Std error	*t*-statistic	Probability
Gender of household head	–354.855	296.714	–1.195	0.233
Age of household head	–13.399	8.368	–1.600	0.110
Education of household head	18.690	36.123	0.517	0.606
Household expenditure on fertilizer (USD)	7.559***	0.375	9.980	0.000
Positive perceptions of market access	470.885	576.498	0.817	0.415
Dummy for province	69.798	339.396	0.206	0.837
Dummy for selling through state agencies	1876.740***	295.679	6.347	0.000
Dummy for hybrid seed usage	36.786	301.521	0.122	0.903
Dummy for access to agriculture credit	–79.449	114.175	–0.696	0.487
Dummy for membership to cooperatives	419.685	278.376	1.508	0.132
Dummy for access to extension services	358.573	282.395	1.270	0.205
Number of males aged between 16 and 65 yrs	220.243**	79.061	2.786	0.167
Number of females aged between 16 and 65 yrs	–38.589	82.419	–0.468	0.640
Dummy for oxen as major means of cultivation	704.384**	328.687	2.143	0.033
Number of cattle owned by household	196.173***	29.710	6.603	0.000
Constant	228.740	956.453	0.239	0.811
Dependent variable: Total production of maize grain (kg) in 2007 season				
Number of observations: 410				
F-statistic: 36.522				
R^2 : 0.581***				

***Significant at 1%, **significant at 5%, *significant at 10%.

plant early. The fact that an increase in number of males aged between 16 and 65 years resulted in 220.243 kg increase in maize output further reconfirms the importance of labour constraints. This is especially true as maize cultivation, especially for commercial purposes, is a preserve for men, while women concentrate on food security crops. The results also show that an increase in the number of cattle owned resulted in an increase in production by 196.173 kg. This is so because, apart from being used as oxen, cattle also produce manure, which is used to fertilize the fields. Households that sold through the state agencies (mainly the FRA) produced 1876 kg more than those that used alternative channels such as private traders. This is explained mainly by the fact that beneficiaries of the FSP are more inclined towards selling their surplus produce to the FRA after paying the loans. Table 10.7 shows that the majority of households (53.5%) sold their maize through the state-operated FRA.

The coefficients on perceptions of market access, education of household head, access to government extension services and farmer group membership all had the correct signs but were insignificant. For example, the variable for cooperative membership shows that households that were members of farmer organizations produced 419.685 kg more than their non-member counterparts.

Although this difference is not significant, households can only access subsidized FSP fertilizer through farmer groups. As such, most of these cooperatives are formed for purposes of obtaining the subsidized inputs, with little or no other services, such as marketing, being offered. Similarly, although the coefficient on the extension variable shows that those with access to the service produced 358.573 kg more than those without, the difference was not statistically significant. This may partly explain the low productivity despite the majority of the farmers reporting having access to extension (Fig. 10.3). Similar findings were reported by Kimhi and Chiwele (2000), who showed that extension services did not have a significant impact on maize production and land devoted to maize. Household head's gender and age were, as hypothesized, negatively related to maize production but also insignificant. Planting hybrid seed had an unexpected sign on the coefficient, with those that planted hybrid seed producing 36.786 kg less than their counterparts who planted recycled seed and open-pollinated varieties. However, this was insignificant. Finally, the provincial dummy also showed the hypothesized sign, with farmers located in the high-potential, high-rainfall agro-ecological Region II[2] (Central Province) producing 69.798 kg more than their counterparts in the drier agro-ecological Region I (Southern Province). However, this also was not statistically significant.

Conclusion and Recommendations

As observed by Salzburg (2008), fertilizer subsidies may not be the best option for addressing the current crisis of high food and fertilizer prices. Significant increases in demand for fertilizer are likely to drive up prices further. Also, the supply response to increased fertilizer use is not assured, given weather and other maize production risks prevalent in most of eastern and southern Africa. Thus implementing large-scale fertilizer subsidy programmes will not guarantee an adequate harvest.

As a tool for increasing overall agricultural productivity, especially for small, poor farmers, fertilizer subsidies have a questionable record. Long experience with input subsidy programmes in Africa is not encouraging on several points: (i) there is very little evidence from Africa that fertilizer subsidies have been a sustainable or cost-effective way to achieve agricultural productivity gains compared to other investments; (ii) there are no examples of subsidy programmes where the benefits were not disproportionately captured by larger and relatively better-off farmers, even when efforts were made to target subsidies to the poor; and (iii) there is little evidence that subsidies or other intensive fertilizer promotion programmes have 'kick-started' productivity growth among poor farmers in Africa enough to sustain high levels of input use once the programmes end (Minde *et al.*, 2008).

In the high-potential areas of Kenya, Zambia and Malawi, many, if not most, households use fertilizer regularly. In less-stable production zones, low or

[2] Zambia is divided into three agro-ecological regions based on length of rainy season, soil types and temperatures.

no fertilizer use by many smallholders is explained not just by credit constraints that limit acquisition but also by the risk of crop failure, with resulting financial losses and consumption shortfalls. The lack of insurance causes inefficiency in production choices (Dercon and Christiaensen, 2007).

The findings of this study show the dominance of maize production, with very little improvement in yields over the period under review. On the other hand, the production of alternative food crops such as sorghum and cassava to mitigate the effects of drought has not shown significant improvement. This is despite earlier efforts by government and NGOs to promote these crops as alternatives, especially in areas where rainfall has become unpredictable due to climate change. The promotion of drought-tolerant crops was done through the provision of free improved sorghum seed and cassava cuttings, among other services, using institutions such as the Program Against Malnutrition. Furthermore, whereas the FSP has resulted in increased fertilizer usage and increased land devoted to maize, the increase in maize productivity has been marginal, raising concerns over the efficiency of the programme in the light of the huge amounts of money being spent.

In addition, as indicated above, the literature shows that the payoffs to fertilizer subsidy programmes could be enhanced by improving the aggregate crop yield response rates to fertilizer application. This requires complementary investment in training for farmers on agronomic practices, soil fertility and water management and efficient use of fertilizer, and investing in crop science to generate more fertilizer-responsive seeds. Some studies indicate that, in some areas, improved management practices may have greater impact on yields than fertilizers alone (Haggblade and Tembo, 2003).

The paper also shows that the presence of the FSP has resulted in reduced participation of the private sector in input marketing. This is contrary to the objectives of the programme, which aims to build capacity of the private sector in input marketing. The operations of the FRA have also been shown to have crowding-out effects on private sector participation in output markets. This is because the FRA can offer above-market prices as they use government resources to cover their operational expenses. This also has implications on the budget, as these recurrent expenditures are using up a huge proportion of the agricultural budget at the expense of other programmes, such as infrastructure development and extension services. At best, the programmes have achieved market distortion, diverting much-needed resources from assisting the poor.

There is a need, therefore, to reorganize the programmes in order to achieve the intended objectives of increasing yields, at least for the staple crop maize. In addition there is a need to consider diversification of the crop portfolio in order to reduce the risk of variability in food supply created from the maize-dominant food supply system in Zambia. Achieving the Millennium Development Goals of halving hunger between 1990 and 2015 still remains a pipe dream and an opportunity missed in Zambia. If only the state actors could reorganize the scarce resources, perhaps at least some renewed hope could emerge. Currently, the government of Zambia devotes at least 60% of their agricultural budget to input and crop-marketing subsidies, leaving relatively little for the long-term investments required for sustainable reductions in poverty and hunger (Minde *et al.*, 2008). A balance is needed between interventions to address short-term supply shortages to avoid widespread hunger versus investments and policies to drive growth and lift poor households out of the poverty trap in which they are caught.

References

ACF/FSRP (2009) Agriculture Consultative Forum breakfast meeting: discussing the 2009 national budget for Zambian agriculture? A presentation made at the 2009 Agriculture Sector Budget Analysis, 4 February 2009, Lusaka, Zambia.

Byerlee, D. and Heisey, P.W. (1996) *Past and Potential Impacts of Maize Research in Sub-Saharan Africa: a Critical Assessment*. Food Policy Paper 21, pp. 255–277.

Chisi, M. (2000) *Sorghum and Millet Breeding in Southern Africa in Practice*. Golden Valley Research Station, Fringilla, Zambia.

CSPR (2005) *Targeting Smallholder Farmers in the Implementation of Zambia's Poverty Reduction Strategy Paper (PRSP). An Assessment of the Implementation and Effectiveness of the Fertilizer Support Programme*. Civil Society for Poverty Reduction, Zambia.

Dercon, S. and Christiaensen, L. (2007) *Consumption Risk, Technology Adoption and Poverty Traps: Evidence from Ethiopia*. Policy Research Working Paper 4257, The World Bank, Washington, DC.

Djurfeldt G., Holmén, H., Jirström, M. and Larsson, R. (eds) (2005) *The African Food Crisis: Lessons from the Asian Green Revolution*. CAB International, Wallingford, UK.

FEWSNET (2009) *Zambia Food Security Update 2009*. Famine Early Warning Systems Network, Zambia.

Foster, K.A. and Mwanaumo, A. (1995) Estimation of dynamic maize supply response in Zambia. *Agricultural Economics* 12, 99–107.

Govereh, J., Jayne, T., Nijhoff, J., Haantuba, H., Ngulube, E., Belemu, A., Shawa, J., Banda, A., Donovan, C. and Zulu, B. (2002) *Developments in Fertilizer Marketing in Zambia: Commercial Trading, Government Programmes, and the Smallholder Farmer*. Working Paper No. 4, Food Security Research Project, Lusaka, Zambia.

Govereh, J., Shawa, J., Malawo, E. and Jayne, T.S. (2006) *Raising the Productivity of Public Investments in Zambia's Agricultural Sector*. Working Paper No. 20, Food Security Research Project, Lusaka, Zambia.

GRZ (2007) *National Adaptation Programme of Action (NAPA)*. Ministry of Tourism, Environment and Natural Resources, Zambia.

GRZ and MACO (2009) *Report on Proposed Reforms for the Zambian Fertilizer Support Programme*. Available at: http://www.aec.msu.edu/fs2/Zambia/FSP_Review_Report_feb_09.pdf (accessed January 2010).

Haggblade, S. (2007) *Returns to Investment in Agriculture: Policy Synthesis 19*. Food Security Research Project, Lusaka, Zambia. Available at http://www.aec.msu.edu/agecon/fs2/zambia/index.htm (accessed January 2010).

Haggblade, S. and Tembo, G. (2003) *Development, Diffusion and Impact of Conservation Farming in Zambia*. Working Paper 8. Food Security Research Project (FSRP), Michigan State University (MSU) and International Food Policy Research Institute (IFPRI). Available at http://www.aec.msu.edu/agecon/fs2/zambia/wp8zambia.pdf (accessed January 2010).

Holden, S.T. (1993) Peasant household modelling: farming systems evolution and sustainability in northern Zambia. *Agricultural Economics* 9, 241–267.

Jha, D. and Hojjati, B. (1993) *Fertilizer Use on Smallholder Farms in Eastern Province, Zambia*. Research Report 94, IFPRI, Washington, DC.

Jorgensen, S.L. and Loudjeva, Z. (2005) *A Poverty and Social Impact Analysis of Three Reforms in Zambia: Land, Fertilizer and Infrastructure*. Social Analysis Paper No. 49, The World Bank, Washington, DC.

Kimhi, A. and Chiwele, D. (2000) Barriers for development in Zambia small- and medium-size farms: evidence from micro-data (unpublished report).

MACO, CSO and FSRP (2008) *Patterns of Maize Farming Behaviour and Performance Among Small and Medium-Scale Smallholders in Zambia. A Review of Statistical Data from the CSO/MACO Crop Forecast Survey – 2000/2008 to 2007/2008 Production Seasons*. Ministry of Agriculture and Cooperatives, Central Statistical Office and Food Security Research Project, Lusaka, Zambia.

Minde, I., Jayne, T.S., Crawford, E., Ariga, J. and Govereh, J. (2008) *Promoting Fertilizer Use in Africa: Current Issues and Empirical Evidence from Malawi, Zambia and Kenya*. Regional Strategic Agriculture Knowledge Support System (Re-SAKSS) for Southern Africa.

NEPAD (2005) *Implementing the Comprehensive African Agriculture Development Program and Restoring Food Security in Africa 'The Road Map'*. New Partnership for Africa's Development, Midrand, South Africa.

Saasa, O.S. (2003) *African Food Crisis – the Relevance of Asian Models: the Role of Policies and Policy Processes*. Report prepared for Lund University of Sweden, Institute of Economic and Social Research, University of Zambia, Lusaka, Zambia.

Salzburg (2008) *Fertilizer Subsidies Convening Synthesis*. Fertilizer Subsidy Meeting, Salzburg, Austria.

Simatele, M. (2006) *Food Production in Zambia: the Impact of Selected Structural Adjustment Policies*. AERC Research Paper 159, African Economic Research Consortium, Nairobi, Kenya.

UNDP and GRZ (2003) *Zambia Human Development Report 2003*. United Nations Development Programme, Zambia.

Wamulume, M. (2003) *African Food Crisis: the Relevance of Asian Models*. Report prepared for Lund University of Sweden, Institute of Economic and Social Research, University of Zambia, Lusaka, Zambia.

Wamulume, M. (2009) *African Food Crisis and the Millennium Development Goals*. Report prepared for Lund University of Sweden, Institute of Economic and Social Research, University of Zambia, Lusaka, Zambia.

Zulu, B., Nijhoff, J.J., Jayne, T.S. and Negassa, A. (2001) *Is the Glass Half-empty or Half-full? An Analysis of Agricultural Production Trends in Zambia*. Food Security Research Project Policy Synthesis No.2, Lusaka, Zambia.

Appendix

Table 10A.1. Variables used and hypothesized relationships.

Variable description	Variable	Hypothesized relationship with quantity produced
Positive perceptions of market access	D2	+
Household is located in Central Province	D3	+
Dummy for market channels	D4	+
Dummy for type of seed used	D5	+
Dummy for access to input credit	D6	+
Dummy for access to extension	D7	+
Dummy for membership to farmer associations	D8	+

Continued

Table 10A.1. Continued.

Variable description	Variable	Hypothesized relationship with quantity produced
Dummy for household head gender	D9	–
Dummy oxen as means of cultivation	D10	+
Age of household head	AGE	+
Education level of household head	EDU	+
Males aged between 16 and 65 years	MAL	+
Females aged between 16 and 65 years	FEM	–
Household expenditure on fertilizer	EXP	+
Number of cattle owned	CAT	+

11 Has the Nigerian Green Revolution Veered Off Track?

TUNJI AKANDE,[1] AGNES ANDERSSON,[2] GÖRAN DJURFELDT[3] AND FEMI OGUNDELE[1]

[1]*Nigerian Institute of Social and Economic Research (NISER), Ibadan, Nigeria;* [2]*Department of Social and Economic Geography, Lund University, Lund, Sweden;* [3]*Department of Sociology, Lund University, Lund, Sweden*

Nigeria is being promoted as the 'Heart of Africa' on account of its geographical location at the intersection of West and Central Africa, as well as on account of its strategic importance as the most populous black nation in the world. With a land area of over 924,000 km^2, more than half of which is arable, a population in excess of 140 million people, many perennial rivers and waterbodies and a most clement climate, Nigeria is greatly endowed. Apart from, perhaps, the Democratic Republic of the Congo, no other country in Africa has the resource base of Nigeria. Nigeria today is democratic, after decades of military dictatorship. The country operates a federal system of governance, consisting of federal government, a federal capital territory, 36 state governments and 774 local government areas.

Agriculture accounts for about 40% of Nigeria's gross domestic product (GDP) as well as employing two-thirds of the workforce. In spite of the importance of agriculture, petroleum dominates the economy, accounting for about 80% of national revenue, over 90% of export earnings and about 23% of the GDP. Nigeria's GDP has grown nearly fivefold in less than a generation, from about US$28 billion in 1990 to about US$140 billion in 2007, as a result of increases in international prices of crude petroleum. Although the GDP is growing at nearly 5–7% per annum currently, this is quite insufficient to address the chronic poverty and employment problems and the challenge of a population rising at the rate of 3.5% per annum. The GDP per capita at purchasing power parity was barely US$1000 at the beginning of the new millennium but has now increased to US$1320 (Central Bank of Nigeria, 2007).

The Nigerian agricultural sector is dominated by smallholder producers, who operate farm sizes of no more than 1–5 ha (NISER, 2002). All the same, the smallholder farmers account for over 90% of agricultural output. The food crops dominate production and include cereals (sorghum, millet, maize and rice), tubers (cassava, yam and cocoyam), vegetables, horticultural products, livestock, fisheries and wild forest products. These are produced in less than

50% of the 70 million ha of available arable land area. The northern part of the country is noted for the production of sorghum, millet, sesame and groundnut. The region accounts for nearly three-quarters of small ruminants, cattle, camels and donkeys. Except for commercial poultry production, the north is also home to the larger populations of domestic poultry (local chicken, guinea fowl, ducks, turkey, etc). The central zone and south-west cultivate mainly roots and tuber crops, maize, plantain and bananas. The south-west and south-east are centres of cash and export production of cocoa, palm produce, rubber and citrus crops. Nigeria leads the world in the production of yam and cassava. It produces nearly 300,000 t of fish per annum.

In spite of its tremendous capacity and potential for livelihood and being the cultural and social centre of rural people, agriculture has performed erratically in recent years. From the slow growth in most of the 1970s and 1980s, agricultural GDP started an upward movement in the 1990s, culminating in an average of 5.6% growth per annum since 2000. This growth rate is above the Africa-wide average, almost achieving the target growth rate of 6% per annum specified under the Comprehensive African Agricultural Development Programme (CAADP). The improved performance of agriculture is uplifting and a credit to policy changes. However, the sustainability of current high growth rates is doubtful. The global food crisis of early 2008, for instance, has not isolated Nigeria from the vagaries of international food shortages and price spirals of the period, thus questioning Nigeria's capability of making food available and accessible at affordable prices to consumers. Nigeria has to rely on food imports to supplement local production and demand. The returns to farmers are declining and farming is not sufficiently profitable as a result of the high costs of production.

The surge in agricultural growth rates experienced in recent years has been powered mainly by expansion in areas planted to staple crops. Productivity has remained flat and yields of most crops have actually declined in the past decade, putting into question the efficacy of public investment in agriculture over the years. Public intervention and investments under the National Food Security Programme were aimed at sparking off a sustainable Green Revolution in the country (FMAWR, 2008). What is more worrisome is that Nigeria may not be able to meet its food production and poverty reduction goals without a significant and sustainable production increase in the agricultural sector. The reason is obvious – more than 70% of the poor reside in the rural areas and depend on agriculture for their livelihood. Agriculture must not only provide cheap food it must also be profitable and income-generating to farmers and rural workers in general to lift them out of poverty. The challenge, therefore, is to have a policy mix that embraces institutional restructuring, strategic investments and coordinated efforts at all levels, to empower farmers and enhance improved conditions in rural areas. Such efforts would focus on rehabilitating degraded rural infrastructure, adopting productivity-enhancing measures and taking steps that will stimulate competitiveness.

However, in order to be able to provide an efficacious and desirable investment plan for the sector, there is the need to undertake a review of the current production configuration of the food subsector of agriculture. Current indicators of food security in the country are pointing towards a declining trend compared with the situation before the implementation of the various food policies and programmes of the government. Though overall agricultural growth rate has been impressive in the last few years, the gap between demand and supply for major staples continues to widen due to rising population and persistent decline in aggregate output and productivity within food production. Access to food in Nigeria is becoming increasingly difficult, particularly for poor households, due to rising food prices, thereby impairing the economic access to food – a state of food insecurity. Thus, a pertinent question that may be asked is what, then, is accounting for the declining situation of food production in the face of increasing public and private investments in the agricultural sector? In order to explain this paradox of growth without improved food security among the rural populace in particular, this paper attempts to engage both secondary and cross-sectional data for the period between 2002 and 2007 to explain factors responsible for the declining condition of food production.

The chapter draws on a number of methods in exploring the apparent contradiction between agrarian policies that, at face value at least, appear to promote the smallholder sector but with little tangible impacts on its productivity. A critical review of the food policy in Nigeria in the pre- and post-2002 periods was carried out, while an assessment of the major agricultural programmes implemented between 2002 and 2007 in the country was done in this study. Using secondary data from the National Food Reserve Agency (NFRA) and the Federal Ministry of Agriculture and Water Resources (FMAWR), the paper also examines trends in production of major food crops between 2002 and 2007. Data collected by the *Afrint* team at the household and village level in 2002 and late 2007 and early 2008 in the states of Kaduna and Osun is used to discuss such trends at the micro level. A comparative approach is employed to assess the drivers of production changes in maize in the Nigerian context, evaluating such trends against the experiences of the other seven countries in the panel (Ethiopia, Ghana, Kenya, Malawi, Mozambique, Tanzania and Zambia). The chapter uses the same econometric modelling strategy as outlined in Chapter 5 (Andersson *et al.*, this volume).

Food and Agricultural Policies and Programmes, 1999 to 2007

The Nigerian agricultural policies between 1999[1] and 2007 were implemented within the framework of the programme of presidential initiatives on arable and tree crops in Nigeria. These were largely commodity-related activities and programmes, in which individual agricultural commodities that were considered extremely important for food security and domestic self-sufficiency were identified and programmes designed to effect accelerated production, increased output and a much higher productivity beyond the existing levels of achievement by the farmers. The presidential initiatives were conceived for rice, cassava,

[1] 1999 marks the beginning of the new democratic dispensation in Nigeria and a radical paradigm shift in agricultural policies and programme planning in the country.

vegetable oil and non-food commodities like rubber and cotton. Another major food-based programme implemented in the country during this period was the Food and Agricultural Organization (FAO)-sponsored Special Programme on Food Security (SPFS).

Specific food policies within the general framework of economic management in the 1999–2007 period took several forms and initiatives. In this section, therefore, a summary of the prevailing agricultural policies and programmes is undertaken, in order to link performance in food crop production to the policies and programmes pursued during this period. The review covers federal government agricultural policies and programmes and, in some cases, state-level policies. In 2004, the National Economic Empowerment and Development Strategy (NEEDS) became the overarching policy framework for agricultural development in Nigeria. This was, however, modified in 2007 and has since translated to the Seven Point Agenda of the Yar'dua Administration. The Seven Point Agenda aims at restoring agriculture to its former status as a leading sector in the economy in terms of its contribution to GDP, supply of raw materials, employment generation, source of exports, provision of staple foods and, hence, food security. The policy thrust includes:

- Provision of the right policy environment and targeted incentives for private investments in agriculture.
- Fostering effective linkages with industry to achieve maximum value added and processing for export.
- Modernizing production and creating an agricultural sector that is responsive to the demands and realities of the economy.
- Reversal of the trend in food imports to stem rising trade imbalance as well as diversifying the foreign exchange earnings base; strive towards food security and a food surplus that could be exported.

Against the backdrop of NEEDS and the Seven Point Agenda the specific policy measures undertaken in the agricultural sector can be broadly categorized into trade, input, fiscal, research and price stabilization policies.

The main thrust of the agricultural trade-related policies is in the form of tariffs. For example, the government announced major increases in import duty on some categories of food and animal products in 2002. The tariff on rice was increased from 50% to 100%, while that on soybean was raised from 30% to 60% in 2002. Also, the tariff on palm oil and its products was increased from 35% to 60%, while that on animal and vegetable fat and oils and related products was raised from 20% to 60%. The main objective of raising the tariff was to discourage imports and induce domestic production of these commodities. A heavy tariff was imposed on rice in 2003, such that the tariff, which stood at 100% in 2002, was raised to 150%. Also in 2003, a ban was imposed on importation of cassava products and export of maize under the food security strategy of the government. Importation of other commodities such as frozen chicken and turkey was banned to encourage home production and protect the domestic producers. In order to promote production, in 2003 the federal government directed that 10% of cassava flour be included in flour milling for the bakery industry. In order to boost domestic production and export of agricultural crops in general, an export subsidy of 10% on agricultural commodities was introduced in 2003. Further interventions came in 2005, in the form of temporary prohibition of the importation of fruit juice, vegetable oil, poultry and related products. The use of fiscal incentives as a major policy instrument used to promote agriculture was given a boost in 2006, through additional import waivers as well as the promotion of increased use of agricultural machinery and inputs through a favourable tariff policy. These were strengthened by the presidential initiatives on various food crops, which were initiated in earlier years. Implementation of the presidential initiatives on rice, cassava, vegetable oil, tree crops, rubber development and tropical fruits, for instance, received a boost in 2005. A total sum of N1.1 billion, including the N687.3 million proceeds from the 10% surcharge on rice importation, was released for the take-off of the crops-related initiatives.

Input policies have focused on chemical fertilizer. Fertilizer policy in Nigeria has continued to be very unstable. For example, procurement and distribution of agricultural inputs including fertilizer started witnessing government intervention by 2001 and 2002, resulting in the re-introduction of a fertilizer subsidy to the tune of 25% (NISER, 2005). This accounted for about N3.5 billion of federal government recurrent expenditure for the year. Available information indicates that out of 163,700 t of fertilizer approved for procurement for the 2002 wet-season farming, only 104,024 t, representing 63.5%, were delivered by government to retailers (Central Bank of Nigeria, 2002). Consequently, most farmers could not access the commodity. Besides, subsidized fertilizer did not reach the intended beneficiaries – the smallholders, particularly in the rural areas. In the financial year 2005, a total of 124,029.5 t of assorted fertilizers, 4200 t of agricultural lime and 56,000 l of micronutrients, all valued at N9 billion, were procured and distributed to the 36 states, the federal capital territory, the River Basin Development Authorities and the National Special Programme for Food Security (NSPFS) at 25% subsidy. Meanwhile between 2002 and 2007 various subsidy rates were adopted by both federal and state governments in Nigeria. While the federal government currently subsidizes fertilizer by 25%, additional subsidy by state governments varies between 25% and 50% across the country. The immediate consequence of the subsidies is shortfall in supply. Shortfall in supply often results in prices being higher than those approved by government. For example, a 50 kg bag of fertilizer in 2007, which is offered at a subsidized price of US$15, was sold in the market at between US$30 and US$35 in most parts of the country (Central Bank of Nigeria, 2008).

In 2006, the National Fertilizer Policy was approved by government to guide and control the production and utilization of fertilizer. In this regard, the moribund National Fertilizer Company of Nigeria was privatized, while the production and utilization of organic fertilizer was being encouraged by government. The federal government also procured and distributed about N250 million or US$2.5 million worth of chemical fertilizer, at 25% price subsidy, to farmers. Government procurement of fertilizer has fallen from 1.3 million t in 1990 to about 245,000 t in 2004. It further fell to 125,000 t in 2005. Moreover, such policies need to be discussed in relation to the large expansion in cultivated areas recorded for many crops (World Bank, 2004).

In terms of credit policy, the merger of the Nigerian Agricultural and Cooperative Bank, People's Bank of Nigeria and the Family Economic Advancement Programme to form the Nigerian Agricultural Cooperative and Rural Development Bank in the year 2000 was designed to ensure financial stability and guarantee credit flow to the agricultural sector. This merger increased the authorized capital of the bank from N1 billion in 2000 to N10 billion in 2001. Thus, in 2002, a sum of N50 billion was earmarked for the bank in order to meet the credit need of small-scale farmers and agricultural processors.

By December 2003, the bank had disbursed about N40 billion to beneficiaries across the country. Although the reforms put in place have redressed the abuse inherent in credit rationing, the issue of inadequate accessibility to credit by small-scale farmers is still pervasive. The interest rate charged remained high and constrained demands for credit by farmers, whose returns have remained very low. In 2004, the federal government established the Agricultural Development Fund. A capital grant of N10 billion was allocated to the fund, to be disbursed over a period of 4 years. Other sources of funds for agricultural development include the 25% appropriation from the sugar development fund (tax accruing from sugar importation) and 1% appropriation from tax accruing to the federal government from the petroleum products pump prices. An Agriculture Credit Support Scheme was established in 2004 to provide credit facilities to all categories of farmers at single-digit interest rate through the initiative of the federal government and the CBN. Under the scheme, the federal government, through the Presidential Committee on Financing of Agriculture, mobilized N50 billion for on-lending to farmers and other agro-allied entrepreneurs nationwide, at an interest rate of 8% for the 2006 farming season.

As for research, a number of breakthroughs were recorded by research institutions, particularly during the period under review. For example, about 43 improved varieties of cassava were introduced in the focal states of the Cassava Enterprise Development Project, comprising Abia, Enugu, Ebonyi, and Imo in the South-East and Bayelsa, Cross River, Edo and River states in the South-South. About 300,000 farm families were expected to benefit from the programme. Another major breakthrough was the development and release of the upland rice variety called NERICA in 2002. The variety is capable of yielding 7 t/ha under intense management. The activities of the NSPFS and the South-South Cooperation (SSC) programme, the Roots and Tuber Expansion Project and the Community Based Agricultural Development Project also assumed significant dimensions in the year 2005. Further, the Ministry of Agriculture, through the 15 agricultural research-related institutions and 12 federal colleges of agriculture, developed and distributed high-yielding and disease-tolerant varieties of sorghum, soybean, rice, oil palm, cocoa and rubber, among others, to farmers nationwide. Government also supported the production and distribution of 429,069 grafted seedlings of mango, capable of planting 4201 ha; 700,000 budded seedlings of citrus, capable of planting 3432 ha; 10,000,000 suckers of pineapple, capable of planting 250 ha; and 117,550 seedlings of avocado, capable of planting 1175 ha. These developments resulted in increased participation of farmers in horticultural activities. Human resource capacity development in the agricultural sector commenced at the Federal College of Horticultural Studies, Gombe State in 2002. The college has the mandate to train students in all fields of horticulture and irrigation technology, and farmers, food processors and other agro-allied groups on vocational skills acquisition.

In terms of price policies the government has attempted to ensure price stability. During the period under review, state governments were expected to store 10% of incremental grain output under a buffer stock scheme, while the federal government was expected to store 5% under the strategic grain reserve scheme. In 2003, the federal government mopped up excess grains of about 75,000 t through the buy-back arrangement, in order to enhance price stability. In order to strengthen the national food security programme and achieve price stabilization, the capacity of the national silo complexes was increased from 100,000 t to 385,000 t in 2006.

Aside from the sectoral policies outlined above, a range of programmes and projects related to food production specifically have been launched during the period under review. For instance, the Policy Coordinating Unit, now the National Food Reserve Agency (2000), showed that the federal government entered into bilateral agreements with a number of international development partners, especially the FAO under the Unilateral Trust Fund in May 2000, to commence the implementation of the SPFS. The programme was planned to be executed over a period of 5 years at an estimated cost of US$45 million. Activities under the programme were spread across 109 locations, with each location selected from each of the 109 senatorial districts in the country. The programme targeted boosting production through cultivation of 500–600 ha of land at each location, involving 250–300 farm families per location and giving the country a total of 23,000 farm families with improved technology and water control. Further, the government initiated the development and rehabilitation of the abandoned irrigation schemes as well as dams in the irrigable areas in all the 36 states of the country, including Abuja. During the period under review, the government collaborated with some states and private sector organizations to rehabilitate selected fish farms and hatcheries for fish and fingerlings production.

The agreement of the Nigeria–France Agricultural Development Project, worth N170 million, which was signed in June 2002, formally took off with the inauguration of the National Steering Committee in 2003. The project was aimed at improving the productivity and access to markets by smallholder farmers in Jigawa, Kano, Katsina and Bauchi states. The project was also designed to promote six counselling and services centres for geographical information system development and to strengthen national expertise in these areas. In 2002, the National Economic Council approved the establishment of three multi-commodity development companies in each state, with a view to boosting agricultural production. The companies were to be floated by the federal government, with N600 million for each state, while each benefitting state was to contribute N25 million of counterpart funds.

Also, the participation of foreign and multilateral agencies in the funding of agricultural activities added further impetus to agricultural production in 2004. The Mobil Nigeria partnership in rice production and the World Bank and African Development Bank support for the Fadama[2] II project contributed substantially to increased output during the year. Efforts on the Second National Fadama Development Programme (Fadama II), spanning 2004 to 2009 and targeting about 120,000 fadama users and 720,000 direct beneficiaries are exemplary. In spite of the achievements of the fadama programme in Nigeria, only a marginal portion of the country's total fadama potential is developed. While under Fadama I, an estimated 55,000 ha of fadama lands were put into cultivation by private smallholders using low-cost motorized pumps, only 12,350 ha or 13% of the country's fadama potential of 950,000 ha were covered (FMARD, 2003).

Policy Outcomes and Recent Performance of Nigerian Agriculture

Nigerian agriculture is currently contributing about 40% to the Nigerian GDP and employing about 70% of the active population, but the performance of the sector is still far below its potential. The growth rate of agricultural GDP was found to have outpaced that of the aggregate GDP in recent times, as shown in Fig. 11.1. Agricultural GDP growth rate rose from 4.2 in 2002 and reached an all-time high of 7.4 in 2007, as against 4.6 and 6.2 for aggregate GDP growth for the same period respectively (Central Bank of Nigeria, several issues). In spite of this, available information (Diao *et al.*, 2009) shows that though this growth rate is well above targets set by the NEPAD CAADP, it is still below the remarkable 10% growth rate set under the National Food Security Programme. Also, this growth rate fell below what is required to achieve the Millennium Development Goal 1 of eliminating hunger and halving the proportion of people in poverty (put at 65.6% in 1996 – the most recent figure available) by 2015 (FOS now NBS, 2005). This, in turn, indicates that Nigeria has not fully tapped its agriculture potential. For example, Nigeria has 79 million ha of fertile land, but only 32 million ha (46%) of these are cultivated. Further, more than 90% of agricultural output is accounted for by households with less than 2 ha under cropping. Typical farm sizes range from 0.5 ha in the south to 4 ha in the north (FMAWR, 2008).

From the *Afrint II* survey, the average area cultivated to maize in Kaduna (north) in the 2006 season was 3.5 ha, while it was 2.5 ha in Osun (south), and for cassava, it was 1.2 ha in Kaduna and 2.7 ha in Osun. A similar situation was observed for rice, with 2.1 ha in both Kaduna and Osun. Though recent growth trends reveal some modest increases in productivity

[2] Fadama is the local name for wetlands where dry-season production can take place.

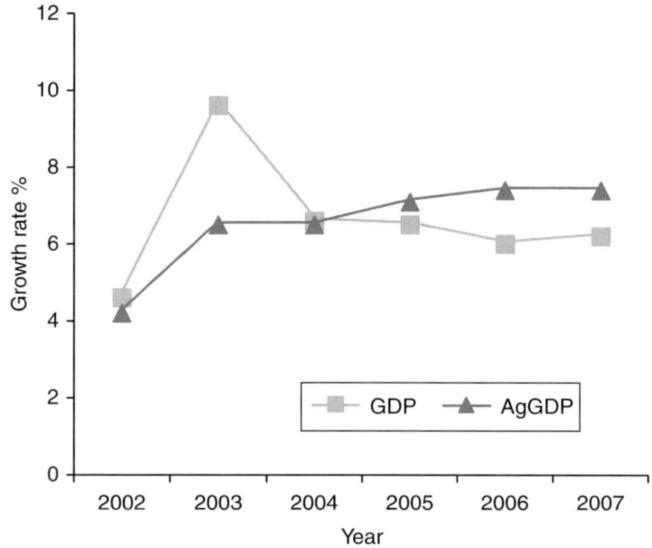

Fig 11.1. Growth rate of gross domestic product (aggregate and agriculture). (Adapted from: *CBN Statistical Bulletin* and *Annual Abstract of Statistics*, various issues.) GDP = aggregate GDP growth rate; AgGDP = agriculture GDP growth rate.

over time, yield levels are generally below potential. This reflects the fact that much of the growth or increase in output has come from expansion in the land area under cultivation. The indication that output growth was accounted for more by expansion in area cultivated than by productivity improvement is reinforced by the significant correlation between output and area harvested compared to the correlation between output and yield (Eboh *et al.*, 2006).

Trends in Food Crop Production

The various policies and programmes highlighted in the section 'Food and Agricultural Policies and Programmes, 1999 to 2007' are aimed at achieving rapid growth within the agricultural sector and reducing poverty. Meanwhile, to meet the 10% overall agricultural growth target set by the government, sector-specific targets were set for major crops and livestock production under the National Food Security Programme (NFSP) (FMAWR, 2008). Table 11.1 presents sector-specific targets for the three major crops covered in this study.

In the attainment of the targets set in Table 11.1, has the Nigerian Green Revolution veered off track? This question can be better understood through critical analysis of production trends for the three major food crops covered

Table 11.1. Crop-specific targets in the agricultural sector. (Adapted from: FMAWR, 2008.)

Commodity	Targets: 2008–2011	% Increase
Cassava	Yield increase from 15 t/ha to 60 t/ha	400
	Production increase from 49 million t to 100 million t per annum	
	Attain 10% cassava flour in breadmaking	104
Rice	Increase production from 2.8 million t of paddy to 5.6 million t paddy rice per annum	100
Maize	Attain 6.5 million tonnes of maize through improved farm inputs and irrigation, from 4.0 million t per annum	62.5

by the study (maize, rice and cassava) using aggregate data between 2002 and 2007.

One major factor accounting for food insecurity is the variability in food production from year to year, which mainly affects the physical availability of food. The historical data employed in this study were derived from the annual crop area yield survey normally conducted by the NFRA and published by the National Bureau of Statistics (NBS). Complementary data were also derived from the Food and Agriculture Organization Statistics (FAOSTAT) web site. Thus, examining historical data on area, output and yield of cassava, maize and rice in the last 5 years, it was noted that not only had area and output been erratic and unstable, the yield level of some of these crops had also declined during most of the period between 2002 and 2007.

Area under crop cultivation

Nigeria covers a total land area of about 92 million ha, out of which about 75% have the potential for agricultural cultivation. However, land area under cultivation is currently put at 59% of the potential arable land area. Of this area, only 0.5% is under irrigation (only 220,000 ha under irrigation, as against the potential of 2–2.5 million ha). Figure 11.2 shows total land area under cultivation for maize, rice and cassava from 2002 to 2007. The total land area cultivated for different crops increased between 2002 and 2007. The introduction of the presidential initiative on cassava in 2003 paid off in the form of increases in area cultivated from 2575 ha in 2004 to 2970 ha in 2005. The unprecedented increase observed in cassava acreage in 2005 was due to government propaganda to export cassava chips and starch to the tune of at least US$3 billion per annum, in addition to making cassava flour compulsory for inclusion with wheat flour for bread baking.

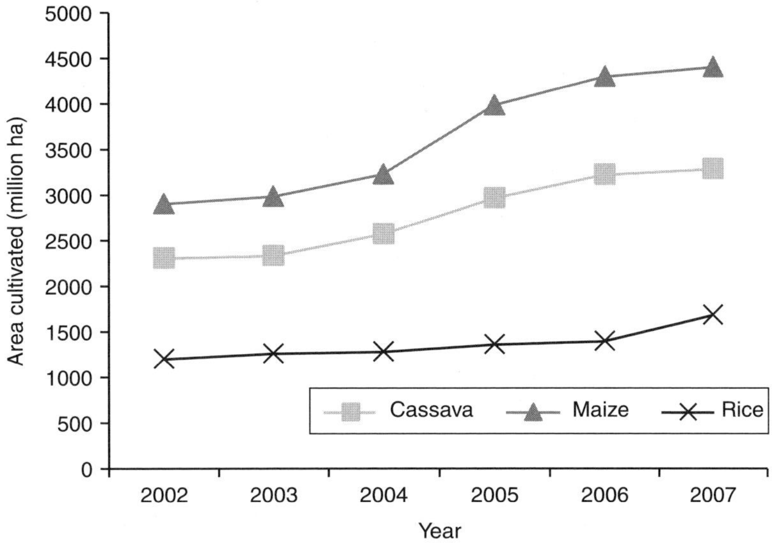

Fig. 11.2. Total land area cultivated to different crops, 2002–2007. (Adapted from: National Food Reserve Agency, 2008 and FAOSTAT data, 2008.)

Like cassava, maize experienced a significant jump in area cultivated between 2002 and 2007, probably due to price increases. In the case of rice, the total area cultivated was relatively small, with low but steady growth.

Structure of production

Analysis of the performance of agricultural output in 2007 shows that crop production grew by 7.51% compared to livestock (6.91%), forestry (6.02%) and fishing (6.58%) (FMAWR, 2009). This performance is consistent with the composition of agricultural output, as dominated by crop production. Over the last 10 years, crop production constituted, on average, about 80% of agricultural GDP. The output of major crops recorded increased growth rates compared to their 2006 levels. Details of the trend in output of cassava, maize and rice between 2002 and 2007 are presented in Fig. 11.3.

The trends presented in Fig. 11.3 show that expansion of cultivated areas resulted in increases in output of these staple crops during the 2002–2007 period. But the output increases appear marginal for maize and rice in particular. The big increase in cassava output may be connected with the fact that cassava has a longer gestation period than maize and rice and, as such, could adapt better to changes in weather conditions. Nevertheless, the false assurance given to farmers that they could export their cassava products in the international market was a dream that went unfulfilled because Nigerian cassava products could not meet international standards. Output of maize rose only marginally as maize farmers could not gain access to sufficient fertilizers at planting time.

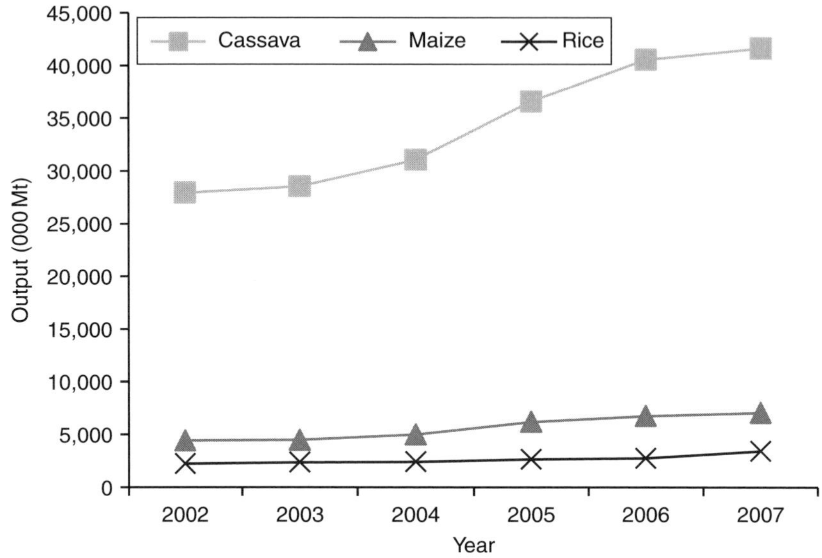

Fig. 11.3. Total output of crops, 2002–2007. (Adapted from: National Food Reserve Agency, 2008 and FAOSTAT data, 2008.)

Productivity (yield)

Aggregate data for major crops shows modest increases in productivity over time; however, the yield levels are far below potential and still less than levels required for global competitiveness in agriculture. Yield levels for cassava, maize and rice either stagnated or only recorded marginal increases between 2002 and 2007, as shown in Fig. 11.4. As a matter of fact, the yield level of cassava declined in 2004 and 2005, despite the implementation of the presidential initiatives on cassava.

Apart from the fact that the current yield levels are not sufficient to meet the various crop-specific targets under the NFSP, the gap between the current yield and the potential yield of these crops are indicative of inefficiencies and low productivity in Nigerian agriculture. While reviewing potential yields for various crops in Nigeria using farm-level data and experimental plots data in 2006, ReSAKSS WA (2009) found that yield gaps ranged from 2.42t/ha for maize and 3.43t/ha for rice to as high as 15.89t/ha for cassava, as shown in Table 11.2.

Agricultural prices

The farm gate price reported in this section is the nominal price and it is the actual price received by farmers for their crop output (see Fig. 11.5). It may be lower than the local market price, for the reason that transportation costs

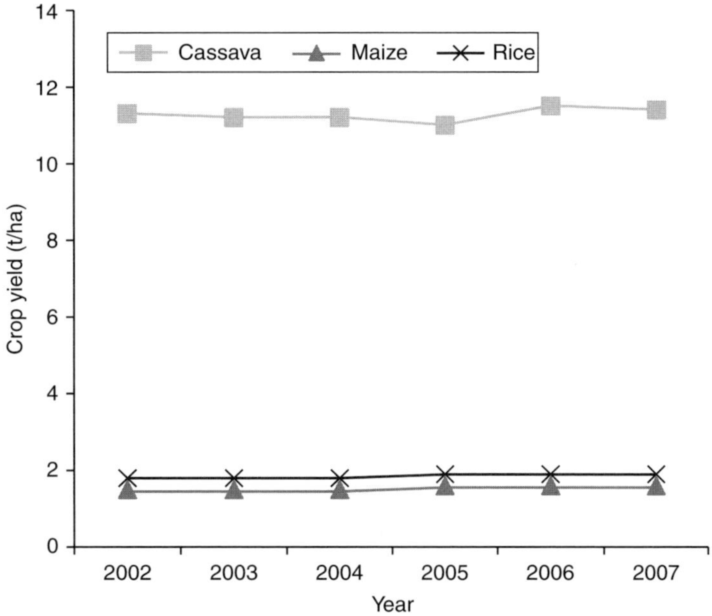

Fig. 11.4. Trend in crop yields in Nigeria. (Adapted from: National Food Reserve Agency, 2008 and FAOSTAT data, 2008.)

Table 11.2. Crop yield gaps in Nigeria. (Adapted from: ReSAKSS WA, 2009.)

Crop	Potential yield (t/ha)	2006 yield (t/ha)	2006 yield yap	
			t/ha	%
Rice	5.40	1.98	3.43	173
Cassava	28.4	12.50	15.89	127
Maize	4.0	1.57	2.42	154

may have been added to the local market price. The farm gate price is used to measure the return to farm enterprises. The farm gate price for the various food crops, except for rice, recorded an increase between 2002 and 2004. Rice prices recorded a decline between 2003 and 2005. In 2005 maize, millet and sorghum also experienced falling prices, although the price of cassava increased dramatically from 2004 to 2005. The sharp decline in the farm gate price for cassava in 2006 may have resulted from the glut arising from the inability of farmers to market the output of the preceding year. In spite of the increase in the farm gate price for rice from N54.0/kg in 2005 to N71.6/kg in 2006, the area cultivated to the crop fell further, from 1.2 ha in 2005 to 1.1 ha in 2006. Probably farmers reduced the area cultivated in 2006 in reaction to the sudden fall in price in 2005 to N54/kg from N62/kg in 2004.

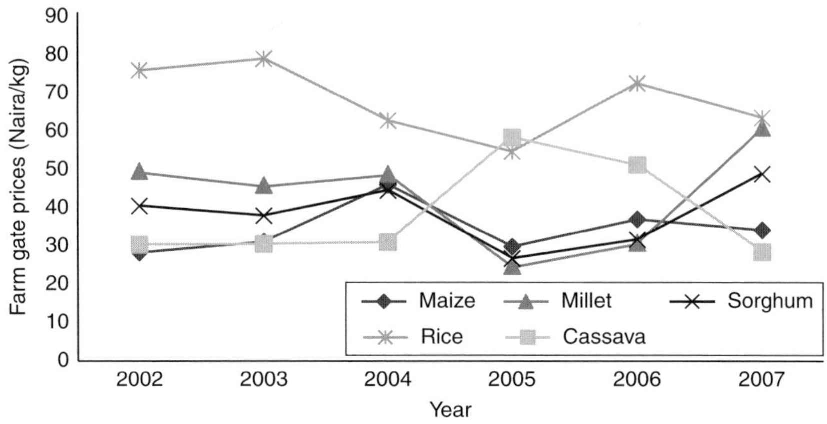

Fig. 11.5. Farm gate prices for staples, 2002–2007. (Derived from NISER Food Crisis Survey, 2008.)

Technology and input use

Available information shows that the rate of farm input used is far from sufficient for achieving potential productivity in Nigerian agriculture. Current fertilizer use is estimated at 0.5 million t/year, far short of the potential of 3–5 million t/year. Government procurement ranged between 69,000 t in 2000 and 76,000 t in 2007 (NBS, 2008). In view of the falling government procurements in fertilizer since the late 1990s, it is not surprising that fertilizer use per hectare arable land (kg of nutrient/ha) decreased from 13 kg in 1989–1991 to 6 kg in 2002 (World Bank, 2004). Similarly, the current use of improved seed/planting materials is put at 12% of potential demand.

Irrigation development

The results of the First Fadama Development Programme (Fadama I), spanning 1993 to 1999 and covering about 55,000 ha, underscore the efficacy of irrigated farming. One of the factors responsible for low competitiveness of Nigerian agriculture is undeveloped irrigation potential, which thus makes reliance on rain-fed farming inevitable. This leads to low productivity, meagre farm incomes and poverty. From Fadama I, widespread adoption of simple and low-cost improved irrigation technologies led to increased farmers' crop incomes, up to 65% for vegetables, 334% for wheat and 497% for paddy rice. Even with Fadama II, only an additional 80,000 ha, representing 8.4% of the fadama potential in the country, was earmarked for development, showing that only about 21% of the country's potential is currently under development (World Bank, 2003).

Production Changes in Maize in Nigeria – a Comparative Analysis

The tendencies outlined in the policy review above are, in some respects, also reflected in the micro-level data gathered as part of the study. This section draws on panel data from the two sampling rounds (2002 and 2008, respectively) to compare the drivers of production in maize with those of the other seven *Afrint* countries. The specificities of the Nigerian experience and explanations of a potentially derailed Green Revolution are contextualized through a comparison using the set of models developed for maize in Chapter 5 (Andersson *et al.*, this volume). Nigeria is compared with the remaining seven countries (Ethiopia, Ghana, Kenya, Malawi, Mozambique, Tanzania and Zambia) in analysing the drivers behind production changes in maize between 2002 and 2008.

The micro-level household survey covered two states, Kaduna in the northern Guinea savannah and Osun in the humid forest. These two states were purposefully selected in 2002 to meet the requirements of the overall objective of the *Afrint* project. The farming system in Kaduna state is cereal based with significant livestock production (particularly cattle and small ruminants), while production in Osun state is predominantly root crop based (mainly cassava), though maize production and rice are equally important in some parts of the state.

As in the other countries, a multi-stage stratified random sampling technique was employed. In Nigeria, each state is divided into Agricultural Development Project (ADP) zones for ease of extension delivery and agricultural development purposes. The sampling procedure comprised the selection of ADP zones after classifying them with respect to their agricultural potential. This was done to ensure dynamism in the areas within each state.

In Kaduna state, five ADP zones were covered, as compared with four zones in 2002. This is because a new zone has been created between 2002 and 2008 and this new zone is now referred to as the headquarter zone. In Osun state, however, all the six ADP zones covered in 2002 were also selected in this current survey. The second stage was the selection of villages, while the third and final stage was the selection of households. In this survey, 14 villages were selected from Kaduna state, as opposed to 24 selected in 2002.[3] Out of these 14 villages, five are new ones, while the remaining nine were covered in the *Afrint I* project. The new villages were added in order to cover some newly created ADP zones. In Osun state, on the other hand, 16 villages were selected against the 25 selected in 2002. Out of these 16, only two villages were new. It should be noted, however, that the selection of villages

[3] The sample design in 2002 was sub-optimal, with too few respondents in each sample village. In the 2008 round it was therefore decided to drop about half of the 2002 villages and increase the number of respondents in each village. This obviously brings down the size of the Nigerian panel.

in *Afrint I* followed the identification of villages along the intensification continuum – early, transition and late. Households were sampled randomly in the villages. The 2002 cross-section included 495 households and the 2008 cross-section includes 434 households. The Nigerian panel data covers 221 households interviewed in both 2002 and 2008. The panel of maize farmers is a subset of the latter, which includes 206 households that grew maize in both 2002 and 2008. The rice and sorghum panels are too small to be analysed on their own, while cassava has its own problem, dealt with in the Introduction to this volume.

Results of the analysis

Generally speaking, Nigeria confirms the overall patterns of production drivers, as outlined in Chapter 5: increased production of maize is more driven by area expansion than by intensification, while commercial drivers emerge as the strongest influence on production increases, especially in the period between 2002 and 2008 (see Tables 11.3, 11.4, 11.5).[4]

One major difference between the seven countries and Nigeria lies in the relative role of technology during the period from the reference year (on average 1982) to 2002. Nigerian producers who used seed fertilizer technology from the outset or adopted it during this period, benefitted less from this than their counterparts in other countries (see Table 11.4). Hence, seed fertilizer technology cannot be shown to be associated with higher maize productivity in the early period of the Green Revolution in Kaduna and Osun. This leads to a critical reflection on an earlier paper by one of the authors (Akande, 2006), where the role of technology in the earlier phases of the Nigerian Green Revolution may have been overestimated. On the other hand, we spot interesting differences in the second equation. Those who used seed fertilizer technology in the reference year had significantly higher production in 2008 than their peers. This is in line with the results for the other seven countries. However, those who have adopted seed fertilizer since the reference year, other things being equal, also enjoyed significantly higher productivity[5] than the non-adopters. There is an evident contrast between the shorter period, from the reference year to 2002, with non-significant effects of seed fertilizer technology, and the longer one, between the reference year and 2008, with stronger effects. The contrast points to a more recent dynamism. This is not enough, however, to yield results in the third equation, which shows that, in Nigeria as well as in the other countries, adoption of seed fertilizer technology apparently has not contributed significantly to increased productivity from 2002 to 2008.

[4] The reader may wish to consult the chapter by Andersson *et al.* (Chapter 5, this volume) in order to understand the modelling strategy, the variables used, etc.

[5] Since the dependent variable in the equations is logged production or logged change in production, and since we control for area in the equations, the regression coefficients can be taken to indicate the impact on *area productivity* of a given independent variable.

Table 11.3. Maize production model: Nigeria compared to seven other countries in sub-Saharan Africa.

	Period 1 (p_1) t_0 to t_1				Period 1 + 2 (p_{1+2}) t_0 to t_2				Period 2 t_1 to t_2			
	Nigeria		Seven countries		Nigeria		Seven countries		Nigeria		Seven countries	
	β	Sig.	β	Sig.	β	Sig.	β	Sig.	β	Sig.	β	Sig.
Constant	7.34	***	5.83	***	6.12	***	6.06	***	–0.57		0.57	***
Controls												
Years since farm established, logged	–0.10		0.02		0.05		–0.09	**	0.15		–0.13	***
Descendant households	0.18		0.09		0.06		0.23		–0.17		0.11	
Area												
Area under maize, logged	0.83	***	0.62	***	0.57	***	0.59	***	0.48	***	0.21	***
Weather												
Drought in 2002	–0.49	***	–0.18	***								
Floods in 2008					–0.36	*	0.46	***				
Fertilizer												
Used fertilizer at the start of the period	–0.14		0.53	***	0.67	**	0.41	***	0.47		–0.07	
Decreased or stopped using fertilizer during the period	0.03		–0.41	***	–0.51		–0.52	***	–0.79		0.14	
Started or increased using fertilizer during the period	–0.10		0.38	***	0.83	***	0.13		0.27		0.12	
Ploughing												
Used ploughing at the start of the period	0.27		0.20	**	1.08		0.60	***	–0.02		0.16	*
Stopped using ploughing during the period	0.35		–0.08		–0.69		–0.31	***	0.24		–0.32	*
Started using ploughing during the period	0.07		0.46	***	0.15		0.55	***	–0.45		0.57	***

Continued

Table 11.3. Continued.

	Period 1 (p_1) t_0 to t_1				Period 1 + 2 (p_{1+2}) t_0 to t_2				Period 2 t_1 to t_2			
	Nigeria		Seven countries		Nigeria		Seven countries		Nigeria		Seven countries	
	β	Sig.	β	Sig.	β	Sig.	β	Sig.	β	Sig.	β	Sig.
Commercialization												
Sold maize at beginning of period	0.14		0.39	***	0.49	*	0.35	***	–0.23		–0.14	
Stopped or decreased selling maize during the period	–0.06		–0.03		–0.41	**	–0.17	*	–0.58		–0.49	***
Started or increased selling maize during the period	0.14		0.57	***	0.34		0.64	***	0.59		0.60	***
Distributional dimensions												
Elite membership in 2002	0.18		0.15		–0.29		0.09		–0.39		–0.16	
Gender of farm manager in 2002	–0.10		–0.09		–0.07		–0.21	***	–0.07		–0.06	
Kaduna	0.64	***			0.09				–0.53	**		
Residual from use of seed-fertilizer technology model									–0.22		0.12	*
Residual from market participation model									–0.03		0.00	
Model info												
No. of cases	159		1158		148		1293		153		1144	
R^2	0.78		0.53		0.51		0.61		0.59		0.29	
Missing cases (%)	0.23		0.28		0.28		0.19		0.26		0.28	

The levels of significance are denoted by: *(10%), **(5%) and ***(1%).

Table 11.4. Fertilizer adoption model for Nigeria compared to seven other countries.

	Nigeria			Other countries		
	β	Exp(β)	Sig.	β	Exp(β)	Sig.
Years since farm established, logged	0.13	1.14		−0.07	0.91	
Descendant households, dummy				0.10	1.36	
Additional land available, dummy	2.44	10.07	**	0.39	1.86	**
Family labour resources increased since 2002	−0.52	0.64		0.36	1.33	**
Increased cattle ownership since 2002, dummy	−0.42	0.73		−0.02	1.10	
Farm management feminized since 2002, dummy	−2.02	0.14	**	−0.10	0.87	
Used fertilizer on maize in 2002	1.05	2.70		1.41	3.03	***
Started or increased sale of maize since 2002, dummy	1.32	3.75	**	0.57	1.93	***
Change in country-level mean nominal producer price of maize, 2002–2008, logged				0.42	5.70	**
Started or increased sale of other food crops since 2002, dummy	0.85	2.35		0.76	2.40	***
Started receiving extension services since 2002, dummy	−0.59	0.55		0.04	0.96	
Constant	−2.21	0.11		−1.23	0.20	***
Valid n (listwise)	147			1083		
Missing cases (%)	0.29			0.32		
Nagelkerke's R^2	0.20			0.23		

The levels of significance are denoted by: *(10%), **(5%) and ***(1%).

In terms of cultivated area there are significant differences in the regression coefficients between Nigeria and the other countries in the first and third equations. Comparing β-value for area in the Nigerian case with that of the other countries, we see that the coefficient for the former country is significantly higher (0.83) than that for the others (0.62). Being a rough indicator of the marginal productivity of land,[6] this would mean that marginal productivity is higher in the Nigerian case, or more specifically in Osun and Kaduna. Moreover, the regression coefficient for area in the third equation, which is an indicator of intensification, is significantly higher in Nigeria than elsewhere. Both these results point to a higher level of intensification in the Nigerian sample, as well as a more dynamic intensification process in Nigeria.

There are few significant differences in the impact of ploughing. Note, however, that, according to equation 5.3, adopters of ploughing did not gain in terms of increased area productivity, unlike their counterparts in the other countries. This could be due to the massive adoption of tractor ploughing,

[6] Cf. Chapter 5, this volume.

Table 11.5. Market participation model for Nigeria compared to seven other countries.

	Nigeria			Seven countries		
	β	Exp(β)	Sig.	β	Exp(β)	Sig.
Years since farm established, logged	−0.19	0.83		0.01	1.01	
Descendant household, dummy				0.70	2.02	
Additional land available 2002, dummy	0.66	1.93		0.74	2.10	***
Maize area increased since 2002, dummy	−0.29	0.75		0.81	2.26	***
Yields of maize increased since 2002, dummy	1.23	3.43		0.83	2.30	***
Used fertilizer on maize in 2002, dummy				0.62	1.86	***
Started using fertilizer since 2002, dummy	−0.82	0.44		0.08	1.08	
Sold or intended to sell maize 2002, dummy	3.70	40.31	**	3.01	20.27	***
Change in country-level mean nominal producer price of maize, 2008 over 2002, logged				−0.18	0.83	
Started or increased sale of other food crops since 2002, dummy	0.32	1.38		0.68	1.98	***
Proxy for elite membership in reference year	−1.06	0.35		0.68	1.97	**
Increased share of maize consumed since 2002, dummy	−2.97	0.05	**	−2.11	0.12	***
Income from non-farm sector 2008, dummy	−0.61	0.54		−0.11	0.90	
Income from sale of non-staples 2008, dummy	0.07	1.07		0.28	1.33	
Constant	0.37	1.45		−2.91	0.05	***
Valid n (listwise)	174			1441		
Missing cases (%)	5			6		
Nagelkerke's R^2	0.37			0.43		

The levels of significance are denoted by: *(10%), **(5%) and ***(1%).

which distinguishes the Nigerian sample from the other countries.[7] Tractor ploughing economizes on labour with marginal effects on yields, while the use of oxen enhances labour productivity as well as yields – the latter through the supply of manure. Thus, the prevalence of tractors in Nigeria may result in lower yield effects of ploughing.

The commercialization indicators behave exactly the same in Nigeria as in other countries. This reinforces the conclusion that commercialization drivers are the most important and they seem at least as strong in Nigeria as elsewhere.

[7] There has been massive adoption of tractor ploughing in the Nigerian maize panel: there are four times as many farmers using tractors in 2008 compared to 2002.

The results reported by Andersson *et al.* in Chapter 5 of this volume pointed to a process where the village elites have partially withdrawn from the maize markets and left space for other and smaller producers to enter. Admittedly, the statistical underpinning of this result is not strong. This is emphasized when we compare Nigeria and the other countries, where we get no significant results. Thus the strong effects of market entry on production increases cannot be shown to have a smallholder profile in the sense of benefitting the non-elite households disproportionately. On the other hand, the strong and significant β value for the Kaduna dummy in the third equation indicates that, during the period 2002–2008, maize production was more dynamic in the forested Osun, where farms, on average, are much smaller than in the Guinea savannah Kaduna.

The above signals, first of all, that most of the drivers identified in Chapter 5 by Andersson *et al.* apply in Nigeria as well. Going by the first two equations in the model presented in that chapter, Nigerian farmers record log production levels that are about 40% higher than the reference case (Kenya). In the reduced form model, however, Nigeria does not stand out as particularly dynamic, in contrast to, for example, Zambia, Mozambique and, to some extent, Ghana (see Andersson *et al.*, Chapter 5, this volume for a discussion of this). The reasons for this may lie at macro level, as we will presently see.

The macro-level variables that are part of the modelling exercise in Chapter 5 earlier in this volume are not included here since they are constant in the one-country regression. It is interesting to compare these indicators for Nigeria and for the other countries in descriptive terms, however (see Table 11.6).

Table 11.6. Comparison of macro-level indicators for Nigeria and seven other countries.

	Nigeria	Seven countries	Total
Government expenditure on agriculture and rural development, 2002 (lagged by 3 years)	1.62	4.32	3.86
Import of maize as per cent of total production 2000–2005	0.02	5.97	3.11
GDP per capita 2001 (constant 2000 USD)	368.71	243.81	255.63
Government expenditure on agriculture and rural development, 2008 (lagged by 3 years Nigeria 2003, Zambia 2004, Ghana 2004, Malawi 2006)	4.80	7.44	7.07
Import of maize as share of total domestic production 2001–2005	0.14	4.95	3.29
GDP per capita 2007 (constant 2000 USD)	473.43	295.08	311.49
Change in budget allocations to agriculture (lagged), 2008 over 2002	2.96	1.72	1.83
Change in import of maize (lagged), 2008 over 2002	7.00	0.83	1.06
Change in GDP per capita 2007 over 2001	1.27	1.21	1.21

Looking first at government expenditure, it is evident that in 2002 Nigeria was much below the mean for the other countries.[8] From that year both Nigeria and the other countries have increased their budgetary allocations to agriculture, in the Nigerian case we record an increase of almost 300%, compared to 83%, on average, for the other countries.

Nigeria had almost negligible imports of maize in the years leading up to 2002. From this scanty numerical base, imports have increased manifold but remain proportionally low.

Being a richer country, Nigeria had a GDP per capita in 2001 which was about 50% higher than the mean for the other countries. Similarly, it experienced 27% growth in GDP per capita over the period 2001–2008, which is considerably higher than the average 21% growth for the other countries. Going by these figures, and the hypotheses we tested in Andersson *et al.* (Chapter 5, this volume), Nigeria should have been poised for a quicker growth in the maize sector than the others.[9] Equation 3 for the maize model suggests that this is not the case.

When we run the subsidiary models (i.e. the Appendix models in Andersson *et al.*, Chapter 5, this volume), we approach the reasons for the sluggish Nigerian Green Revolution (see Tables 11.2 and 11.3).

As is clear from these tables, there are no significant differences between Nigeria and the other countries. This is true both for the fertilizer adoption model and for the one on market entry. The drivers for adoption and entry seem to be the same as for the other countries.

Adoption of seed fertilizer technology is, in all countries, associated with commercialization and having access to set-asides or fallow land. An ad hoc interpretation of this can be that increased maize production on the whole relies on extensive growth and draws on set-asides that can be made productive by adding fertilizer. Recall, however, that the maize production model indicated a slightly more intensive growth pattern in Nigeria than in the other countries.

As can be seen, two variables are excluded from the Nigerian model of fertilizer adoption: (i) descendant households; and (ii) nominal price change of maize since 2002, which is a country-level variable (thus being constant in a regression for a single country). The exclusion of the former is due to its low variance: there are only six descendant households.

Nigeria experienced an increase in producer prices for maize of 28% if maize prices are taken at the two points of 2002 and 2007, respectively, and measured in USD at 1999/2001 USD value. In local currency the increase was 23% during the same period (FAOSTAT). In the *Afrint* database, however, a nominal price increase of 15% over the period was recorded, which is clearly below the inflation rate. This is probably an important explanation for the slow

[8] The Nigerian figure here is from another source than those used for the other countries, so the comparability may be problematic.

[9] Remember that the maize model in Andersson *et al.* (Chapter 5, this volume) predicts a 2% growth for every percentage of growth in GDP per capita.

adoption rate for technology and for a consequently sluggish growth in production and may indicate that the Nigerian Green Revolution has lost steam due to non-remunerative prices. If this is true it is slightly ironic: while the government seems to have done everything to follow the recipe for a Green Revolution, one important prerequisite, i.e. cost-effective technology packets, are either lacking or supplied at non-remunerative prices, if they are supplied at all. The earlier discussion of fertilizer procurement indicated procurement has been going steadily down, and since there is a virtual state monopoly on trade in fertilizer, this leads to black markets with higher prices and, most importantly, leaving most smallholders without access to fertilizer.

Looking finally at the market participation model, the same variables that had to be excluded in the previous model are excluded here, along with the lagged fertilizer adoption variable. As in the previous model, the drivers of market participation appear to be largely similar to those in the other countries: set-asides, area- as well as yield-intensive growth and market participation in 2002. We also note that the association between self-provisioning and market entry is negative in Nigeria, as in the other countries.

It is tempting to interpret the above as adding up to at least a tentative diagnosis of the Green Revolution in the Nigerian maize sector, not as having veered off track but having failed to gain steam. The growth potentials are not, as yet, fully exploited, although there are some indications of significant effects of seed fertilizer technology on maize production. Development is entirely driven by commercial incentives, but these tend to be sluggish, partly due to non-optimal state intervention (in fertilizer markets). This is also in line with the finding for the other countries.

Thus, our overall conclusion must be that although the government, on the face of it, has gone for sensible policy packages, a crucial factor is missing at farm level, i.e. more vigorous incentives, in the form of remunerative prices, supply of technologies which are already standard (i.e. fertilizer) and cost-effective cropping technologies, more generally. These seem to be the most important reasons for the slightly discouraging results of agricultural policies in the maize sector so far. Unfortunately, data for the other staple crops are too scanty to allow checking if this applies to other staples as well.

References

Akande, T. (2006) Food Policy in Nigeria: Analytical Chronicle. New World Press, Ibadan, Nigeria.

Central Bank of Nigeria (2002) Annual Report and Statement of Account. Abuja, Nigeria.

Central Bank of Nigeria (2007) Economic Report for the First Half of 2007. Abuja, Nigeria.

Central Bank of Nigeria (2008) Annual Report and Statement of Account. Abuja, Nigeria.

Xinshen, D., Nwafor, M., Alpuerto, V., Kamiljon, T., Akramov, T., Salau, S. (2009) Agricultural growth and investment options for poverty reduction in Nigeria. Draft report. IFPRI and IITA, Nigeria.

Eboh, E.C., Larsen, B., Oji, K.O., Achike, A.I., Ujah, O.C., Oduh, M., Uzochukwu, S.A. and Nzeh, C.C.P., (2006) Renewable Natural Resources, Sustainable Economic Growth and

Poverty Reduction in Nigeria. AIAE Research Paper 1. African Institute for Applied Economics, Enugu, Nigeria.
FAOSTAT (2008) FAOSTAT data. Available at: http://faostat.fao.org/ (accessed December 2008).
FMARD (2003) Second National Fadama Development Project (FADAMA 2). Project Implementation Manual Vol. 1. Federal Ministry of Agriculture and Rural Development (FMARD) PCU, Abuja, Nigeria.
FMAWR (2008) National Food Security Programme. Federal Ministry of Agriculture and Water Resources (FMAWR), Abuja, Nigeria.
FMAWR (2009) Agricultural Development Efforts in Nigeria and Alignment with CAADP: Stock Tacking Report. Federal Ministry of Agriculture and Water Resources (FMAWR) Abuja, Nigeria.
National Food Reserve Agency (2000) Special Programme on Food Security (SPFS). Project document. Abuja, Nigeria.
National Food Reserve Agency (2008) Data and Statistics on Agricultural Production, Land Area and Productivity in Nigerian States. Abuja, Nigeria.
National Food Reserve Agency (2008) Report of Annual Crop, Area, Yield Survey. Abuja, Nigeria.
NBS (2005) Poverty Profile for Nigeria. National Bureau of Statistics (NBS), Abuja, Nigeria.
NBS (2008) Annual Abstract of Statistics. National Bureau of Statistics (NBS), Abuja, Nigeria.
NISER (2002) Annual Survey of Crop Production Conditions in Nigeria. A publication of NISER Annual Monitoring Research Project (NAMRP), NISER, Ibadan, Nigeria.
NISER (2005) Public–Private Partnership in Nigeria Development: NISER Review of Nigerian Development 2003/2004. NISER, Ibadan, Nigeria.
NISER (2008) The Global Food Crisis: Impact and Policy Implications in Nigeria. Research report submitted to the National Fadama Development Office, Abuja, Nigeria.
ReSAKSS WA (2009) Review of potential crop yield in Nigeria. Draft. Regional Strategic Analysis and Knowledge Support System West Africa (ReSAKSS WA), International Institute of Tropical Agriculture, IITA, Ibadan, Nigeria.
World Bank (2003) Second National Fadama Project: Project Implementation Manual. Federal Ministry of Agriculture and Rural Development. Abuja, Nigeria.
World Bank (2004) Nigeria: value and supply chain study. First draft report prepared by Consilium International Inc., Washington, DC.

12 Addressing Food Self-sufficiency in Tanzania: a Balancing Act of Policy Coordination

Aida C. Isinika[1] and Elibariki E. Msuya[2]

[1]*Institute of Continuing Education, Sokoine University of Agriculture, Morogoro, Tanzania;* [2]*Department of Agricultural Economics and Agribusiness, Sokoine University of Agriculture, Morogoro, Tanzania*

After Structural Adjustment (1986–1994), Tanzania moved from an era of heavy state involvement in agriculture to full liberalization, when all direct and indirect subsidies were removed (Isinika, 2003; Skarstein, 2005). There are others who argue that economic liberalization in Tanzania, as is the case in some other African countries, was never complete, being partial due to emphasis on price liberalization, uncoordinated timing and sequencing, lack of local commitment and ownership, and weak institutional capacity (Kherallah *et al.*, 2000; Cooksey, 2003). Pressure for economic liberalization came from bilateral and multilateral donors, led by the International Monetary Fund (IMF) and the World Bank, which used different techniques, including withdrawal of donor funds, which led to a significant decline in foreign aid (Havnevik *et al.*, 1988). Economic reforms were considered necessary to liberate the private sector and to get prices right so that they would reflect relative scarcities of resources for more effective allocation to achieve static efficiency. Many African countries were forced to agree with SAP prescriptions because they were desperately in need of foreign exchange to service outstanding debts. It was assumed that government withdrawal from market operations would enable farmers to respond to factor and product price signals, leading to innovation, specialization and accumulation (Skarstein, 2005).

The reform process continued during the 1990s, when Tanzania, like many other African countries, undertook more extensive reforms designed to turn around declining growth rates and reverse balance of payment deficits. In agriculture the reforms aimed to eliminate bias against the sector by removing price controls, deregulation of agricultural markets (which had been achieved by 1990) and closure of state-owned monopolies, which was completed in 2007 (Isinika, 2009). In 2001, Tanzania developed an Agricultural

Sector Development Strategy (ASDS), which defined the role of the state to be policy making, regulation and provision of public goods. The use of subsidies as policy instruments was ruled out (Skarstein, 2005), only to be reversed a few years later.

While some initially hailed withdrawal of the state from market operations as a move in the right direction and that economic transformation was on course (IMF, 1986, 1995; World Bank, 1992), there are many other studies, however, that portray a different picture. A simulation analysis based on poverty reduction rates for the period 1992–2002 shows that growth rates attained until then were not enough to meet the Millennium Development Goals (MDGs) (Demombynes and Hoogeveen, 2004). Analysis of the post-Structural Adjustment Programmes (SAP) period up to 2000 shows that the outcome of the reforms fell short of expectations for agriculture in general but especially in relation to food production, forcing some countries, including Ghana, Malawi, Nigeria and Tanzania, to undergo policy reversal on fertilizer subsidies. Several countries also restored other forms of government intervention in the input sector (Kherallah *et al.*, 2000).

Thus the policy and institutional environment of the new millennium (post 2000) represents a relatively liberalized agricultural sector with some level of government intervention. Despite a strong push from international finance institutions, urging many African governments to pursue a hands-off policy in markets, it is common knowledge that many governments intervene in their food and agriculture sectors in a variety of ways, using subsidies, taxes, credit, price stabilization programmes and expenditure programmes to provide incentives or to achieve income transfer for equity or to stimulate economic development (Stiglitz, 1987; Giovanni and McCalla, 1995). It is none the less correctly argued that such interventions should be carefully managed to minimize efficiency loss.

This calls for a careful balancing act so that, in addition to promoting policies that optimize static efficiency in resource allocation, the policies also enhance dynamic efficiency, such that technical progress and growth of land and labour productivity moves on a path of dynamic efficiency in the long term (Uma Lele, 1989; Rune, 2005). On this basis, careful phasing of subsidies over time has been recommended for India and Pakistan, in order to discourage inefficient use of fertilizer, water and electricity, and reverse escalating government spending (Vaidyanathan, 2000). In addition, the literature suggests that the marginal opportunity cost of spending on subsidy or government transfer programmes is more likely greater than unity, to the tune of up to 50% higher (Alson and Hurd, 1990).

For poor countries such as Tanzania, the option to restore subsidies is a tough one, considering that there are many competing ends to use the same scarce resources (health, education, roads, etc.). Economic liberalization policies should therefore aim at achieving efficiency by maximizing returns, equity for distribution of income and food security. Realizing these policy objectives may be limited by supply constraints (resources, technology, relative prices and management), demand constraints (population, income, taste and relative prices) and world prices constraints through export and import (Monke and Pearson, 1989). Hence, governments may from time to time be required to make trade-offs between efficiency, income distribution and food security using different policy instruments.

This chapter examines the performance of food production and productivity in Tanzania since 2000, in relation to post-SAP policies. This discussion assumes that individual households in Tanzania strive to achieve food security through their own production as well as purchases from the market. Meanwhile, the government strives to meet national food self-sufficiency of the main staples (maize, rice and cassava) from the local production, implying that individual farmers must produce a surplus, which is then marketed efficiently so that everybody can access sufficient and good-quality food at all times at affordable prices. Any change in the policy environment changes the opportunity set and hence the choices individuals make, which in turn shapes the aggregate performance of economies over time (North, 1993). It is in this context that the analysis in this paper looks at the performance of food production and marketing, at the micro and macro levels, during the post-SAP period in Tanzania, as influenced by preceding and prevailing policies and institutions, in particular focusing on the magnitude and direction of change. The discussion is guided by several questions. Is there any change happening in food production? What is driving that change? Can the change be sustained? What is the role of supporting institutions, markets and governance in directing this change?

Reinforcing the Market Reforms

The timeline in Tanzania shows that while the thrust of the economic reforms during the 1980s was on markets – to get the prices right – the focus during the 1990s shifted to institutions. Tanzania, like many other African countries, followed the bandwagon of institutional reforms to consolidate market reforms that began in 1986. Specific for agriculture, there was a land policy in 1995 (Shauri, 1995; Kaduma, 2005), followed by the land laws of 1999, which became operational in 2001, with amendments in 2003. Although the president still holds all land in trust on behalf of the people of Tanzania, the new policy recognizes that land has intrinsic value, and hence can be marketed (URT, 1994; Shauri, 1995; Kaduma, 2005), which represents a major departure from the socialist past. In 1997, the agricultural policy was approved, recognizing the private sector as a key player for agricultural transformation, especially in relation to input supply, value addition and service delivery (Yoshida, 2005).

Considering the need for a sector-wide approach, Tanzania undertook further analysis of agriculture to determine how to foster accelerated sector transformation for wealth creation and poverty reduction. This was preceded by macro-level poverty reduction strategies, and followed by the ASDS, completed in 2001 whose salient features are presented in Figure 12.1. The ASDS was designed to conform and contribute to the National Strategy of Growth and Poverty Reduction – more commonly known as MKUKUTA,[1] which has set targets in three clusters for: (i) achieving growth and poverty reduction; (ii) improving the quality of life and social wellbeing; and (iii) good governance. The ASDS

[1] MKUKUTA is the Kiswahili acronym of *Mpango wa Kukuza Uchumi na Kuondoa Umaskini*, which is equivalent to National Strategy for Growth and Poverty Reduction (NSGPR).

ASDS purpose: to stimulate and facilitate agricultural sector growth and reduce rural poverty.
ASDS strategic objectives:
(i) Create enabling and favourable environment to improve agricultural productivity and profitability; and
(ii) Increase farm income to reduce rural poverty and ensure household food security.

ASDS is in line with the Comprehensive Africa Agriculture Development Program (CAADP) and Millennium Development Goals (MDGs).

ASDP Phase one: 5 years (2005–2010)
ASDP Phase two: 4 years (2011–2014)

Seventy-five per cent of funding goes to local component for LGA to finance District Agricultural Development Programmes (DADPs).
Twenty-five per cent of funding for national component (Ministries).
Uses basket funding from government (75.6%), donors (21.7%) and farmers (2.6%).
Coordination and funding of research and extension services designed to improve and involve more stakeholder participation in co-funding and decision making.

Fig. 12.1. Salient features of the ASDS and ASDP.

is operationalized through the Agricultural Sector Development Programme (ASDP), which requires coordination between five agricultural sector led ministries[2] as well as Local Government Authorities (LGAs). These are responsible for coordinating programme implementation at the local level.

In relation to food security, the ASDS aims to support regions and LGAs (districts, wards and villages) to plan and implement effective District Agricultural Development Programmes, such that they meet food security needs of vulnerable groups through assured input provision, training for skills upgrading, regular monitoring and strengthening the capacity of smallholder farmers as well as service providers to organize and have a strong voice in markets as well as other local institutions that affect their livelihoods (URT, 2001, 2005). These institutional reforms were expected to change the incentive structure in accordance with North (1993), which would in turn induce a change in choices available to actors in agriculture, hence translating into different technical measures and changes in farm practice, and hence improvement in farm productivity and production (Gibson and Knoontz, 1998).

However, as institutional reforms proceeded, the experience of many African countries on food production and productivity during the post-SAP period did not live up to such expectations. In Tanzania, analysis of data for the period 1986–2000 shows that while total output of main staples may have been increasing, productivity, however, was declining, especially for maize and rice (Kherallah *et al.*, 2000; Isinika *et al.*, 2005). Rice was hailed by the World Bank as the fastest-growing crop during the 1990s (World Bank, 1994), but such production

[2] The five agricultural sector led ministries are: Ministry of Agriculture, Food Security and Cooperatives (MAFSCO), Ministry of Livestock Development and Fisheries (MLF), Ministry of Water and Irrigation (MWI), President's Office – Regional Administration and Local Government (PO-RALG). The exact name of the ministry may change from time to time but basic functions generally remain the same.

Table 12.1. Labour productivity in major food grains. (Adapted from: Skarstein, 2005.)

	Maize			
Period	Labour productivity	Kg per capita of total population	Five major food crops[a]	
1976–1986	−0.66	+0.25	+1.08	+0.66
1985–1998	−1.99	−2.35	+1.35	−1.80

[a]Five major food crops include; maize, rice, wheat, sorghum and millet. They accounted for 59.7% of food tonnage (1995/96–1997/98).

growth came from area expansion (Isinika *et al.*, 2003; Skarstein, 2005). Productivity declined for both land and labour, and for all major food staples. Skarstein (2005) noted that per capita maize productivity (land and labour), and even agricultural gross domestic product (GDP), actually fell by 2.5% in the interval 1986–1998, while the trend growth rate of maize production declined by 1.1%. The analysis by Skarstein shows further that maize productivity performance post-SAP was worse than before structural adjustment was introduced (Table 12.1), contrary to earlier positive prognosis (Delgado *et al.*, 1999).

A Declining Trend

Following the commencement of SAP policies in 1986, fertilizer use fell steeply after the removal of fertilizer subsidy, reaching only 63,000 tonnes in 1998/99, from a peak of over 100,000 in 1990 (World Bank, 2000; Isinika *et al.*, 2003). The proportion of farmers who used fertilizer fell from 27% in 1990/91 to only 10.5% in 1997/98. Maize farmers used rates below recommended levels (Hawasi *et al.*, 1999; Isinika and Mdoe, 2001) because of high prices and fertilizer not being available as reported by 47% and 27%, respectively, of the farm holding according to the Expanded Agricultural Survey (URT, 1998). Fertilizer prices increased up to the point where, in some parts of the country, correlation between the price of maize and fertilizer became negative (Bilame, 1996). In remote regions such as Ruvuma and Rukwa, use of inorganic fertilizer on maize became unprofitable, changing the spatial distribution of maize-producing areas in the country as regions in the central part of Tanzania (e.g. Dodoma) gained prominence due to their competitive advantage in marketing, being close to main consuming areas (Kherallah *et al.*, 2000; Isinika *et al.*, 2005). Similarly, regions in the north (such as Manyara and Arusha) resumed prominence in maize production due to their comparative advantage of natural fertility, such that maize can be produced using less fertilizer (Skarstein, 2005). Declining fertilizer use was reinforced further by soil fertility decline due to soil mining. Farmers responded by rolling back to subsistence production and diversification out of agriculture – that did not amount to specialization. While fertilizer use is the most documented input, the use of other inputs (improved seed and agrochemicals) also fell. The persistent use of the hand hoe by more than 60% of farming households also continued to be a limiting factor (URT, 2006).

The rate of innovation uptake is a function of several factors, including availability of technologies and the means by which farmers can access and use those technologies. Lack of credit has also been mentioned as a serious bottleneck to technology uptake among smallholder farmers. The weak link between farmers, extension services and research has also been blamed for low uptake of many agricultural innovations, especially in Africa, where both extensions and research services are very weak. In 1997 extension services in Tanzania were decentralized, relegating powers for planning and delivery of these services to LGAs, which fall under the President's Office – Regional Administration and Local Government. The ministries responsible for agriculture and livestock development retained the mandate for policy making, advisory and technical backstopping when called upon to do so (Isinika, 2000, 2003). Within most LGAs, extension services were relegated to the back seat, receiving low priority on resource allocation and in technical upgrading of staff through training, which was compounded by staff attrition, the outcome of a freeze on staff recruitment since the early 1990s (Isinika, 2002) and decimation from HIV and AIDS (Arndt and Wobst, 2002). Thus agricultural extension and research increasingly comprised an ageing personnel, who had very low motivation due to a multitude of factors. All these changes, plus other institutional constraints, culminated in declining productivity of the main staple crops. Something had to be done to reverse the situation.

How did food markets behave? The immediate effect of economic liberalization was to increase the number of private traders, especially in the product market. According to Kherallah *et al*. (2000) the impacts of market reforms in several African countries such as Tanzania have included expansion of private traders, even where parastatal organizations are still active. However, further expansion is constrained by lack of credit and uncertainty about government commitment to the reform (Cooksey, 2003). Bigsten and Danielsson (1999) attribute such uncertainty to Nyerere,[3] who they argue never fully supported the reforms, and he continued to have influence even after his retirement in 1985. It is consequently argued that the reforms were partial, emphasizing price liberalization. Timing and sequencing was not well coordinated; commitment and ownership was low and institutional capacity was weak. Cooksey (2003) none the less admits that maize market liberalization has been successful, and the availability of maize has kept pace with demand.

In general, the market impacts of the policy reforms have improved market integration as vertical linkages with traders and exporting firms have facilitated financing of crop purchased, especially for rice. None the less, the level of their investment in food markets has been low, with little evidence of specialization in service delivery (such as storage) to facilitate marketing. Transport is often a bargained-on-the-spot market. The analysis by Kherallah *et al*. (2000) for the

[3] Julius Nyerere was the first president of Tanganyika, which gained independence in 1961. In 1964 Tanganyika formed a union government with Zanzibar to form Tanzania. Nyerere retired as president of Tanzania in 1985.

post-SAP period showed that markets, not only in Tanzania but in many African countries as well, remained risky, personalized and cash based. There were numerous traders but many of them lacked experience, and oversupply forced many of them to exit from distribution.

On a positive note, the reforms reduced inflation from over 30% during the early 1990s to single digit by 1999 (7.9%), declining further to 4.6% in 2001 (Ratasitara, 2004). But inflation has since crept back to double digits, being 12.2% in December 2009. Other positive impacts are: reduced fiscal burden, improved timing and delivery of inputs, and facilitated regional trade in food crops. There has also been some increase in farm prices but with reduced marketing margins, especially for food crops (Kherallah *et al.*, 2000). Isinika *et al.* (2005:210)[4] also reported that 54% of the respondents considered maize prices to have improved since their households were formed, but the study found little evidence of market integration happening for food crops in general, except in accessible areas. In the case of rice, however, the majority (58%) of respondents from the same study reported improved prices and market integration following upgrading of transport infrastructure, but profitability had decreased as input prices had gone up.

Policy Reversal

By the end of the last millennium (1990s), food production, especially on a per capita basis, was stagnant or declining. The market reforms did not induce smallholder farmers to specialize or to use improved technologies as envisaged. Nor did the reforms solve the underlying problems of credit availability and poor infrastructure for transportation, communication and irrigation, confirming the post-Washington Consensus that macroeconomic stability, trade liberalization and getting the price right is not enough (Stigltz, 1998a,b). During the budget of 2003/04, the government announced the intention to restore subsidies for fertilizer. Maize and sorghum seed have also been subsidized since 2005 (Isinika, 2009). Tanzania joined several other African countries (Ghana, Malawi, Nigeria, Zambia) that have taken similar steps. In 2008 parliament passed the Fertilizer Act to provide for more effective regulation of the fertilizer industry, including promoting more effective private sector participation while ensuring quality and adherence to standards.[5] Other countries, including Kenya and Zimbabwe, are also reported to exercise varying degrees of market intervention policies (Minot and Benson, 2009), such as marketing boards, development programmes and projects (Cooksey, 2003). There is agreement that the span of 10 or 15 years is probably too short for the first generation of reforms, focusing on prices, to have their full impacts felt through the economy, especially in Tanzania, where the economy is still at the pre-industrial stage.

[4] This paper derives from the *Afrint I* microstudy for Tanzania, for cross-sectional data collected in August–September 2002.

[5] http:/www.bunge.go.tz/POLIS/BTS/general/GENERAL_FR.asp?fpkey. Accessed 20 December 2009.

By the Abuja declaration (2006), African policy makers resolved that member states should grant targeted subsidies in favour of the fertilizer sector (AU, 2006). Thus, the number of African countries resorting to restore subsidies is likely to increase. Subsidies are being justified, first, on efficiency grounds. It is argued that, following the decline of fertilizer use in many African countries after structural adjustment, subsidies can help farmers to reach optimum rates, such that additional farm income exceeds the cost of subsidy programmes. Second, on equity grounds, it is argued that subsidies may be the most effective way of reaching the poor (Minot and Benson, 2009). In Tanzania, the process of policy reversal seems to be continuing. In October 2009, the government passed a bill to establish a board, which will handle mixed crops – mainly food crops. This board is expected to play a role similar to the defunct General Agricultural Products Export Corporation (GAPEX), which collapsed during the 1980s, along with other agricultural parastatal organizations.

There are differing points of view on whether such policy reversal is the right or wrong move. Cooksey (2003) argues that patronage, cronyism and rent seeking, as well as the desire of governments to go back to the project mode, motivated the reversal. Meanwhile, others (Kherallah *et al.*, 2000) argue that what is needed is not state withdrawal from the market but an accountable and determined developmental state that walks a balanced line to pursue a portfolio of instruments which stimulate long-term dynamic growth while minimizing negative distributional impacts. For example, governments can use input and output price ratio as a policy instrument – not to be determined exclusively by the market. Other policy options could be construction of roads and irrigation infrastructure, storage facilities, providing cheap credit, supporting cooperatives and establishing other supporting institutions. In the next section, we look at how the production and productivity of food in Tanzania has performed, following policy reversal, which restored some direct intervention of the government in the market.

The Impact of Policy and Institutional Change

The period of policy reversal (post-2002) also covers a period when the government of Tanzania is expected to conform to and meet targets set by other regional and global frameworks to which Tanzania is a signatory. Under the Comprehensive Africa Agriculture Development Program, which is coordinated by the New Partnership for African Development, the target is to achieve 6% annual growth rate for agriculture. To achieve this, countries are expected to have reached at least 10% annual budgetary allocation by 2010. Agriculture is defined to include crops, livestock, forestry and fishing.[6] The government strives to align national policies and strategies to the MDGs. In relation to food security it is MDG1 that is most relevant, whose aim is to reduce by half, between 1990 and 2015, the proportion of people living under extreme poverty (less than US$1/day). This goal forms an important component of the

[6] http://www.africa.union.org

National Strategy for Growth and Poverty Reduction, where the target is to reduce the proportion of people living below basic needs from 38.6% in 1990 to 10.3% by 2015, and to reduce the proportion of people living below basic food needs from 21.6% in 1990 to 10.8% in 2015. Levels of achievement by the year 2000 were 35.7% and 18.7%, respectively (Volker, 2005). Looking at how Tanzania has fared in terms of resource allocation from the public sector in support of agriculture, it is obvious that there is a need to leverage more resources from the private sector so that agricultural transformation and poverty reduction can happen at the intended pace, as envisaged.

Has Tanzania attracted more investment into agriculture from local and external sources? A study by ESRF (2008) notes that the current public financial support to agriculture is low compared with regionally and globally. In 2004 spending on agriculture as a share of total public spending was 2.3%, compared with over 10% in transforming countries during the 1980s, when they experienced their agricultural growth spurt (World Bank, 2008 cited by ERSF, 2008) as shown in Table 12.2. Isinika (2009) similarly reported low levels of spending, especially in real terms. Spending on agriculture as a share of agricultural GDP is equally low; being 1.3% in 2006, compared with 4% in other developing agriculture-based African countries, including Kenya and Uganda (4.1%), Malawi (7.4%), Zambia (8.3%) and Zimbabwe (9.3%).

In Tanzania, the ASDS is facing a financing gap, being funded at less than 50% of the original plan since its commencement in 2005. Future financing for ASDP looks equally grim, facing a gap of up to 52.5% of the government commitment over the life of the programme up to 2014 (ESRF, 2008; Isinika, 2009). Funding from development partners, who are expected to cover 21.7% of the ASDP cost, has been equally lagging and is actually under threat, as some donors opt to switch from the initial sector-wide funding framework back to the programme/project mode. In principle, public funding is expected to leverage private sector investment from local and external sources, such that in the medium and long run the private sector drives agricultural transformation, but this is not happening at a desirable rate.

Table 12.2. Public spending in agriculture-based countries. (Adapted from: World Bank, 2008.)

	Agriculture-based countries		Transforming countries		Urbanized countries		Tanzania	
Category of spending	1980	2004	1980	2004	1980	2004	2004	2006
Public spending on agriculture as a share of total spending (%)	6.9	4	14.3	7	8.1	2.7	2.3	1.9
Public spending on agriculture as a share of total GDP (%)	3.7	4	10.2	10.6	16.9	21.1	1.4	1.3
Share of GDP in agriculture (%)	28.8	28.9	24.4	15.6	14.4	10.2	26.1	25.9

In order to attract more private sector investment, therefore, in 1997 the government established the Tanzania Investment Centre under the Investment Act, to unify and streamline investment incentives (Pigato, 2000). Although Tanzania is listed among African countries which have attracted a fair share of foreign direct investment (FDI), such investments have generally gone to extractive industries and tourism, which have limited backward linkages compared to agriculture (Volker, 2005; Msuya, 2007). In agriculture, most investments have gone into traditional cash crops rather than food crops. Although the government has set up incentive packages including tax holidays,[7] economic processing zones and privatization,[8] investments into agriculture have been hampered by weak physical infrastructure (transportation, communication and energy), low quality of labour, accentuated by deteriorating education and health, and lack of back-up services for enterprises. Other factors, also found in other African countries, are corruption, lack of access to global markets, lack of access to finance, high cost of doing business, excessive taxation and weak tax regulatory framework and policy uncertainty (Asiedu, 2003; Garbe-Madhin, 2006).

In the case of Tanzania, it has been argued that the current combined levels of public and private investments are too low compared to Asian countries during the 1970s, at the height of their Green Revolution, where expenditure on agriculture was up to 20% of government spending, on average, in some countries (ESRF, 2008). Some African countries have exceeded their 10% commitment to meet the Maputo Declaration but Tanzania is lagging behind. By 2008 Tanzania had reached only 6.2%, rising to 7.1% in 2008/09 and promising to reach the 10% target in the next budget (2010/11). Despite the shortfall, the current level of government commitment to support agriculture represents an improvement compared to the past. The challenge remains – can these trends be sustained?

An assessment of the first generation of reforms has been summarized by Kherallah *et al.* (2000) as having reduced the fiscal burden, increased competition and improved timing and delivery of inputs in accessible areas. But the reforms did not overcome the underlying problems of credit. The authors go on to suggest that there is a need to address other reasons for the low level of fertilizer use in Africa, including the low volume of imports resulting in high cost, insurance and freight cost of fertilizer, high distribution costs due to poor infrastructure and low population density, low levels of irrigation at less than 5% of the planted area, and lack of credit. In the next section, we assess how these challenges have been addressed in Tanzania, during the post-SAP period and how the interactive effect of markets, institutional and governance reforms

[7] In many cases the tax holidays have been assessed to be generous to recipients but costly for African countries (Pigato, 2000).

[8] The privatization of parastatals began in 1992 as part of economic reforms. By 2007, when the process was completed, 270 companies had been disposed through divesture or disposal of non-core assets. Out of 28 agricultural companies, 21 were privatized to Tanzania nationals, while 7 went to foreigners (Isinika, 2009).

have played out in terms of input and service delivery and their ultimate impact on productivity, production and food security at the household level and self-sufficiency at the aggregate national level.

Access to and Use of Resources

Land use and land tenure

Land is a key input into any agricultural production process in Tanzania. While land that is available to smallholder farmers has not changed since the early 1990s, utilization has increased significantly, imposing pressure on land. This is consistent with findings from the *Afrint* panel study (Ashimogo *et al.*, 2003; Msuya, 2009), where data from Iringa and Morogoro regions show that the average area under maize decreased by 13%, from 1.033 ha/household in 2002 to 0.874 ha/household in 2008 (Fig. 12.2). Meanwhile, area under rice decreased by 5%, from 1.02 ha/household in 2002 to 0.92 ha/household in 2008. The same applies to cassava, where the area was 0.267 ha/household, on average, in 2002, falling to 0.22 ha/household in 2008, representing a 17% decline. Considering the population growth and competing use of land, the trend towards land scarcity should be expected to increase. Whether this trend will encourage more agricultural intensification remains uncertain due to partial implementation of the Land Act no. 5 of 1999.

The Land Act no.5, which governs the administration and management of village land, requires all villages to be mapped and titled before villages can issue customary titles within their boundaries. Since 2003, when the land laws became operational, only a few villages have been demarcated and mapped in most regions, and even less have title deeds (Kaduma, 2005; Ashimogo, 2008). This has encouraged encroachment into village land by serious and speculative investors. There are some investors who have been invited by local governments to

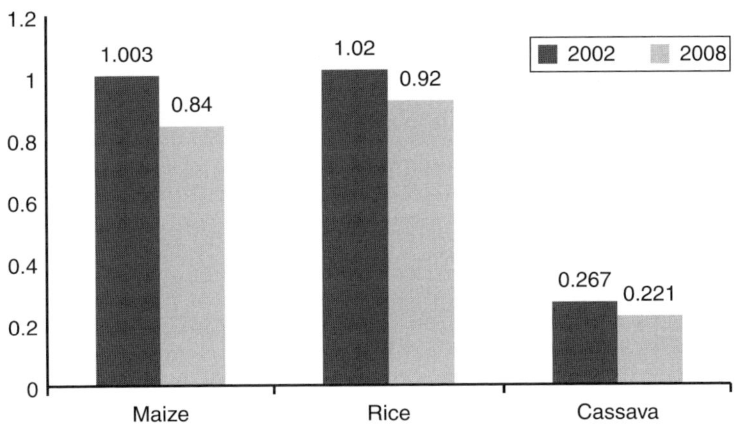

Fig. 12.2. Average area under crops (ha/household). (Adapted from: Msuya, 2009.)

invest in the production and processing of biofuels, as happened recently in Rukwa region (Kiwele, 2009). Other investors have been invited to participate in food production. For instance, Saudi Arabia has requested to acquire 500,000 hectares of land to produce food for exporting to their country. The Food and Agriculture Organization of the United Nations (FAO), among others, has warned developing countries of the dangers of such land-grabbing (Braun and Meinzen-Dick, 2009). It is evident that, as competition for land intensifies, the land tenure system is not robust enough to ensure availability of land for food self-sufficiency.

Tools and implements

The absence of key productive assets such as draught animals and implements has been identified as another major constraint for agricultural productivity (Winters *et al.*, 2004). In Tanzania, the hand hoe is the most dominant agricultural tool, used by the majority of smallholders for cultivation and weeding, accounting for 56% of the planted area, followed by oxen (32%). Tractors account for only about 4%, while 8% of the planted area falls under no-till. Animal-drawn technology (ADT) use is most common in Shinyanga region, where about 65.4% of the planted area was cultivated using ADT. Use of oxen or donkeys is low in Morogoro (9%) and moderate in Iringa (35.6%), but hand cultivation is common in both regions, being used by over 60% of the households. Data for *Afrint II*, which was collected in 2008, shows little improvement (Msuya, 2009). About 75% of the maize farmers used the hand hoe during the most recent harvest (2007), while 22% used ox ploughs and only 3% used tractors. More respondents (96%) used the hand hoe in Morogoro, compared to 59% in Iringa. In the case of rice production, 81% of the farmers used hand hoes and 19% used tractors, while all farmers who planted cassava used hand hoes. There was high and significant positive correlation between households that practised lowland irrigated rice and use of tractors for land preparation (Msuya, 2009).

The hand hoe and use of other tedious and taxing farm processes have been blamed for luring rural youths away from farming. Consequently, the farming population is fairly old. The average age of respondents from *Afrint II* was 45 years (Msuya, 2009), having only 5 years of schooling. While the years of schooling have not changed compared to a similar study 5 years earlier (Ashimogo *et al.*, 2003), the average age has increased by 3 years, which is consistent with the observation that younger people often do not choose to engage in farming.

The government has expressed the desire to replace the hand hoe with more modern technology, especially for land preparation. Tools for weeding, especially in rice production, are also being promoted. Under the ASDP, district councils have been encouraged to increase the number of power tillers available to farmers, who can buy them in groups or as individuals. Groups can pay 20% of the value to receive an 80% grant from the District Agricultural Investment Fund. The challenge is how to manage a group facility that requires regular maintenance and care. Past experiences under the *Ujamaa* regime provide many examples of failed efforts involving group ownership and management of facilities. The policy of giving priority to groups to acquire jointly owned tools may require re-evaluation before implementation goes too far.

Transport and communication infrastructure

As noted earlier, private investments in agriculture or any other sector are attracted by low cost, which is a function of good transport and communication infrastructure. Data from TANROADS[9] show that the length of trunk and regional roads that are considered to be in good condition has improved from 4081 km in 2002 to 14,764 km in 2005, representing 14% and 51% of all trunk and regional roads for the two periods respectively. Conversely, roads considered to be in a poor condition are reported to have decreased from 14,052 km (49%) in 2002 to 6440 km (22%) in 2006. None the less, lack of suitable infrastructure remains a major limiting factor to development in Tanzania. It has been reported that Tanzania dropped four notches as a favoured destination for foreign direct investment largely due to poor infrastructure and low education of the labour force (*Daily News Tanzania*, 2009). Funds allocated for road maintenance reached a peak in 2002/03 but have declined since then, in both nominal and real terms. Funds for railways and harbours also declined between 2002 and 2006 (Isinika, 2009). Although market access, which is strongly influenced by the condition of rural roads, remains a limiting factor, more respondents of the *Afrint II* micro-study indicate that market access has improved in 2008 compared to 2002 (Msuya, 2009). The proportion of respondents reporting carrying luggage on head loads decreased from 51% to 41%, while those using bicycles increased from 40% in 2002 to 57% in 2008. An insignificant proportion of respondents used donkeys or motorized transportation (Isinika *et al.*, 2005; Msuya, 2009). These data are consistent with most of the farmers selling their crops at the farm gate or within village markets.

Communication by mobile phones is also emerging as an important means of transmitting information into rural areas, especially on marketing. The mobile phone sector has shown significant growth, while the fixed line sector, a monopoly of the state-owned Tanzania Telecommunications Company Limited Company, has remained stagnant since 2000. Mobile telephone penetration currently stands at 30%, growing by 10% in the interval 2006–2009, and destined to grow even faster. Estimates show that a 10% increase in penetration will lead to a 1.2% rise in per capita GDP. Platforms such as Nuru SMS are emerging. This will provide an opportunity for information-sharing for various purposes, including marketing, health and technology, similar to Sokoni SMS of Kenya. Other specific uses for agriculture include making more

[9] TANROADS is a government agency that has the mandate to undertake regular maintenance of all regional and trunk roads. TANROADS is represented in every region. Local government authorities are responsible for maintaining the district and village roads.

efficient crop forecasts and more accurate surveys of commodity and input demand. This new development is expected to open up uncharted opportunities for farmers and traders in agriculture. In Tanzania only 34% of the mobile phone lines are currently used for business, compared to 85% in Egypt and 89% in South Africa.[10]

Irrigation infrastructure

Water is the most limiting factor for food crop production in Tanzania, since agriculture is largely rain-fed. Only 2.7% of the total planted land was irrigated during 2002/03, translating to 211,872 ha on the mainland, of which 77% was irrigated during the long rains and the remainder during the short rains. The number of smallholders practising irrigation was about 240,721, having changed little compared to 1995/6. However, Morogoro and Kigoma regions had experienced significant increase in the number of irrigating farmers, while Dodoma and Manyara experienced the most decline. Comparing with data from the *Afrint* micro-studies (Ashimogo *et al.*, 2003; Msuya, 2009), it seems that the area under irrigation has decreased since 2002. About 48% and 7% of the respondents from Iringa and Morogoro, respectively, reported to have irrigated at least 25% of their maize farm during 2002, compared to only 10% of the maize farmers in 2008. In the case of rice, however, the farmers who practised irrigated lowland rice production increased from 1% in 2002 to 16% in 2008. However, only 1% among them grew more than one crop per year (Msuya, 2009). Considering the current threat posed by climate change, as a result of which some regions of Tanzania are expected to have subnormal rains while others expect to get above-normal rains (Agrawala *et al.*, 2003), developing irrigation is strategically important. After 2 years of implementation, the ASDP review reported that the area under irrigation had increased by 25,000 ha (0.9%), from 264,000 to 289,000 ha from 2006 to 2008 (Mlaki, 2008), representing a 36% increase (from 211,872 ha in 2002). None the less, this new level represents less than 4% of the total planted land.

Services

Besides inputs, tools and implements, farmers also need quality services (extension, research, information, business development, marketing and others) in order to optimize technology use as well as market opportunities. Farmers also need to be involved in planning for their development at various levels so that their input contributes to making the services relevant for them. However, there has been failure in general to integrate research and extension as complementary services, especially at the district level. Districts have been slow to widen the

[10] http:www.telecomsmarketresearch.com/reseach/TMAAAQUQ-Tanzania (accessed 2 December 2009).

scope of service providers by inviting other non-state service providers (private sector, non-governmental organizations (NGOs) and community-based organizations) to participate in service provision through competitive bidding, as required under the ASDP (Ashimogo, 2008; Development Associates, 2008; Matee *et al.*, 2008; Mlaki, 2008). It requires a mindset transformation among local government staff so that they perceive non-state service providers as complementing the limited capacity of the government services rather than competing with public extension staff. The current number of village extension agents represents only 22.4% of the requirement (Isinika, 2009). The government has set up a crash programme to train 3000 additional agricultural technical staff (at certificate and diploma level) within 3 years (2009–2011), who will be hired by local government authorities. This would increase the number of beneficiaries who access agricultural extension services from the current 35% to 45% (Ashimogo, 2008). Meanwhile, studies have consistently shown that traders, input suppliers and neighbours are the most common source of production and marketing information among farmers (Isinika and Mdoe, 2001; Ashimogo, 2008).

Availability of financial services has been another limiting factor. Since 1994, the government established a revolving agricultural input trust fund to fill the vacuum following the collapse of cooperatives during the 1970s and failed attempts to revive them during the 1980s. However, the fund has not lived up to its expectation. Available data show that between 2002 and 2006 only 1130 loans were issued, being less than 300 loans per year, and the beneficiaries have often been well-connected government and political leaders as well as business owners (Isinika, 2009). Other avenues for smallholder farmers to access credit have included Savings and Credit Cooperative Societies (SACCOS), various microfinance institutions, the presidential empowerment fund, local government supported project funds and banks. However, lending from commercial banks has not specifically targeted agriculture (Fig. 12.3). Moreover, interest rates remain too high (15–22%) for most agricultural investments to benefit.

The tax regime and inflation have also had their toll on agriculture. At one time the sector faced 55 taxes, compared to 7 in Zambia, 4 in South Africa and 25 in Morocco (Ashimogo, 2008). The government has removed a number of taxes that undermined agriculture. In addition, there are some agricultural activities that have been zero-rated for Value Added Tax (VAT).[11] These include all unprocessed agricultural produce (but not for local market), industries producing agricultural inputs (fertilizer, fishing gear, pesticides) and a VAT rebate for small agricultural exporters (through cooperative unions or associations). Most smallholder farmers, however, cannot benefit from a VAT rebate since they are not registered (URT, 2007). Further reforms are necessary to liberate the sector, because the tax burden remains relatively high. It has been reported that, despite the tax reforms, agriculture pays 17 times more tax than industry (5% compared to 0.3%). A 10% reduction in taxes to agriculture would raise annual economic growth rate by 0.43% (URT, 2007).

[11] In accordance with the VAT Act of 1997.

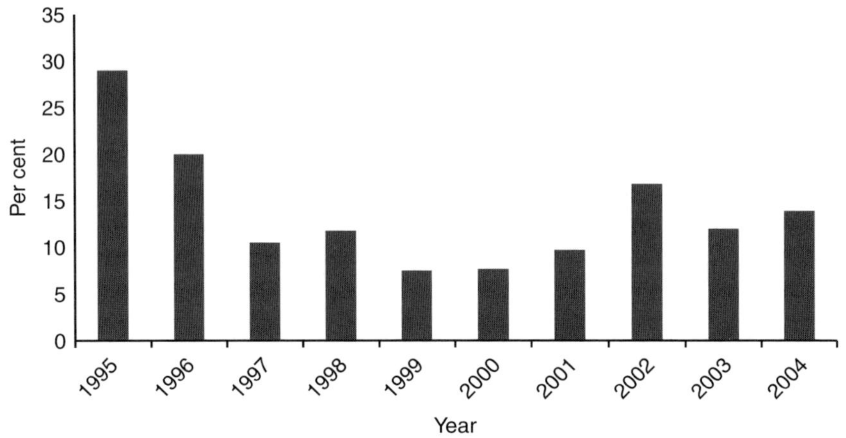

Fig. 12.3. Commercial banks lending to agriculture (1995–2004). (Adapted from: Machude and Nkini, 2005, cited by Ashimogo, 2008.)

Inflation is another vice that must be addressed. Until 2003, inflation had been successfully reduced to below 5%, from an all-time high above 30% during the 1980s and early 1990s. From 2005, inflation began creeping up, reaching 14.8% in June 2008, influenced by rising food and fuel prices globally but also due to increasing borrowing by the government. Although inflation since declined to around 10.3% in 2009, it rose again to 12.2% by December of the same year. This level is too high for healthy economic development. In addition, rising food staple prices have the potential to choke off growth from demand-side linkages (Delgado *et al.*, 1999).

Fertilizer use and other inputs

In Tanzania, the percentage of households using inorganic fertilizer remains very low but is improving in the case of maize. Currently only 9 kg are used per hectare, compared to 27 kg in Malawi, 53 kg in South Africa, 16 kg for SADC and 279 kg in China (URT and TBC, 2009). Although the supply of fertilizer has increased since 2004, following government's deliberate efforts to enhance fertilizer use by restoring the subsidy, supply still lies in the range of 50–80% of what is required (Fig. 12.4).

Results from the *Afrint II* micro-study (Msuya, 2009) show that about 21% of the sampled farmers used artificial fertilizer for maize production during the most recent season, representing an improvement from 2002, especially in Iringa region, where significantly more respondents (16%) indicated the amount of fertilizer used has increased (Table 12.3), probably reflecting the effect of the fertilizer subsidy programme introduced by the government since 2002/03. The fertilizer subsidy has increased tenfold in the last 5 years (since 2005). In the case of rice, not much has changed since 2002, as 88% reported not using any artificial fertilizer

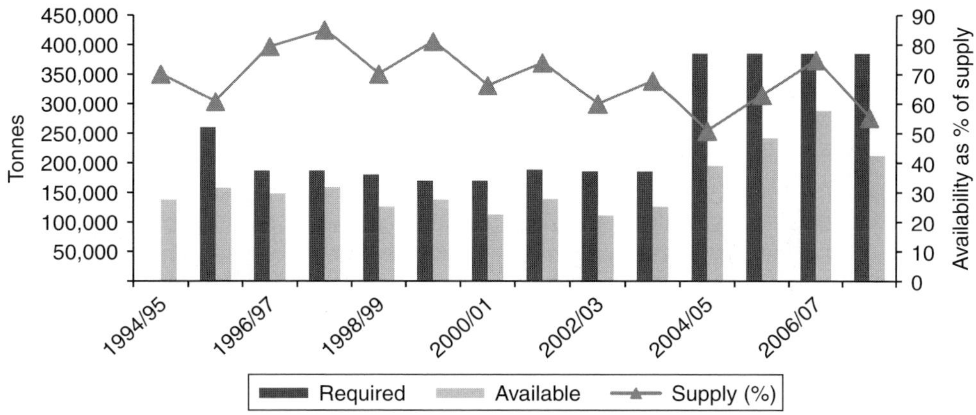

Fig. 12.4. Supply and availability of fertilizer (mainland Tanzania), 1995–2008. (Adapted from: Ministry of Agriculture and Food Security, unpublished data.)

during both periods. No artificial fertilizer was used for cassava production, but manure use increased from 19% in 2002 to 25% in 2008 (Msuya, 2009).

The use agrochemicals is much lower, being applied to only about 9% of the planted area in mainland Tanzania, and even lower for fungicides (2%) and herbicides (2%). However, insecticides are used more often (72% of the applied area) than fungicides (15%) and herbicides (13%) in mainland Tanzania. Iringa and Shinyanga regions had the highest planted area to which agrochemicals were applied, probably due to production of permanent cash crops (cotton and tobacco) (URT, 2006). From the *Afrint II* micro-study (Msuya, 2009), overall, 43% of the respondents used pesticides on maize during the most recent season (Table 12.3), being significantly higher in Iringa (73%) compared to Morogoro (3%), which has not changed since 2002, when corresponding figures were 72% and 3%, respectively. Meanwhile, there seems to be a marked increase in pesticide use for rice production. About 71% of the households who cultivated paddy rice applied pesticides in 2008 compared to less than 33% in 2002. Also one farmer used pesticides for cassava compared to none in 2002. This could indicate rising awareness and availability of this input or a rising trend of pests forcing farmers to look for solutions.

Use of improved seed represents only 7% of the demand (URT and TBC, 2009). Results from *Afrint II* (Msuya, 2009) reflect the continued dominance of traditional seed, which was used by 73% of the maize farmers, 95% of the

Table 12.3. Use of selected improved inputs (Msuya, 2009 and Ashimogo *et al.*, 2003).

	Fertilizer on maize	Improved maize seed	Pesticides on maize	Pesticides on rice
2002	16	12	<10	33
2008	21	27	43	71
% Change	7	15	>33	38

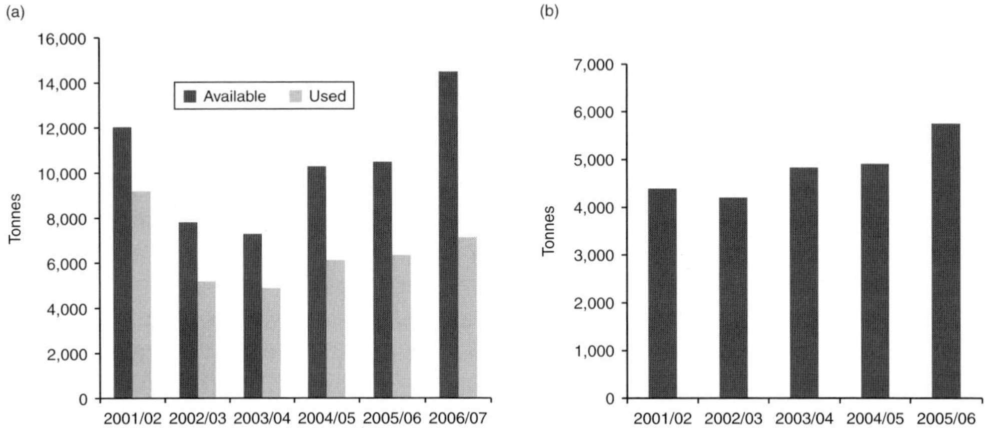

Fig. 12.5. Availability and use of improved seed (2001–2006). (a) All improved seed: quantity available and used; (b) quantity of improved maize seed used. (Adapted from: Ministry of Agriculture and Food Security, unpublished data.)

rice farmers and 20% of the farmers planting cassava. Conversely, only 27% of the respondents planted improved maize seed, being higher in Iringa (37%) than in Morogoro (13%). This represents an increase compared to 2002, when only 12% and 3% of the maize farmers in Iringa and Morogoro, respectively, used improved seed. The trend towards using more improved maize seed has been increasing since 2001/02 (Fig. 12.5), and it should be expected to accelerate further, since distribution of improved seed now also benefits from the transport subsidy, which was extended to cover seed as well since 2006/07.

The use of improved rice seed has also been increasing since 2004/05, but according to Msuya (2009) only one farmer out of 194 in Morogoro region[12] planted NERICA or NERICA descendants. During 2008, the majority of farmers (64%) acquired maize seed from their own stock; 12% obtained seed from neighbours; 21% bought from the market; and 3.2% got maize seed from NGOs. Likewise the main source of rice seed was own stock (76%), followed by other farmers (13%), marketplace (7%) and purchased from extension agents and NGOs (4%).

Production and Food Security Response at Macro Level

Crop production

Considering the policy and institutional environment, let's now look at how farmers have responded in terms of production of the main food crops. Maize is the main staple crop, followed by rice and cassava. Sorghum and millet are important in drier parts of the country. Maize is grown in all regions of mainland

[12] Only one and two farmers reported growing rice in Iringa during 2008 and 2002, respectively.

Tanzania, but regions in the south (Iringa, Mbeya, Rukwa and Ruvuma – the big four) dominate, producing about 40% of the maize in 2005. Production of maize has been increasing very gradually since 2002/03, the combined effect of area expansion and yield increase (Fig. 12.6). The trend in Iringa mirrors the national aggregate, reflecting the impact of fertilizer subsidy and the impact of the big four regions on the national supply of maize. Effective from 2008, Morogoro and Kigoma regions have been added to regions that are focal for food production. It would be expected that the level of food production will improve to match the policy ambition of Kilimo Kwanza (Agriculture First) giving the sector priority, such that Tanzania becomes a net exporter of maize to neighbouring countries (URT and TBC, 2009).

Rice, which is mainly produced in five regions (Morogoro, 19.7%; Shinyanga, 18.5%; Mwanza, 13.6%; Tabora, 10.2%; and Mbeya, 8.5% in 2002/03), has also shown increasing production, largely from area expansion. The yield of rice has not increased consistently for 2 consecutive years, reflecting annual variation of rainfall and the low level of improved technologies (seed, spacing, fertilizer), as discussed earlier. The average yield of rice in 2005/06 was 1.3 t/ha, 34% lower than that obtained in 2001/02 (1.96 t/ha). The production trend for Morogoro region mirrors the national aggregate (Fig. 12.7), reflecting this region's influence, which produces about one-fifth of the rice national supply.

Cassava has similarly exhibited rising production since 2004 (Fig. 12.8), largely attributed to productivity gain. The area under cassava has changed very gradually, growing at 6% annually. Meanwhile, cassava yield has increased from 1.5 t/ha in 2002/03 to 2.1 t/ha in 2005/06, probably reflecting recovery from the devastating attack of cassava mosaic virus during the 1990s, the yield recovery being due to introduced varieties that are resistant to the virus. The production and yield of sorghum has remained almost stable nationally since 2002/03,

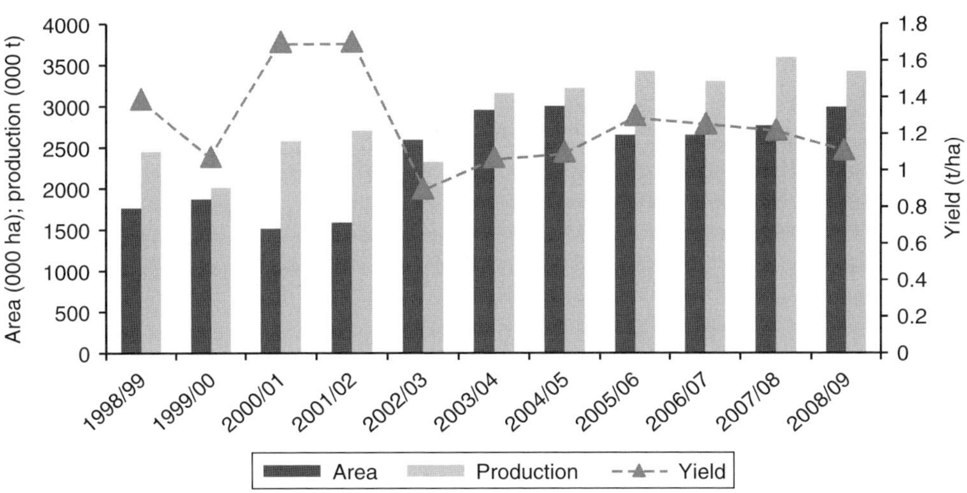

Fig. 12.6. Production and yield of maize in mainland Tanzania. (Adapted from: World Food Program – Tanzania, unpublished data.)

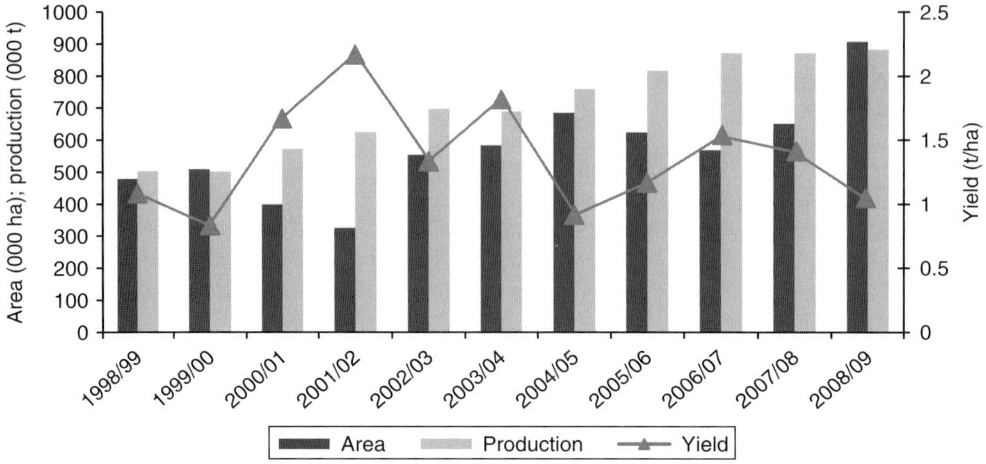

Fig. 12.7. Production and yield of rice in mainland Tanzania. (Adapted from: World Food Program – Tanzania, unpublished data.)

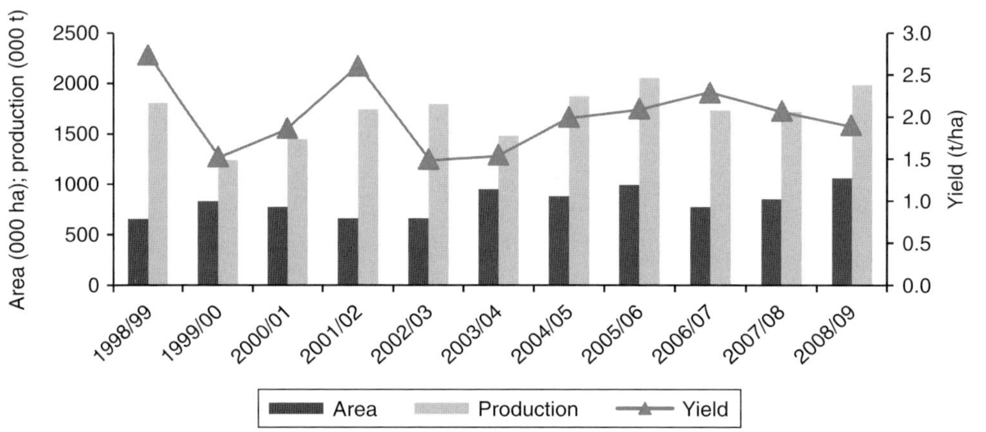

Fig. 12.8. Production and yield of cassava in mainland Tanzania. (Adapted from: World Food Program – Tanzania, unpublished data.)

but there has been a significant yield increase in Morogoro region (Msuya, 2009), where the lead research station for sorghum and millet is located. It is safe to assume that uptake of improved sorghum seed would be higher here.

Food self-sufficiency

Despite the observed gradual production increase, and in some cases a decline, production of the main food staple crops has, in general, kept pace or is slightly ahead of the population growth rate: 2.8% compared to growth of main staple crops (6.4% for maize and 7.3% for rice). Overall, Tanzania is self-sufficient for maize during most years, which is the most important food crop, contributing about 31% of the food supply, according to crop estimates for 2005/06. Cassava comes second (19%), followed by potatoes (13%), sorghum and rice rank fourth (7%) (URT, 2006).

Analysis of data for the period (1994/95–2007/08) shows that food self-sufficiency in Tanzania was achieved in 9 out of 14 years when the self-sufficiency ratio (SSR) was between 102% and 118%, with a gradual declining trend since 1996/97 until 2003/04, then gaining gradual momentum since then (Fig. 12.9). Pockets of food shortage continue to exist in about 8 regions (38%) and 37 districts (33%). During the 8-year period 2001/02–2007/08, the number of regions which experienced food deficit ranged between 5 and 14, while the number of districts was between 13 and 62, being lowest in 2002/03 and highest in 2003/04, which was a dry year. The north-eastern part of the country had also been hit hard by a severe drought during this year (2009), but the southern part of the country had a good crop, ameliorating the effect of the drought (Appendix 12.3).

Is this performance good enough to meet household food security needs and national self-sufficiency? Will this trend in production and productivity enable Tanzania to live up to the ambition of being a net food exporter, as proclaimed under Kilimo Kwanza? While there are indications of a gradual improvement in the macro-production of all main food crops (maize, rice and cassava), the rate of production and productivity growth is not enough to meet set development targets. Let us now examine the response of smallholders farmers to the above-mentioned policy and institutional changes, drawing evidence from the *Afrint I* (Ashimogo *et al*., 2003) and *Afrint II* studies (Msuya, 2009), specifically focusing on three major agriculture transformation constraints: (i) the subsistence nature of markets (measured in terms of percentage of marketed produce, whereby a high degree of subsistence exists if more than 50% of produce is for own consumption); (ii) transaction costs (defined as the total cost of transforming products through space, form and time, along with the costs of arranging

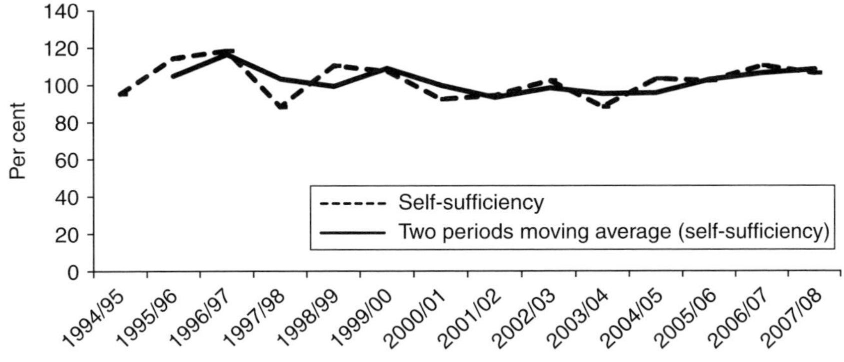

Fig. 12.9. Food self-sufficiency ratio in Tanzania (1995–2008). (Adapted from: URT, 2006 and unpublished data.)

transactions in complete agricultural systems); and (iii) missing market (measured by limitations farmers face in accessing market institutions).

Subsistence nature of maize, rice and cassava markets

Participation of farmers in markets is necessary for structural transformation from subsistence agriculture to an economy based on specialization, exchange and technological innovation. For the *Afrint II* micro-study (Msuya, 2009), overall, 53% of total harvested maize was used for home consumption, while 38% was sold. Cassava, however, was produced mostly for subsistence. Although the proportion of households participating in cassava markets decreased from 21% in 2002 to only 8% in 2008, market participation had changed very little, involving 46% and 47% of the respondents over the two periods, respectively, 85% selling at the farm gate and 10% at the village market. On the other hand, rice was mostly produced for the market. Fifty-five per cent of the harvested paddy was sold. The remainder was used for home consumption (35%), paying hired labour (5%) and others uses (5%). The percentage of paddy sold has also increased, from 49% in 2002 to 55% during the most recent season.

In the case of maize, 52% of the farmers sold maize following the most recent harvest (2007), representing an increase relative to 2002 (Ashimogo *et al.*, 2003), but with regional differences. Actually, the proportion of households from Morogoro region selling maize has dropped from 49% in 2002 to 39% in 2008, but it has increased for Iringa, from 42% to 56% over the same interval. Meanwhile, the proportion of households participating in paddy trade has remained above 70% in the past three recent seasons. Compared to maize, paddy production was more commercially oriented. Of those who sold maize, less than 10% were net sellers. There was a positive and significant ($P > 0.001$) correlation between average per cent of staple food crops sold by households and total household income. Overall 60% of households indicated sale of food staples generated most cash in the course of the last year, followed by sale of other food crops (13%) and micro-business (11%). Sale of food for cash income was more pronounced in Morogoro region, where 81% of respondents said sale of food staples was the major source of income, compared to 39% of households in Iringa, implying more options for diversification in Iringa, probably due to better accessibility for most of the villages within the sample. These findings differ from the conclusion by Kherallah *et al.* (2000), that reforms have been more beneficial to export crops.

As was the case for maize, paddy was mostly sold at the farm gate, as indicated by 77% of households that sold paddy. Fifty per cent sold paddy at the village market, while only 8% sold in markets outside the village. After harvesting, farmers collect and store the maize at home and sell it only when they need cash. Depending on the urgency of household requirements, farmers would sell at any price just to cover the immediate cash needs. There are many factors constraining the participation of farmers in markets but the most important is poorly functioning markets, which squeeze them out (domestic and regional).

In theory, subsistence agriculture is seen as just an early stage of development, which will perish once Ricardo's comparative advantages are perceived and result in wealth-generating trade (Abele and Frohberg, 2003). This does not seem to be happening in Tanzania due to poorly functioning markets.

Transaction cost (institutions, infrastructure and information)

High transaction cost is another major problem facing farmers, often due to high transport costs along with limited market information, lack of product standards and low competitiveness of markets. According to Msuya *et al.* (2009), maize farmers receive about 53% of the final price when a sack of maize (100–120 kg) is sold within the same region and about 45% when it is sold across regions. What is wrong with this is that, first, high transport and handling costs are passed on to consumers, who pay high food prices. Secondly, there is no value added whatsoever along the chain. Limited market information and lack of product standards compound the transaction cost problem.

Limited flow of information also makes market coordination difficult and inefficient. For example, most farmers do not know the selling price before making production decisions. Information asymmetry between sellers and buyers creates room for dishonest traders to take advantage of farmers' lack of price information. If emerging SMS platforms for information sharing, as noted earlier, are encouraged and supported, this problem should decline as mobile phone technology reaches deeper into rural areas. The main source of price information for both traders and maize farmers includes friends and neighbours (Msuya *et al.*, 2009), similar to findings of another study 10 years earlier (Isinika and Mdoe, 2001). Cross-checking with many middlemen is another popular source of information for farmers, even though it is well known that middlemen often collude to offer lower prices to farmers. The public market information system is the least used means of price information because it is often unreliable and inaccessible (Msuya, 2009).

Sometimes, farmers opt to take their produce to markets directly to avoid being cheated by middlemen. Given the high cost of transport due to poor infrastructure, small amounts of produce to be sold and unreliable product markets, the whole exercise is largely inefficient. Meanwhile, the number of middlemen is still increasing, thus adding the squeeze on what the smallholders receive, and when they collude to offer low prices, they effectively operate as a private monopoly/monopsony, thereby nullifying the whole purpose of liberalization (Winters *et al.*, 2004). Farmers are also squeezed on account of quality, when traders do not pay a premium for quality improvement and farmers, in turn, do not invest to present the best-quality products in the market. In the case of the *Afrint II* study (Msuya, 2009), maize quality had minimal impact on the price offered by traders. About 60% of farmers indicated that traders did not pay a lower price for their produce as a result of postharvest quality deterioration. Only 6% of sampled farmers who sold maize indicated they received a lower price for most of their produce due to postharvest quality deterioration. Quality control by rewarding higher prices for better quality is an important incentive for quality improvement. There was minimal quality differentiation in the case of cassava, as 25% of farmers indicated that traders paid a lower price for some produce due to postharvest quality deterioration. Paddy markets, however, were more differentiated by quality. Up to 43% of farmers received much lower prices from traders as a result of postharvest quality deterioration.

Lack of standards, which is quite prevalent in the maize marketing system in Tanzania, distorts the market in two ways. First, as one price is paid for different grades of maize, it removes the incentive for farmers to produce high-quality maize. Smallholders are undermined again, since most of the procurement at lower levels (village) uses volume measures (tins and buckets) instead of weight (kg). These findings differ from the assertion by Rweyemamu (2003), who argued that markets can assure product quality, hence there is no need for commodity boards to issue export permits or register growers, which creates market barriers. This would have nullified the proposed move by the government to establish another commodity board for mixed crops. It should be noted, however, that there are many other countries that maintain regulatory boards for different purposes. For example, the Farm Products Council of Canada has the mission to oversee the national supply and management of poultry and eggs and national promotion of research agencies to ensure an efficient system works in the balanced interest of stakeholders, from producers to consumers. Similar agricultural and marketing boards are also found in many other countries. Often, these boards are formed by stakeholders, to whom they are accountable, even though they may receive subsidies from the government.

Missing markets

Missing and thin markets are common in many African countries due to failure of public good, access failure and transaction failure (Doward, 2005). In addition to problems of poor infrastructure alluded to earlier, missing or thin markets for credit, labour and information on potentially tradable commodities have been cited as constraints to market integration in Africa (Asharf *et al.*, 2008). High contract risks, lack of credit facilities, high price and unavailability of inputs in the staple food crops subsector are signs that input and credit markets are missing in the current market set-up. Price uncertainties remain very high in the maize market. Without contractual agreements farmers are not assured of next season's price and thus tend to produce just enough for subsistence.

For the *Afrint II* study (Msuya, 2009), only 3% of sampled farmers grew maize on the basis of a prearranged contract with private traders. None of the households that sold cassava had a prearranged contract with private traders, while less than 2% of the paddy farmers had contractual agreements with private traders. This is despite efforts by the government to promote and encourage this form of market arrangement as a solution to linking smallholder farmers to markets, while also working to improve farmers' collective voice through cooperatives and improving the policy environment, as discussed earlier. While the contract-farming model involving smallholders has worked somehow for traditional cash crops (sugar, tea, sisal) and commodities that require central handling and/or processing (horticulture, dairy), it is yet to be developed for annual food crops such as maize and rice. The warehouse receipt system has been tested for rice and cashew nuts but the tendency of contracting parties to cheat (both farmers and traders) remains high, largely attributed to failure to enforce contracts. The warehouse receipt system law, which was enacted in 2008, has attempted to tighten such loopholes (Isinika, 2009). It remains to be seen if this will improve contract enforcement in farming as a model for smallholder farmers.

The number of farmers accessing credit is also low; a sign of missing credit markets, limiting in turn the use of inputs. Only 17% of farmers had obtained agricultural inputs on credit in the most recent season (Msuya, 2009), being higher than the national average of 3% in 2002 (URT, 2006). This is a result of eliminating support prices and grain marketing boards (under SAPs), together with a weak private sector. Informal lending institutions, which tend to have very high interest rates, have now become the major source of credit for both traders and farmers of major staple food crops. Even with the reintroduction of fertilizer subsidies in the Southern Highlands zone, farmers find it difficult to access inputs due to very high prices, pushing farmers further towards subsistence. In 2002, Morogoro region had twice the number of households categorized as very poor, compared to Iringa region (Ashimogo *et al.*, 2003). During 2008 the number of households categorized as very poor is almost the same for Iringa and Morogoro regions (Msuya, 2009), probably reflecting the higher dependency on purchased inputs for farmers in Iringa (see Fig. 12.10).

Although the implementation of SAPs increased competition and reduced marketing costs in many cases, its overall impact on farmers has, in general, been negative (Msuya *et al.*, 2009). According to Ponte (2002) and Gabre-Madhin

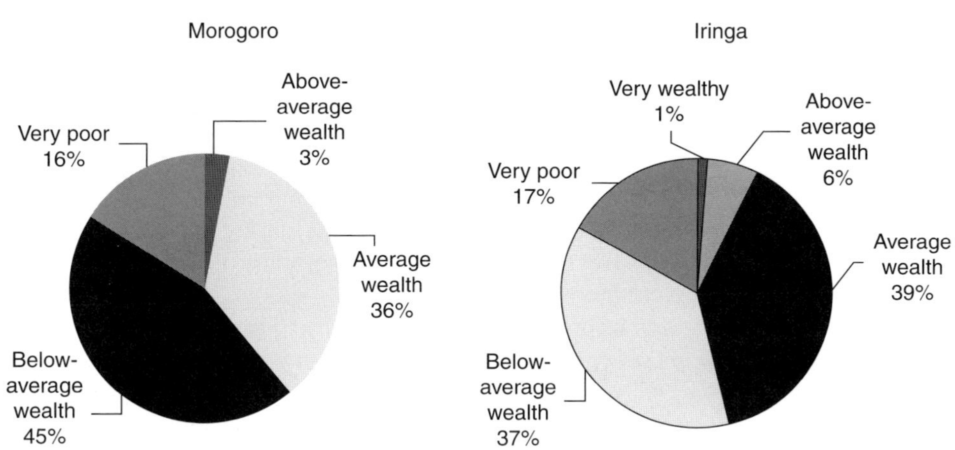

Fig. 12.10. Wealth status of households with respect to assets ownership 2008. (Adapted from: Msuya, 2009.)

(2006), the gap left by the state in secondary distribution and credit provision has not been adequately compensated by the private sector, and both these markets, together with output markets, are altogether missing in many parts of sub-Saharan Africa, including Tanzania. The development of a wide range of private marketing institutions is important for smallholders to improve market access, which will then induce a stronger production response of food and other crops.

The Way Forward: Need for Balanced Reforms

For Tanzania to be a net exporter of food, her aggregate self-sufficiency ratio has to exceed 120% consistently over time, which has not been achieved since 1995. Moreover, per capita production of food has been declining. According to data from the FAO (FAOSTAT), by 1999 per capita food production in Tanzania stood at 108 t, compared to 135 t for Africa and 343 t for the world as a whole. The percentage of irrigated land stood at 3.3%, compared to 3.8% for Africa and 18.3% for the world; intensity of fertilizer use was 5 kg/ha, compared to 12 kg/ha for Africa and 94 kg/ha for the world, and average daily per capita calorie supply, at 1940, remains below the average for Africa and the world. While the growth rate of agriculture has improved, from less than 2% in 1997 and 1999 to around 4% since 2002 (see Fig. 12A.1 in Appendix 2), it remains below the target of 6% set by the Maputo Declaration (AU, 2006; Minde *et al*., 2008). Since agriculture remains the largest sector in the economy, accounting for 24.6% of total GDP by 2007, poor performance of this sector also pulls down overall economic performance. If food production is to play a leading role in poverty reduction, therefore, more needs to be done to improve the performance of agriculture, especially food crops production, which is the largest subsector within agriculture (BOT, 2008).

The first generation of reforms in Tanzania had a strong focus on prices, but it has since been demonstrated that getting prices right is not enough; market development should remain on the reform agenda. For example, fertilizer prices are only one of several factors affecting use (Kherallah *et al*., 2000; Skarstein, 2005; Minot and Benson, 2009). Well-functioning markets, defined by adequate infrastructure, functioning market institutions and better incentives, are vital for agricultural transformation to take place. According to Pingali (1997), for a smooth transformation of agriculture there should be long-term strategies, including investment in rural markets, transportation and communications infrastructure, to facilitate integration of the rural economy. Likewise, to complete the reforms, governments should, in addition, promote good governance and improve the state's capacity to monitor market development in order to encourage market participation and competition, and contract enforcement, as well as property governance, to avoid channelling investments to rent-seeking groups (Pingali, 1997). Other aspects of the reform should include encouraging farmers to diversify, with a focus on specialization, addressing problems of vulnerable groups in remote areas, where price transmission is often poor, and continuing to institute credible macroeconomic policies. Equally important, governments should constantly monitor whether markets exist at all, especially to meet the needs of the poor. According to Winters *et al.* (2004), extreme adverse poverty shocks are often associated with the disappearance of markets, while strong poverty alleviation can arise when markets are created for previously untraded or unavailable goods or services.

With functioning markets in place, several benefits emerge, including rising investment in agriculture and subsequent farm growth, organizing farmers to strengthen their position in the markets to benefit from economies of scale, and improvements of other institutions such as policies that foster trade and reduce transaction costs. All these are expected to have a higher chance of producing desired results when markets are efficient. Currently, agricultural markets are undergoing rapid changes due to globalization, among other things. It is obvious from the preceding discussion that market institutions are the key missing link in government's efforts to transform agriculture. Building market institutions is a long-term strategy expected to reduce investment risk and decrease transaction costs for both farmers and traders by clarifying property rights, enforcing contracts, ensuring quality control and establishing rules of market conduct, among other legal concerns. While incentives and infrastructure components can be spearheaded by the public sector, building of market institutions is a role championed by the private sector. However, for smallholder farmers to benefit from such developments they need to be better organized.

Before liberalization (implementation of SAPs) smallholder farmers were mostly organized under cooperatives. The economic functions of these cooperatives included distribution of subsidized inputs on credit as well as bulking of farm produce. Primary cooperative societies were the main vehicles for assembling produce at the farm gate, while second-tier structures such as the cooperative unions were responsible for intermediate processing and marketing, usually to the parastatal marketing boards (for either export or domestic distribution). The cooperatives enforced quality standards and assured farmers of a market outlet and predictable prices (URT, 2005). However, after liberalization, cooperatives were mostly marginalized and completely abandoned in some parts due to mistrust by farmers regarding government motives and poor governance by managers (URT, 2005). This led to inefficient markets, which in turn forced farmers to act independently in production and marketing of produce, and eventually many cooperatives collapsed, having the most negative impact on the production and marketing of food crops. By improving marketing efficiency, marginal farmers can again participate in the market. Reducing fixed marketing costs or reducing farmer-specific marketing costs, especially for smallholders who are currently not participating in the market, will improve marketing efficiency.

For the *Afrint II* study (Msuya, 2009), only 17% of households (20% in Iringa and 14% in Morogoro) were members of farmer associations. Although the number of SACCOS has increased recently, they have little to do with staple food crops' production and marketing. Rising urbanization and growing consumer power exerts a growing influence on food production and marketing systems. On one hand, demand for processed convenient foods is rising, creating new market opportunities for high-value products as well as staples. None the less, these developments impose new constraints to the conventional markets. Generally, changes in these markets create significant access challenges for farmers, including more stringent quality and standards, increased variability in prices and bulking difficulties, which limit regularity of supply of economic volumes by smallholders as well as resulting in increased transaction costs (Sautier and Biénabe, 2005). The future and prosperity of farmers thus depends largely on how they are organized to overcome such challenges. Given this reality, reorganizing smallholders is crucial for them to benefit from market institutions being developed by the private sector. Organizing traders who still play an important role in local markets is equally important. Social capital remains a significant barrier to entry in wholesale and external trade as well as in transportation. Markets are risky, personalized and cash-based (Kherallah *et al.*, 2000).

According to Msuya (2007), integrated producer schemes designed to develop the capacities of smallholders through extensive provision of extension services and close monitoring of production and quality control are a better form of producer association, especially those focusing on specific value chains, compared to conventional and multipurpose cooperatives (which were mostly politically motivated). It has been observed that creation and development of market institutions is easier for crops whose farmers are well organized. Institutional innovations, such as contract farming, credit associations, group lending and the warehouse receipt systems, are being developed by different actors, including NGOs and development projects (Ashimogo, 2008).

Integrated producer schemes introduce a competitive environment by making prices a public good. In other words, contracts between the two parties will include price information, and such prices will be available to all farmers in the area as a benchmark for decision making. Given such interventions, smallholders will have certainty on prices. Secondly, farmers are motivated to improve product quality if they are rewarded with higher prices. As prices are certain, farmers can now concentrate on lowering transaction costs by achieving economies of scale. This becomes sustainable if smallholders are well organized. Establishment of wholesale markets (auctions) in major buying areas would probably create the same impact by making prices public. Therefore, efforts to foster integration and creation of strong bonds between smallholders and private sector actors within value/supply chains through integrated producer schemes can increase market participation and productivity and hence improve food security.

Conclusion

This discussion set out to assess the impact of policy and institutional reforms for agricultural transformation in Tanzania. Tanzania, like many other African countries, was forced to accept donor prescription for economic reforms during 1986, in order to address declining economic trends in all sectors of the economy. Expectations were raised that the economic downturn would be reversed if recommendations were followed according to the Washington consensus – focusing on macroeconomic stability, market liberalization and getting prices right. Many of the prescriptions for African countries have borrowed heavily from the Asian experience, despite many contextual and temporal differences, such as the effect of globalization.

The first generation of economic reforms were followed by institutional reforms during the 1990s, which covered a number of aspects, including land reforms, local government reforms, tax reforms and other institutional reforms. Specific to agriculture, the government developed a sector strategy (ASDS) and programme (ASDP) to guide transformation. While these reforms brought partial success to realign macroeconomic stability during the 1990s, empirical evidence points that the gains are not strong enough to bring about significant transformation as expected.

The immediate aftermath of the reforms was to increase the participation of actors from the private sector. In agriculture these actors sought opportunities in the provision of inputs (fertilizer, pesticides and farm implements). There have also been improvements in credit availability. However, the grace period did not last long. As soon as all direct and indirect subsidies were removed in 1994, the country experienced a declining trend in food production and productivity. Use of purchased inputs declined, coinciding with reduced opportunities for fallowing as pressure on land increased, creating room for further soil fertility decline due to soil mining. Investments in agriculture were not increasing at the expected rate, thus limiting the follow-on of public goods (roads, irrigation, research, extension, etc.) and private goods (value addition, communication, transport, etc.). It is now evident that, while some success was recorded, the reforms were not enough to unlock prevailing problems of thin, weakly integrated and missing markets for credit, inputs and outputs. The reforms could also not respond in the time of price volatility emanating from globalization since the 1990s. Critics have blamed such failure on half-hearted partial adoption of the reforms. Others point to inadequate time for the full impact of the reforms to play out.

Considering the gravity of the declining production threat, something had to be done. The government of Tanzania joined several other African countries to reverse earlier hands-off policies. First, a partial transport subsidy has been restored for fertilizer and improved seed since 2001. Secondly, marketing boards have been retained and more are being formed to oversee coordination of key subsectors within agriculture. Proponents of market reforms lament that such reversal is motivated by the rent-seeking interests of a few at the expense of economic efficiency. There is a counterargument, however, that hands-off is not an optimum solution. What is required is a developmental state that will pursue market mediation in a balanced manner so that private sector participation is supported and enhanced by providing a conducive policy and institutional environment, and necessary public goods and services. This essentially calls for a balancing act to ensure sustainable dynamic growth of agriculture and hence the economy.

Evidence from the post-2000 data shows that, following the policy reversal, something positive is happening. There is improvement in agricultural input availability; some gains are seen in production and productivity, especially for maize, and food self-sufficiency remains marginally stable. There is a rising trend of credit availability and use; some gains are observed for area under irrigation and there are government efforts to increase investments into agriculture, including leveraging resources from the private sector. The analysis shows, however, that the trend rates of all these changes are still too weak to bring about visible sector-wide transformation and sustained dynamic growth.

For these reasons, some recommendations are made in order to sustain the positive gains that have been attained thus far. Considering the importance of food production for poverty reduction, efforts to support agriculture should also focus on supporting food production. Partial reforms have been blamed for the weak results observed until now. It has therefore been recommended that continuing efforts on the reforms should foster long-term dynamic growth so that actors benefit from improving access to technologies along with improving capital goods, economies of scale and competition induced by fully functioning markets.

To overcome the limitations of subsistence production, characterized by autarchy, it is recommended that government should pursue complementary policies, which target small farmers to accumulate assets that will enable them to benefit from opportunities availed by the ongoing economic and institutional reforms. Essentially, efforts should be directed at improving market coordination, including reducing the cost of coordination, enforcement of contracts, enhancing collective action and reducing the risk of all actors in the market. As concluded by Garbe-Madhin (2006):

> the potential for harnessing markets for smallholder agricultural development depends on both market development and addressing challenges of scale, location, assets and power. Building institutions requires tailoring to a country context and product nature, capturing linkages between institutions rather than a piece-meal approach.

References

Abele, S. and Fohberg, K. (eds) (2003) Introduction to subsistence agriculture in central and eastern Europe. How to break the viscous circle? In: *Studies in Agriculture and Food Sector in Central and Eastern Europe*. IAMO Vol. 22, pp. I–VI.

Agrawala, S., Moehner, A., Hemp, A., van Aalst, M., Hitz, S., Simith, J., Meena, H., Mwakfwamba, S.M., Hyera, T. and Mwaipopo, O.U. (2003) *Development and Climate Change in Tanzania: Focus on Mount Kilimanjaro*. Development Cooperation Directorate. A working party on global and structural policies. OECD, Paris.

Alson, J.M. and Hurd, B. (1990) Some neglected social cost? Government spending in farm programs. *American Journal of Agricultural Economics* 72, 149–156.

Arndt, C. and Wobst, P. (2002) HIV/AIDS and labour markets in Tanzania. TMD discussion papers, International Food Policy Research Institute, Washington, DC.

Ashimogo, G.A. (2008) Analysis of the role of the private and public sector to support rural enterprises. Unpublished consultancy report prepared for Oxfam GB – Tanzania.

Ashimogo, G.A., Isinika, A.C. and Mlangwa, J.E.D. (2003) Africa in transition: micro study, Tanzania. Final research report for the *Afrint* research project. Sokoine University of Agriculture, Tanzania.

Ashraf, N., Giné, X. and Karlan, D. (2008) Finding missing markets (and disturbing epilogue): evidence from an export crop adoption and marketing intervention in Kenya. *American Journal of Agricultural Economics* 91(4), 973–991.

Asiedu, E. (2003) Foreign direct investment to Africa: the role of government policy, governance and political stability. Department of Economics, University of Kansas, Lawrence, Kansas.

AU (2006) Abuja declaration on fertilizer for the African Green Revolution. Declaration of the African Union special summit of the heads of state and government, Abuja, Nigeria. Available at http://www.africafertilizersummit.org/Abuja%20Fertilizer%20Declaration%20in%20English.pdf (accessed 30 November 2009).

Bigsten, A. and Danielsson, A. (1999) Is Tanzania an emerging economy? A report for the OECD project 'Emerging Africa'. Available at: http://www.oecd.org/dataoecd/ 40/30/2674918.pdf (accessed 7 November 2009).

Bilame, O.S. (1996) Performance of maize during structural adjustment programmes in Tanzania. MA thesis, University of Dar es Salaam, Tanzania.

BOT (2008) Bank of Tanzania annual report, 2007/08. Bank of Tanzania (BOT).

Braun, J. and Meinzen-Dick, R. (2009) Land grabbing by foreign investors in developing countries – risks and opportunities. IFPRI Policy Brief No. 13, IFPRI, Washington, DC.

Cooksey, B. (2003) Marketing reform? The rise and fall of agricultural liberalization in Tanzania. *Development Policy Review* 21, 67–91.

Daily News Tanzania (2009) Former *Nyumbu* factory to produce power tillers – prime minister. *Daily News Tanzania*, October 2009.

Delgado, C.L., Minot, N. and Courbois, C. (1999) *Agriculture in Tanzania since 1986: Follower or Leader of Growth?* IFPRI, Washington, DC.

Demombynes, G. and Hoogeveen, J.G.M (2004) *Growth, Inequality and Poverty: Simulated Growth Paths for Tanzania. 1992–2002.* World Bank Policy Research Working Paper No. 3432, World Bank, Washington, DC.

Development Associates (DASS) (2008) A review of policies, strategies and institutional arrangements in support of agricultural transformation in Tanzania, focusing at the local level. Unpublished report prepare for Oxfam GB - Tanzania.

Doward, A. (2005) Rural livelihood and making markets work for the poor. A PowerPoint presentation, presented at the Centre for Environmental Policy, Turin, Italy, 20 September 2005.

ESRF (2008) *Agriculture Sector Development Programme Financing Study*. Economic and Social Research Foundation (ESRF). Submitted to Food and Agriculture Organization of the United Nations (FAO), Rome.

FAO (2004) The State of Food and Agriculture 2003/2004 – Agricultural Biotechnology Meeting the needs of the poor? Food and Agriculture Organization of the United Nation's (FAO), Rome. Available at: http:// faostat.fao.org (accessed 24 November 2009).

Gabre-Madhin, E. (2006) *Building Institutions for Markets*. IFPRI, Stockholm, Sweden.

Gibson, C.C. and Knoontz, T. (1998) When community is not enough: institutions and the value in community based forest management in southern Indiana. *Human Ecology* 26, November 4, 621–647.

Giovanni, A. and MacCalla, A. (1995) Assessing the impact of agricultural technology improvement in developing countries in the presence of policy distortions. *European Reviews of Agricultural Economics* 22, 5–24.

Havnevick, K.J., Kærby, F., Meena, R., Skarstein, R. and Vuorela, U. (1988) *Tanzania Country Study and Norwegian Aid Review*. Centre for Development Studies, Bergen, Norway.

Hawasi, F.G.H., Mdoe, N.S.Y. and Turuka, F.M. (1999) Efficiency in fertilizer use among smallholder farmers in Mbinga District. *AGREST Proceedings*. Sokoine University of Agriculture, Morogoro, Tanzania, pp. 72–86.

IMF (1986) Statement by the IMF representative. Paper presented at the meeting of the Tanzania Consultative Group, Paris, 10–11 June.

IMF (1995) Statement by the IMF representative. Paper presented at the meeting of the Tanzania Consultative Group, Paris, 27–28 February.

Isinika, A.C. (2000) Mechanisms for contracting out extension services to different agents. Unpublished consultancy report for the Ministry of Agriculture and Cooperatives, Tanzania.

Isinika, A.C. (2002) Agricultural education in Tanzania. In: Ngugi, D., Isinika, A., Temu, A. and Kitalyi, A. (eds) *Agricultural Education in Kenya and Tanzania (1968–1998)*. Regional Land Management Unit (RELMA) Technical Report No. 25, Regional Land Management Unit (RELMA), Swedish International Development Agency (Sida), Nairobi, Kenya, pp. 56–102.

Isinika, A.C. (2003) Tanzania macro report: addressing national food self sufficiency. Unpublished report for *Afrint I* research project.

Isinika, A.C. (2009) Tanzania macroeconomic report: addressing national food self sufficiency. Unpublished macro-report for *Afrint II* research project.

Isinika, A.C. and Mdoe, N.S.Y. (2001) *Improving Farm Management for Poverty Alleviation: the Case of Njombe District*. REPOA research report 01.1. Mkuki na Nyota Publishers, Dar es Salaam, Tanzania.

Isinika, A.C., Ashimogo, C. and Mlangwa, J.E.D. (2003) Tanzania macro report: addressing national food self sufficiency. Unpublished report for *Afrint I* research project.

Isinika, A.C., Gasper, C.A. and Mlangwa, J.E.D. (2005) From Ujamaa to structural adjustment – agricultural intensification in Tanzania. In: Djurfeldt, G., Holmén, H., Jiström, M. and Larsson, R. (eds) *The African Food Crisis: Lessons from the Asian Green Revolution*. CAB International, Wallingford, UK, pp. 197–217.

Kaduma, F.E. (2005) Public awareness and attitudes on individualized land tenure: A case study in Njombe district in Tanzania. Unpublished manuscript.

Kherallah, M., Delgado, C., Minot, M. and Johnson, M. (2000) *The Road Half Travelled: Agricultural Market Reform in Sub-Saharan Africa*. IFPRI policy report, IFPRI, Washington, DC.

Kiwele, P. (2009) Bioenergy policy implementation Tanzania context. Compete policy conference, 26–28 May, Lusaka, Zambia. Available at http://www.compete-bioafrica.net/events/events2/zambia/Session-2/2-4-COMPETE-Conference-Lusaka-Kiwele-Kenya.pdf (accessed 15 December 2009).

Matee, A.Z., Ngetti, M.S. and Rwambali, E.G. (2008) An assessment of the performance of agricultural extension services delivery under ASDP: a case study of Kilosa and Kilombero districts. Unpublished report for the Embassy of Ireland – Tanzania.

Minde, I., Jayne, T.S., Crawford, E., Ariga, J. and Govereh, J. (2008) Promoting fertilizer use in Africa: current and empirical evidence from Malawi, Zambia and Kenya. Report prepared for the Regional Strategic Agricultural Knowledge Support System (Re_SAKSS) for Southern Africa, based at the International Water Institute, Pretoria, South Africa.

Minot, N. and Benson, T. (2009) Fertilizer subsidies in Africa. Are vouchers the answer? IFPRI Issue Brief 60. Available at: http://ideas.repec.org/p/fpr/issbrf/60.html (accessed 25 November 2009).

Mlaki, H. (2008) *The Agriculture Sector Development Programme (ASDP): Framework and Implementation Status*. A report presented at the annual learning event of the Agricultural Non-State Actors (ANSAF) Forum, Morogoro, Tanzania.

Monke, E.A. and Pearson, S.R. (1989) *The Policy Analysis Matrix for Agricultural Development*. Cornell University Press, London.

Msuya, E.E. (2007) Analysis of factors contributing to low FDI in the agriculture sector in Tanzania. Proceedings of the 10th International Conference of the Society for Global Business and Economic Development. *SGBE* IV, 2846–2865.

Msuya, E.E. (2009) Afrint II microstudy – Tanzania. Unpublished research report.

Msuya, E.E., Hisano, S. and Nariu, T. (2009) An investigation into commercialization constraints facing smallholder farmers in Tanzania. *Journal of Agricultural Economics Society of Japan*, Proceedings of the 2009 Agricultural Economics Society of Japan, pp. 551–558.

North, D.C. (1993) Institutional change: a framework of analysis. In: Sjöstrand, S.E. (ed.) *Institutional Change. Theory and Empirical Findings*. M.E.Sharpe, London.

Pigato, M. (2000) Foreign direct investment environment in Africa. Unpublished paper. Available at: http://www.ifc.org/ifcext/flas.nsf/AttachmentsByTitle/FDI_in_Africa (accessed on 25 November 2009)

Pingali, L. (1997) From subsistence to commercial production systems: the transformation of Asian agriculture. *American Journal of Agricultural Economics* 79, 628–634.

Ponte, S. (2002) *Farmers and Markets in Tanzania*. Mkuki na Nyota Publishers, Dar es Salaam, Tanzania.

Ratasitara, L. (2004) *Exchange Rate Regimes in Tanzania and Inflation*. African Research Consortium Research Paper 138. The Regal Press Kenya Ltd, Nairobi, Kenya.

Rweyemamu, D. (2003) Reforms in the agricultural sector: the Tanzanian experience. Economic and Social Research Foundation – Tanzania. Unpublished research report submitted to the Global Development Award Competition, 2003.

Sautier, D. and Biénabe, E. (2005) The role of small-scale producers' organizations in addressing market access. In: Almond, F.R. and Hainsworth, S.D. (eds) *Beyond Markets Work for the Poor*. Proceedings of International Seminar. CPHP, Westminster, London, pp. 69–85.

Shauri, V. (1995) Village titling and its legal ramifications in development of rural land tenure in Tanzania, LLM thesis, University of Dar-es-Salaam, Tanzania.

Skarstein, R. (2005) Economic liberalization and smallholder productivity in Tanzania. From promised success to real failure, 1985–1998. *Journal of Agrarian Change* 5, 334–362.

Stiglitz, J.E. (1987) Some theoretical aspects of agricultural policies. *World Bank Research Observer* 2, 43–60.

Stiglitz, J.E. (1998a) Redefining the role of the state. Paper presented at 10th anniversary of the MITI Research Institute (Tokyo, Japan).

Stiglitz, J.E. (1998b) *More Instruments and Broader Goals: Moving Towards the Post-Washington Consensus*. Presented as the WIDER Annual Lecture, at the World Institute for Development Economics Research in Helsinki, Finland.

Uma, L. (1989) Sources of growth in East African agriculture. *World Bank Economic Review* 3, 119–144.

URT (1994) *Report of the Presidential Commission of Inquiry into Land Matters. Volume 1, Land Policy and Land Tenure Structure*. United Republic of Tanzania (URT) Ministry of Lands, Housing and Urban Development, Government of the United Republic of Tanzania and the Scandinavian Institute of African Studies, Uppsala, Sweden.

URT (1998) *Expanded Agricultural Survey 1996/97, Mainland Tanzania. Volume II, Main Report*. United Republic of Tanzania (URT) Ministry of Agricultural and Cooperatives Development and National Statistics Bureau, Dar es Salaam, Tanzania.

URT (2001) *Agricultural Sector Development Strategy*. United Republic of Tanzania (URT), National Printing Company (KIUTA), Dar es Salaam, Tanzania.

URT (2005) *Poverty and Human Development Report 2005*. United Republic of Tanzania (URT), Mkuki and Nyota Publishers, Dar es Salaam, Tanzania.

URT (2006) *National Sample Census of Agriculture (2002/03): Small Holder Agriculture, Volume II: Crop Sector – National Report*. United Republic of Tanzania (URT) National Bureau of Statistics, Dar es Salaam, Tanzania.

URT (2007) Impact assessment of tax reforms on agriculture in Tanzania. United Republic of Tanzania (URT) unpublished report.

URT & TBC (2009) Towards a Tanzanian Green Revolution: policy measure and strategies. United Republic of Tanzania (URT) and Tanzania Business Council (TBC). Unpublished policy document.

Vaidyanathan, A. (2000) Agricultural subsidies. *Agricultural Situation in India* 57, 261–265.

Volker, T. (2005) *Tanzania Growth Progress and Success in Reducing Poverty*. IMF Working Paper wp/05/35, IMF, Washington, DC.
Winters, A., McClloch, N. and Mackay, A. (2004) Trader liberalization and poverty: evidence so far. *Journal of Economic Literature* XLII, 72–115.
World Bank (1992) A vision for sustained growth in Tanzania. Paper presented at the Tanzania Consultative Group Meeting, Paris, 29–30 June.
World Bank (1994) *Tanzania Agriculture*. World Bank, Washington, DC.
World Bank (2000) *Tanzania Agriculture since 1986 – Follower or Leader in Growth*. World Bank Report 20639. World Bank, Washington, DC.
World Bank (2008) *World Development Report*. World Bank, Washington, DC.
Yoshida, M. (2005) Land tenure reform under economic the liberalization regime: observations from Tanzanian experience. *African Development* 4, 139–149.

Appendix 1

Table 12A.1. National food security statuses. (Adapted from: URT, 2006 and unpublished data.)

Measure of food security status	2001/02	2002/03	2003/04	2004/05	2005/06	2006/07	2007/08	Average
National level (SSR%)	94	102	88	103	102	110	106	101
Number of food-deficit regions	6	4	14	6	10	5	5	7
Number of food-deficit districts	15	13	62	41	41	50	–	37

Appendix 2

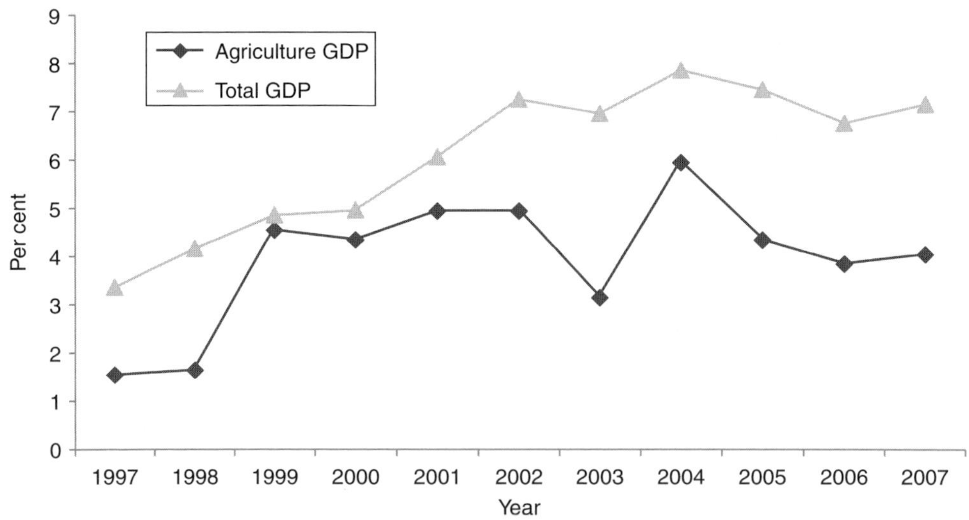

Fig. 12A.1. Agricultural and total GDP growth rate. (Adapted from: BOT, 2008.)

Appendix 3

Food insecurity severity scale

Generally food secure

Moderately food insecure

Highly food insecure

Fig. 12A.2. Tanzania food security outlook: October 2009–March 2010. (Adapted from: Famine Early Warning Network System Network FEWSNET, September 2009.)

13 Focusing on the Majority – Rethinking Agricultural Development in Mozambique

Peter E. Coughlin

EconPolicy Research Group Ltd, Maputo, Mozambique

Unlike most African countries, Mozambique possesses a huge coastline, vast tracts of virgin arable land and no landless peasants, but, despite these advantages, it suffers extreme poverty. Colonialism trained exceptionally few Africans and left an infrastructure appropriate for serving the metropolis but, in large measure, inappropriate for the economic development of an independent state. The revolutionary war and, in their last days, the flight of most Portuguese (including manufacturers, merchants and commercial farmers) were quickly followed by a civil war, abetted initially by Rhodesia and later by apartheid South Africa. The warfare systematically destroyed the rural commercial and transport networks and educational and health systems and killed or displaced millions, thus aggravating the economic disruption. During this period, until 1987, government policies encouraged huge, inefficient state farms, crop price and movement controls, and Agricom's (the centralized agricultural marketing board) centralized crop purchasing, all of which greatly discriminated against small farmers. Given the violence, bloodshed, massive flight of refugees and repeated crop failures due to mismanagement and a prolonged drought, many districts suffered starvation and gaunt, naked poverty. The socialist experiment with state management of agriculture had failed. Breaking with the past, the Economic Rehabilitation Program started in 1987, to drop price and purchase controls and to liberalize the agricultural sector (Alfieri *et al.*, 2007:5). The economy responded: annual gross domestic product (GDP) growth averaged 5.4% between 1987 and 1989 and 'inflation fell from 160% in 1987 to 35% in 1991' (Cravinho and George, 2007:802).

Finally, with peace in 1992, most refugee farmers headed back home, and between 1992 and 2004 real annual agricultural output averaged 6.2% and growth in gross national income averaged 8.1% per annum between 1993 and 2003, 7.2% in 2004, 7.7% in 2005, 10% in 2006, 7.4% in 2007 and 6.8% in 2008, the latter despite the global financial crisis and economic recession (INE, 2008:10). But, besides the resettlement of refugees, the huge tasks of rehabilitating, reforming and, eventually, expanding or creating systems, both physical and human, also began. The process has been heavily state and donor driven. Within the framework of structural adjustment, the state rapidly privatized state enterprises, withdrew from directly productive activities and eliminated nearly all controls over prices and markets. It also guided an evolving process of investment in infrastructure and systems, including decentralization, capacity building and progressive reform, gradually enabling the country to merit confidence and support and to attract investment, first in the richer south and, more recently, in the poorer central and northern parts of the country. Overall, growth is now quick and the economic possibilities are interlinked and fast evolving.

Annual inflation in consumer prices fell from 56.5% in 1995, averaging 12% from 1996 to 2004, and, December to December, was 13.1% in 2005, 8.1% in 2006, 12.1% in 2007 and 11.8% in 2008. Thus, throughout this period, monetary and fiscal policy has kept growth strong and inflation moderate and fairly steady.

Though pressed, in 2008 and early 2009, by soaring and, later, slightly abated prices of food and chemical fertilizers and hit by falling export earnings, Mozambique has maintained – despite the crisis – a high rate of GDP growth. Bolstered by investment and continuing aid inflows and donor confidence in its macroeconomic policies, the government has used redistributive policies (via tax reductions, minimum wage increases, direct relief) to mitigate temporarily the impact of international market fluctuations on the poor. But these external shocks have distributive and strategic implications for Mozambique and beyond.

This shifting dynamic confronts farmers and defines the possibilities and limits of agricultural development, as reflected in their productive and technological choices. Blessed with abundant land but squeezed between low farm-gate prices and high input costs, the majority remain subsistence farmers, selling little or nothing to the market, and those who produce for sale do so mostly without the benefits and risks of modern inputs. Indeed, though agricultural intensification is occurring among the farmers participating in the rapidly expanding contract farming schemes encouraged by the government's policies and infrastructural investments, few farmers outside these zones use pesticides, fertilizers and hybrid seeds. And, except for fallowing, crop rotation and improved seeds or varieties (e.g. for maize, cassava, sunflower and sweet potatoes), even the use of pre-industrial methods of intensification is limited, e.g. composting, manuring, small-scale irrigation, use of nitrogen-fixing crops and integration between land and animal husbandry (Coughlin and Givá, 2009:8, 16). Though extension services now reach all but one of the country's 128 districts, in 2007 only 10% of farmers received information or advice from an extension officer (MAG, 2007b). Moreover, the farmers are cash strapped and fearful of risks, and their response to the extension officers' messages has been disappointing. For most food crops, productivity per hectare has improved little. Except for the initial surge in production due to the onset of peace and stability and the resettlement of refugees in the 1990s, the impact of the government's policies and programmes on farm size and productivity has been small, except for farmers operating within concessionaire schemes that furnish inputs and extension services and guarantee crop purchases. Except for efforts to improve seeds or cultivars, the vast majority of farmers remain outside the scope of any – governmental or non-governmental – programmes aimed at enhancing farm outputs.

The situation, however, is far from static. Roads, electricity, communication and education are expanding. Competition among input suppliers is also growing; agricultural price and supply information is more readily available; and, as primary, secondary and feeder roads are built, traders are penetrating deeper into the countryside, initially as lone buyers and later in more open competition. Despite the problems of monopsony, false measurements and lack of information and negotiating power, the farmers are being gradually enticed by the market, especially for the production and sale of cash crops. And, though nascent and far from uniform, there are initial signs that those investments are affecting both the scope and intensity of the farmers' activities. Moreover, the government has encouraged investments in large-scale multinational sugar plantations, targeting both the domestic and European markets and, more recently, in large irrigated rice schemes organizing numerous small farmers. With this push, rice production grew from 190,000 t in 2008 to 260,000 t in 2009, and the government has plans to produce 400,000 t by 2011, to cut or eliminate its import dependence on rice (Banda, 2009). The government is also betting on vast biofuel projects, especially on semi-arid or previously little-occupied land. Nevertheless, even with all these investments in infrastructure and projects, most farmers will remain untouched: tiny, traditional and impoverished. For example, between 2002 and 2008, though average annual rural income increased from US$890 to US$1019, the median fell from US$339 to US$287. 'Most people became poorer but the best off became richer' (Hanlon, 2010 based on preliminary data from TIA, 2008). Thus, for the majority, hope is eclipsed.

Might that change? Are circumstances shifting to permit them, too, to be more easily and effectively covered by programmes aiming to improve the technologies they use, thereby increasing their productivity and, in many cases, farm size? Might the world financial crisis and economic recession benefit many small farmers? High prices for food and chemical fertilizers and, once the recession is over, for petroleum will inevitably imply more costly imports plus a redistribution of income from urban dwellers to most farmers who, on net, sell food. The yet faster rise in the prices of inorganic fertilizer and petroleum will also squeeze the profits of farmers who use these inputs (Connolly and Athenry, 2008; Skalsky, 2008:1164). Together with higher short- and medium-term crop prices, this will improve the relative profitability and decrease the risk of investments in alternative farm technologies (e.g. organic fertilizers, animal-drawn transportation, small-scale irrigation and improved seeds and storage techniques).

Low profitability, capital scarcity and risk aversion have long been the major impediments for farmers to accept and implement extension messages, even when those are deemed technologically and economically appropriate. Might much better profitability and greatly decreased risks spur a paradigm shift in the way the government's extension workers function? In this context, farmers may become far more receptive to enhanced agricultural techniques that depend little on imported inputs (e.g. organic fertilizers, animal traction, small-scale irrigation, improved seeds and better storage), especially if extension workers were firmly backed by numerous local- or district-level projects to extend credit to organized groups of farmers, much in the way that the cotton and tobacco concessionaires operate, albeit on a smaller scale. With the significantly higher profitability of agricultural investments, might it also be time to have a much more integrated, pro-active and dramatically better-funded approach for the development of agricultural value chains, including improved farm techniques and storage, crop marketing by farmers' associations and agro-processing, along with supportive clusters?

Might crisis bode opportunity?

Research Focus and Methodology

'As part of a ten-country study examining the trends in agricultural intensification among small farmers and searching out what works well or badly in Africa and why' (including some understanding of the driving forces behind technological change), the present study for Mozambique combines an extensive review of the literature with results of surveys in 2005 and 2008 from ten villages: four in the north, four in the centre and two in the south, an area hit by prolonged drought.

> Considering their agro-ecological characteristics, the villages had an agricultural potential ranging from low to high. The selection criteria deliberately excluded disastrous and extremely successful examples, preferring instead to choose districts and villages representative of the gamut of the most common experiences, the best of which might serve as useful models for widespread adaptation.
>
> (Coughlin and Givá, 2009:1)

In each of the ten villages, approximately '40 households were selected and administered a structured questionnaire while another questionnaire was used to interview village leaders' (Coughlin and Givá, 2009:1; see also Coughlin, 2006; Mole, 2006). The 2008 study attempted to revisit the same households that had been interviewed in 2005, although, if for any reason (e.g. death, migration) 'a household could not be contacted, another was selected to replace it', preferably with a household descended from the absent one. 'The district agricultural directors or, later, the chief economic officers as well as the local extension workers were interviewed to understand better the policy, infrastructural, climatic, commercial and other agricultural factors shaping the context in which the villages operate' (Coughlin and Givá 2009:2).

The Micro, Macro and Tax Environments Affecting Agricultural Production and Productivity

Despite abundant and fertile land, Mozambique's farmers typically maintain tiny plots and achieve low productivity. Snared in poverty, few escape. The explanations, in part, can be found at both micro- and macroeconomic levels, including the recent grave impacts from the international environment and the way that tax reform affects agriculture.

Micro-level performance

Rare for a sub-Saharan country, Mozambique has vast tracts of unused forests and cultivatable land, with most regions getting more than 800 mm annual rainfall and few being semi-arid. Seventy-eight per cent (62 million ha) of the land has forest vegetation;[1] 46% is cultivatable, though only 10% is cultivated, and 97% of that by smallholders (World Bank, 1996, 2003; Issufo, 2003:1; MADER, 2003a:14). Theoretically, that could be 12–13 ha for each farm family, as opposed to the actual average of 1.4 ha (World Bank, 2005a:17). This potential is, however, far from utilized. The farmers' productivity is low and projects to improve their yields or cultivated acreage often prove unsustainable.

Performance: farm-level evidence

Despite the general availability of agricultural land, farms are small, and most farmers are impoverished because few use agrochemicals, improved seeds, animal traction, water management, manuring and composting, or other enhanced farming techniques like micro-dosing[2] or conservation farming, and, consequently, their per farm and per hectare productivity is mostly stagnant and quite low (see Fig. 13.1 and Tables 13.1 and 13.2). Moreover, on average, many farmers use inputs very inefficiently and the gap between efficient and inefficient farmers is great. For example, Zavale *et al.* (2005:21) calculate that, if farmers became more efficient in the way they use inputs (labour, seeds, fertilizer, agrochemicals), they would need 70% less inputs to get the same output. As a result:

> over the past decade, agricultural growth was almost entirely driven by factor accumulation, with little technological improvement. The main sources of growth in the agricultural sector have been expansion of cultivated land area and an increase of the rural labour force. The improvement of agricultural technologies

[1] Including 19 million ha classified as 'valuable for timber production' (Issufo, 2003:5).

[2] Micro-dosing means the 'application of a small quantity of inorganic fertilizer, whether applied directly in the hole during planting, mixed with seed before planting, or applied after the plant emerges'. Experiments by the International Crops Research Institute for the Semi-Arid Tropics 'showed placed application of 3, 5, and 7 kg/ha of P led to significant productivity gains of 72%, 81%, and 88%, respectively' (Pender *et al.*, 2008:8).

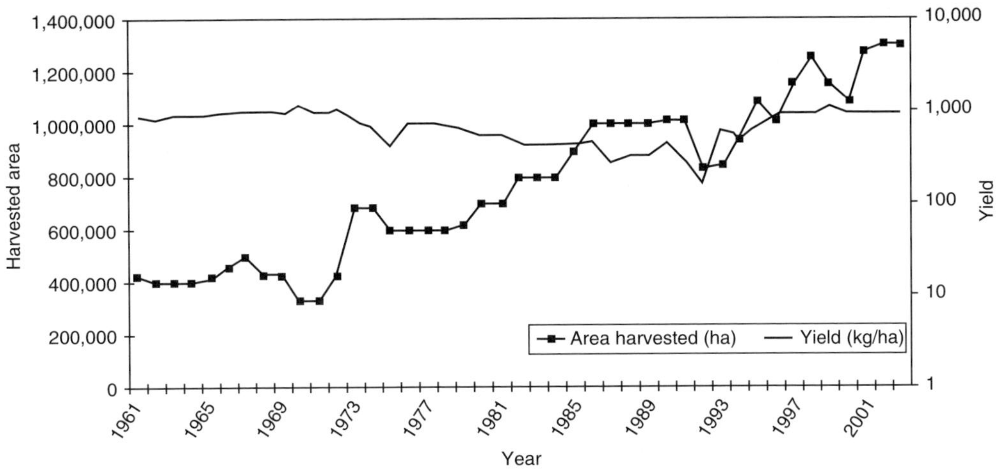

Fig. 13.1. Production and yield of maize from 1961 to 2003 in Mozambique. (Adapted from: FAOSTAT data, 2004, cited in Zavale *et al.*, 2005:2.)

Table 13.1. Rural household technology and characteristics (% of farmers using). (Adapted from: MADER, 2003b, 2006; MAG, 2008.)

	Cropping season		
Technology	2001/02	2004/05	2006/07
Fertilizer use (% of users)	3.7	3.5	4
Pesticide use (% of users)	6.7	5.1	5
Animal traction use (% of users)	11.2	8.6	12
Improved maize seeds (% of growers)	n.a.	6	10
Improved rice seeds (% of growers)	n.a.	6	10
Improved large groundnuts (% of growers)	n.a.	2	6
Improved small groundnuts (% of growers)	n.a.	4	9
Cattle vaccination (% of producers)	11	9	12
Chicken vaccination (% of producers)	n.a.	3	5
Irrigation (% of farmers using)	10	6	13

n.a., not available.

Table 13.2. Yields for some crops in Mozambique for small and medium holdings (kg/ha). (Adapted from: MAG, 2007b.)

	2005	2006	2007
Maize	538	839	681
Sorghum	314	497	435
Cowpeas	156	203	167

> has only played a minor role. Access to and use of improved crop technologies remains very limited, and there is evidence that crop yields are stagnant. Thus, if appropriate action is not taken, agriculture growth will slow down and rural poverty will remain widespread.
>
> (World Bank 2005b:1)

The yield gap between sub-Saharan Africa and other countries has greatly widened over the last half-century (Fig. 13.2). Currently, for example:

> Mozambique lags behind all other East and Southern African countries in maize productivity. In 2004, its maize yield averaged 960kg/ha compared to 1,500kg/ha for Kenya, 1,100kg/ha for Malawi, and 2,600kg/ha for South Africa (FAOSTAT, 2005). These low yields are a reflection of Mozambique's limited use of irrigation and ... yield-enhancing inputs such as fertilizers and improved seeds.
>
> (Uiaene, 2006:1)

Why so? Various studies in Mozambique have revealed little or no impact of extension services on farmers' technological choices. Walker *et al.* (2004:vii, 49) argue that, in Mozambique, 'agricultural extension had no measurable impact on either net crop income or livestock sales' though they later acknowledge that 'households ... [that] received information from extension agents had somewhat

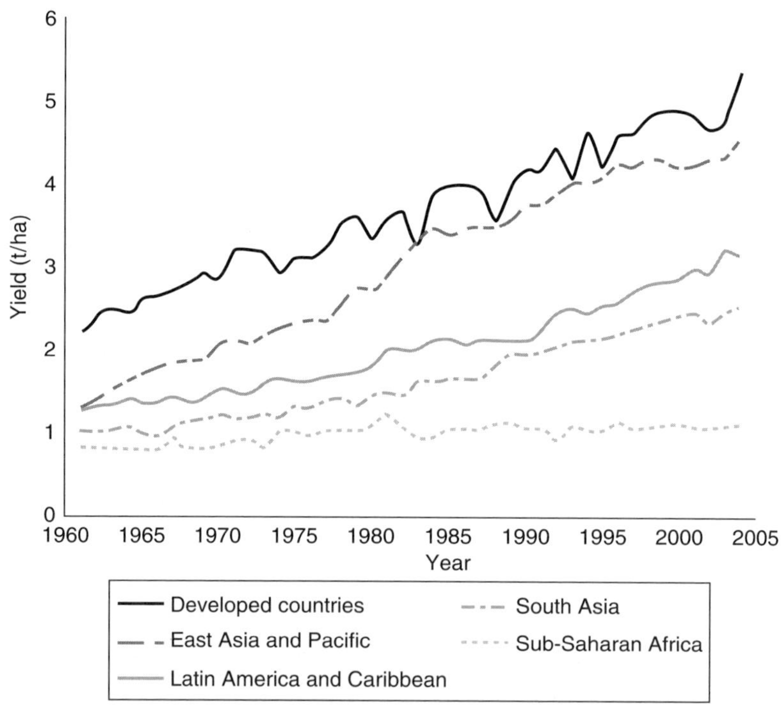

Fig. 13.2. Yield gap for cereals between sub-Saharan Africa and other regions has widened. (From: http://faostat.fao.org, assessed June, 2007.)

higher (5% with borderline statistical significance) net crop income than other households' and suggest that this may, in part, be due to 'constraints on access to improved inputs and to more location-specific adapted technologies'.[3] A recent econometric study by Uaiene *et al.* (2009:19) found that 'extension appears to only influence the decision to adopt animal traction', though they admit that, if extension services work indirectly through associations and leaders, their model would have missed its impact. A model by Zavale *et al.* (2005:11) even found access to extension negatively and significantly associated with the adoption of improved maize. On the other hand, Perumalpillai-Essex (2005:48) argues, on the basis of his model, that 'access to rural extension increases farm production by about 8.4%', largely by promoting improved seeds, natural pesticides and soil conservation.[4] Nevertheless, for all these studies, the returns are, at best, low.[5] By contrast, a study for Zimbabwe found that 'receiving one to two visits

[3] 'The National Directorate for Rural Extension Services and Sasakawa Global 2000 ... were conscious that the same type of technological package (one for maize and one for rice) was not adequate for all different agro-ecological regions. However, at that time, ... [they were] reasonable packages' since the research institutes were, in general, unable 'to recommend specific fertilizer application levels for specific crops and locations'. This observation suggests a major line of inquiry necessary to improve significantly the productivity of the agrochemicals used. IIAM (Institute of Agrarian Research of Mozambique) and its constituent research bodies have been working to identify various input packages appropriate for maize and rice for specific agro-ecological zones, though this is a process that will take some years. Similar research for other crops would be beneficial but this depends on priorities in the face of scarce resources. Letter of 17 August 2006 from Hélder Gemo, National Director for Rural Extension between April 2000 and July 2006, plus a follow-up interview on 4 September 2006.

[4] The analysis finds that extension works mainly through the introduction of new crop varieties. According to the survey, 43% of respondents introduced new varieties in the last 5 years if they had received advice. Only half as many (21%) introduced new varieties if they had not received advice. Farmers that introduce new varieties can count on a significantly higher probability of reporting an improvement in living conditions. The extension service also works by encouraging new techniques, in particular, by 'promoting natural pesticides' and soil conservation (World Bank, 2005b:104).

[5] Various studies also examined the annual internal rate of return (IRR) from investment in extension services. A review of 27 studies in Africa revealed that 21 had internal rates of return exceeding 12% (Oehmke *et al.*, 1997:5). A meta-review of 19 studies of the average IRR of agricultural extension yielded 80% if the unit of observation was farms, and in five studies where the focus was aggregate, the IRR was 75%. For Africa, six studies had a 90% average IRR on investment in extension and 35% for research (Evenson, 2001:80). An earlier review of 11 studies in Africa revealed an average IRR of 40% for investments in agricultural technology development and transfer, with most rates falling between 21% and 60% (Oehmke and Crawford, 1993:5). Another meta-analysis of 281 studies between 1953 and 1997 revealed that, worldwide, 'the medium of the rate of return estimates was 48.0% per year for research, 62.9% per year for extension services, 37% for studies that estimated the returns to research and extension jointly, and 44.3% for all studies combined' (Alston *et al.*, 2000:ix). In Kenya, Evenson and Mwabu (2001:24) estimated elasticities and found that for all crops, on average, a 10% increase in the intensity of extension efforts raised production by 1.3% and, specifically for maize, the main food crop, by 2.9%. The intensity of extension efforts was measured by the 'number of extension workers per farm in a given cluster', which presumably reflects both the extension workers' own training and their effectiveness in training farmers (Evension and Mwabu, 2001:4, 5).

per agricultural year raises the value of crop production by about 15%' (Owens *et al.*, 2003:356).[6]

If, in Mozambique, extension services are having little impact, might a more fundamental cause be at work? Uaiene (2006:12) argues that low profitability – not risk aversion and capital scarcity – is the main reason farmers are slow in using improved maize and agrochemicals. His model reveals that, if maize is sold at harvest time when prices are low, improved technologies are less profitable than traditional methods (Uiaene, 2006:12). By contrast, if grain is stored well and sold at typically high prices during the 'hungry season', improved farming techniques are far more profitable than traditional methods.

> The use of new improved cultivars and fertilizers can be accelerated if farmers can exploit for their benefit the seasonal price variation by selling when the prices recover. Pooling, storage and inventory credit are part of the strategy. The model results indicate that if inventory credit were available, new technology would be adopted with a consequent increase in farm income. The high returns to capital invested, with a shadow price of capital of 82%, indicates a further potential dynamic effect for farmers to reinvest their increased profits in new technologies in the following crop year.
>
> (Uaiene 2006:vi)

Significantly, in Mozambique, an 'analysis of improved agricultural technology adoption indicates that households with access to credit and extension advisory services as well as members of agricultural associations are more likely to adopt new agricultural technologies' (Uaiene *et al.*, 2009:18). The synergies that occur when extension workers are working with farmers who are organized in associations and have access to credit and, if possible, assured outlets for their crops are critical. 'Making credit accessible to farmers would increase [the] adoption and intensity of use of improved maize varieties by 24% (15% being the probability of adoption and 8% the intensity of use of the varieties)' (Langyintuo and Mekuria, 2005:1).

This conclusion was certainly borne out in our micro-study of ten villages (Coughlin and Givá, 2009:17).[7] In eight of the ten villages studied, hardly anyone used advanced techniques. By contrast, the two villages in Gaza province in southern Mozambique had irrigation projects with an extension worker assigned to each village. These projects had dramatically improved the output of participating farmers, mostly women, by teaching them to plant high-value crops and to use agrochemicals to improve yields. In the other eight villages, the efforts of extension workers, when available, revealed minimal impact, except for the development of 43 tiny fish ponds, built mainly with sweat equity, in Nacocolo, a village in Nampula province in northern Mozambique.

[6] Although Langyintuo and Mekuria (2008:165) found that, in Mozambique, extension contact would increase the probability of adoption of improved maize seed by 18.5%, they did not convert this to a comparable increase in a farmer's net income.

[7] Due to irrigation and the use of chemicals, the participating farmers got three crops per year – 4500 kg/ha/season for maize and 1250 kg/ha/season for beans – rates far higher than with traditional farming methods.

There, 'each pond is maintained by individual families, sometimes diligently, sometimes, lackadaisically. The fish harvest is small and mainly consumed by the farmers' families' (Coughlin and Givá 2009:8). Even there, except for the introduction of a pedal pump to irrigate tiny vegetable plots near the river, farm technology was entirely traditional. Analytically, the critical difference between the two classes of villages was that those in Gaza obtained not only extension advice but also a significant, albeit temporary, infusion of capital to initiate the irrigation scheme.

The majority, however, have no access to credit, even through their associations. Consequently, few will improve their technology or farm size, and, with the present snail's pace of improvement in agricultural productivity, most farmers will stay below the poverty line. Although, between 1996/7 and 2002/3, absolute poverty fell from 69% to 54%, between 2002 and 2005, 'though mean household income per adult equivalent increased, ... the lowest two income quintiles experienced a decrease in household income' (Cunguara, 2008:75). Whereas:

> the use of fertilizers, pesticides, and animal traction has increased among households moving out of poverty,... households moving into poverty have experienced a decrease in the use fertilizers, pesticides, and animal traction.... The use of seed and improved cultivation practices is higher among the non-poor, and reduction in the use of such technologies is correlated with moving into poverty.... Animal traction is used to cultivate relatively large areas and households cultivating such areas are more likely to be non-poor.
>
> (Cunguara, 2008:80)

The conclusion is manifest: to avert and overcome poverty, farm sizes must increase and productivity must improve. But, for farmers, that requires loan capital, not just advice.

Sustainability of projects to improve farm productivity[8]

Sustainability is a grave problem for some projects that strive to improve farmers' agricultural systems and practices. Many initiatives to build farmers' technological capability and performance collapse after the projects end. The case of the Foundation Against Hunger (Fundação Contra Fome) in Nhamantada suggests the need for serious reflection. The foundation organized farmers into groups or associations and provided them with improved seeds (cowpeas and other beans such as bambara nuts, pigeon peas, sorroco, groundnuts, sorghum), field assistance by an agricultural extension agent, and a rotational credit and savings programme. According to the farmers we interviewed, the extension agent taught improved agricultural practices, including better planting methods (e.g. use of appropriate plant spacing and alignment), use of botanic pesticides produced with available local plants, better land preparation, soil improvement techniques and introduction of fruit trees

[8] I thank my colleague, Nicia Givá, for permission to use excerpts from our joint report (Coughlin and Givá, 2009:17–18 and 23–24) in this subsection and in the final conclusions to the present chapter.

(oranges, papaya, litchis and avocado). During the project, the farmers in the association were reportedly happily engaged in these activities. But after 3 years, the project stopped and the group could neither sustain the activities nor adopt the disseminated techniques in their fields. According to the village leader, the programme for agricultural credit stopped with the project, though the rotational credit and savings scheme for small businesses survived and still has 30 members, divided into seven groups.

The organizational collapse after the project stopped and the subsequent demoralization of the farmers happened for two interrelated reasons: (i) the project's two extension workers stopped orienting, mobilizing and encouraging the farmers; and (ii) the project had not been structured to build up the farmers' capital and strengthen their association so that, after the project ended, they would have sufficient funds and organizational ability to continue to purchase and utilize improved inputs and farming practices.

What is the lesson? It seems that adoption of improved practices must occur together with a steady, programmed improvement in the farmers' investment capacity (capital). Without that, when the project ends, impoverished farmers will necessarily revert to traditional, low-input, low-technology farming systems. How can projects avoid such a relapse? Coughlin and Givá (2009: Box 2 and Annex 1) propose a low-tech, easy to manage and highly profitable solution valid for some circumstances (Table 13.3). As opposed to traditional storage methods or, alternatively, the use of central or village-level grain storage facilities, with their concomitant managerial difficulties and risks, they suggest low-volume, hermetically sealed grain storage bins needing no chemicals, initially managed as an inventory credit scheme but quickly transitioning the farmers into bin owners and savers instead of borrowers (Ferizli *et al.*, 2001; Giovannucci *et al.*, 2001; Villiers *et al.*, 2006). As savers, they avoid treatment costs, interest charges and the worst consequences – a huge drop in net income – if hungry-season prices fail to rise above postharvest prices. Moreover, having converted the farmers into savers, the project runs far less risk of collapse after external support terminates.

A sustainable improvement of agricultural productivity can be achieved if farmers adopt yield-increasing inputs and significantly improve managerial

Table 13.3. Comparative impact on total net income per hectare per year due to choice of storage methods for maize and grain, assuming average irrigated yields. (From: Coughlin and Givá, 2009.)

Type of technology	Maize	Beans	All three seasons (2 beans, 1 maize)
No storage ($/ha/year)	175	1005	1180
Traditional storage ($/ha/year)	456	1649	2105
Hermetically sealed bins ($/ha/year)	764	2411	3175
Average % increase over no storage	337	140	169
Average % increase over traditional storage	68	46	51

practices while market efficiency is also improved (MADER, 2004). However, although improved crop varieties have been released in many sub-Saharan Africa countries, less than 10% of farmers use them. Indeed, in our sample, only 3.3% of the households have been using improved maize seed, despite various promotional efforts. In the south, only the few farmers in the irrigation projects use improved seeds, although, according to Chibuto's administrator,[9] the improved seed programme is a district-wide priority. In the north, only 4% of the sampled farmers use improved or hybrid seeds, and in the centre, only 6%.

These low usage rates for improved seed reflect the very slow rate with which farmers use improved inputs and techniques, a result that Uaiene (2006) argues arises because, at harvest-time prices, such technologies are suboptimal or even make a loss. For example, using prices from 2004, if maize was sold soon after the harvest, the improved input package lost US$25.47/ha, as opposed to a gain of US$27.35/ha with traditional seeds and technology. Only if the crop is sold at a high price during the hungry season does the improved package earn more (US$86.71/ha) than traditional inputs (US$57.95/ha). Based on this analysis, Uaiene advocates inventory credit schemes, better storage and delayed sales.

Macroeconomic and international environment

As for the macroeconomic and international environment, when international petroleum and food prices soared in 2008, the government needed to mitigate the impact on low-income groups. To do so, it suspended VAT (Value Added Tax) on wheat and petroleum for public transportation providers and, though the metical appreciated nominally by only 1%, it was allowed to appreciate 'by 22% in nominal terms against the euro during the year through February of 2009, and by 27% against the rand. ... [In real terms,] the metical appreciated by 17% ... against the euro during the year through January of 2009, and by 37% against the rand. Meanwhile, the metical appreciated by only 6% in real terms against the dollar, reflecting nominal bilateral stability' (Vitek, 2009:4–5) (Fig. 13.3). Aside from these measures, international price fluctuations were allowed to pass through to the domestic economy (Arndt *et al.*, 2008:3).

> A variety of indicators suggest that Mozambique has recently lost external price competitiveness with respect to its major trading partners. Consistent with these indicators, an exchange rate assessment based on the macroeconomic balance, equilibrium real exchange rate, and external sustainability approaches indicates that the metical is overvalued by 26% to 41% in real effective terms.
>
> (Vitek, 2009:16)

[9] Interview with Zacarias Souto, Chibuto district administrator, 6 May 2008.

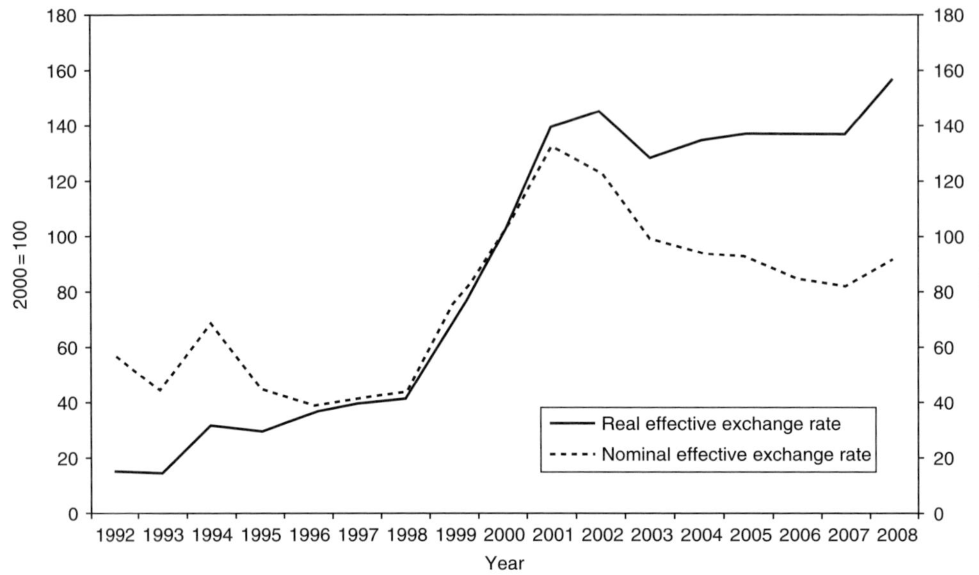

Fig. 13.3. Real effective exchange rate versus nominal effective exchange rate. (Adapted from: Vitek, 2009:4.)

In the short term, this favours consumers, though it draws down international currency reserves. If, however, overvaluation persists, it will curtail exports, encourage imports, hurt local producers and lower personal incomes.

Soaring fuel, fertilizer and food prices in 2008 sucked resources out of the national economy and redistributed incomes. According to a computable general equilibrium model by Arndt *et al.* (2008:14), this would cause GDP to fall by 1.2% and absorption (C + I + G) by 5.1%.[10] Since recurrent government expenditure is assumed to be fixed and investment declines by only 1.2%, household consumption bears the bulk of the adjustment, declining by more than 7.0%. As for incomes, since '74% of rural households are net food sellers, whereas 76% of urban households are net buyers', food price increases favoured small, typically low-tech, non-mechanized farmers, though a minority suffered as net food buyers, selling after harvests and buying even more during the hungry season (Arndt *et al.*, 2008:6, 15).

Since fuel comprises 12% of imports and food 5%, the impact of fuel price increases is preponderant (Arndt *et al.*, 2008:6, 8). The biggest winners were the farmers in northern Mozambique, whereas most of those in the drier south are net food buyers and therefore lost when food prices rose. On average, urban dwellers lost regardless of their income level.

The increases in food prices between 2006 and 2008 were sharp and, though followed by a fast fall, few analysts believe that they will fall back completely and resume earlier historical trends.

[10] C = consumption; I = investment; G = government expenditure.

> The consensus outlook is for world agricultural prices to remain high and volatile, and global economic conditions add considerable uncertainty to that prediction. Commodity prices are now linked more strongly via biofuels demand and depend strongly on exchange rate adjustments and macroeconomic outcomes.
>
> (Abbot, 2009:49)

According to the February 2009 USDA projections for grain and oilseed prices:

> long-term growth in global demand for agricultural products, in combination with the continued presence of U.S. ethanol demand in the corn sector and EU biodiesel demand for vegetable oils, holds prices for corn, oilseeds, and many other crops well above their historical levels, although season-average annual prices are not projected to reach the record highs seen in the first half of 2008.
>
> (USDA, 2009)

OECD–FAO (2009:10) (see Fig. 13.4) opines that, though:

> the situation varies by commodity, ... average prices in real terms (adjusted for inflation) for the next 10 years are still projected at or above the levels of the decade prior to the 2007–08 peaks. Average crop prices are projected to be 10% to 20% higher in real terms relative to 1997–2006, while for vegetable oils real prices are expected to be more than 30% higher.

On top of the traditional sources driving food demand (growing populations and incomes), the growing, often mandated, demand for biofuels has

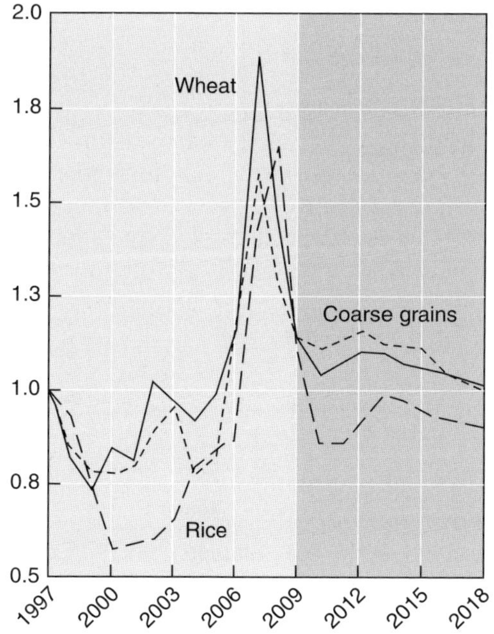

Fig. 13.4. Outlook for real world crop prices to 2018 (1997 = 1). (From: OECD–FAO, 2009.)

lifted demand and prices, implying a permanent shift in the long-term trend. Nevertheless, 'biofuels will struggle to compete with relatively low fossil fuel prices as long as crude oil prices remain in the USD 60–70 range', biofuel's threshold profitability point (OECD–FAO, 2009:10).

Pushed by rising energy prices and demand for food crops (partly for biofuel), fertilizer prices also soared and then plunged back, though not fully. They 'peaked in September 2008 at more than seven times their 2002 value, a much steeper increase than for grains and oilseeds, and even energy. They subsequently fell to varying degrees by type of fertilizer, depending on market structure, and in many retail markets remain well above historical levels. In early 2009, the price of urea in the Ukraine remained at about 2.5 times the 2002 price level' (Fig. 13.5) (Abbott, 2009:9). Rising energy prices pushed fertilizer prices principally through rising transportation costs and gas prices, the latter because gas is used to produce ammonia, the main input for all nitrogen fertilizers. Energy costs also impact on transportation expenses, a significant part of the cost of fertilizer. For example, transportation for ammonia shipped to the US Gulf Coast represents 22% of total costs if coming from Trinidad and Tobago and 50% if from Russia (Huang *et al.*, 2009:30). By mid-2009, petroleum prices were around US$70/barrel, but, as we come out of the current recession, they will very likely head upward, putting pressure on transportation, gas and fertilizer prices.

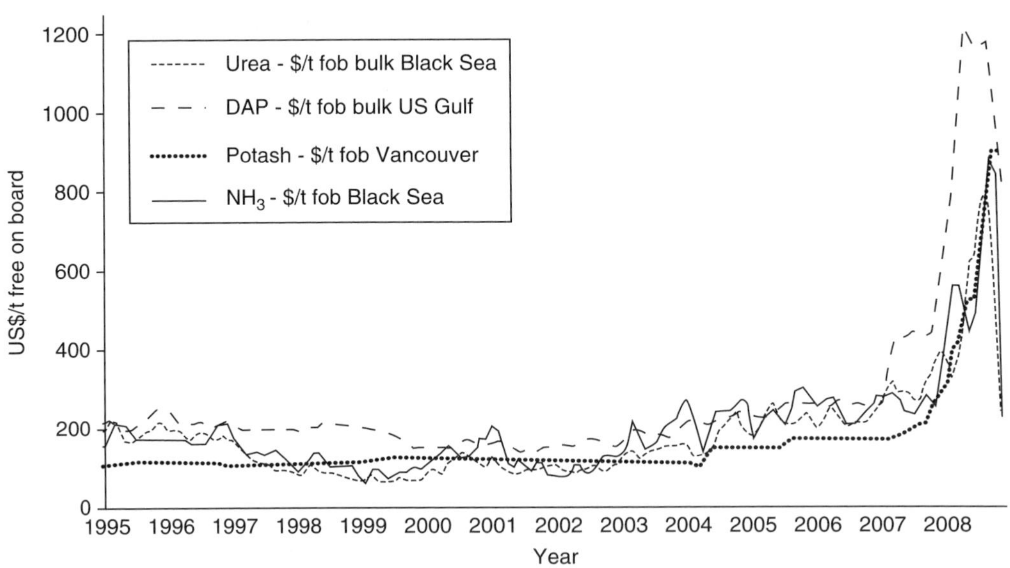

Fig. 13.5. Fertilizer prices, 1995–2008 ($/t). Adapted from: FertEcon.com (a subscriber service) cited by National Corn Growers Association, available at http://www.ncga.com/files/pdf/kenJohnson11-14-08.pdf. Note: By 25 February 2010, urea bulk had fallen to $273.75/t, DAP to $422/t and NH_3 to $382.50/t, according to FertEcon's weekly price list. Even these more recent prices are well above the long-term trend.

Since only 4% of Mozambican small farmers use fertilizers, the rise in their prices did not affect most, at least not immediately. However, since improved plant varieties perform best if given fertilizer, the long-term rise in fertilizer costs will squeeze profit margins and hamper that transition. A partial, but only partial, solution is for farmers to intensify their use of a cheaper, local solution: more organic fertilizers and nitrogen-fixing crops.

The prices for fertilizer and pesticides are unnecessarily high due to the inflated c.i.f. costs of imported chemicals, high local transport costs and the exaggerated margins charged by some suppliers, especially when they have a regional monopoly as, for example, in the contract farming schemes.[11]

> Given the small market demand for fertilizer, dealer/distributors are generally unable to negotiate a discount. For example, minimum order for fertilizers from Saudi Arabia is 10,000 tonnes per order. At this volume, the delivered price of urea in Beira is approximately $295/tonne. Given that even the largest dealer/distributors in Mozambique only order between 3,000 to 7,000 tonnes of urea per year, local companies are generally unable to purchase fertilizers at competitive prices. As a consequence, dealer/distributors have little choice but to purchase fertilizers from South Africa, at prices as high as $415/tonne delivered in Maputo.
>
> (GDS, 2005:6)

These costs are passed on to farmers. For example, 'purchased seed and fertilizer make up 68% to 80% of total maize production costs (exclusive of family labour) in the three regions [Ribáuè, Malema, and Monapo/Maconta]. . . . [Hence,] even small reductions in the farm-gate cost of fertilizer and seed (e.g. by reducing transport and other marketing costs) could significantly increase farm profits.' For example, a 25% reduction in agrochemical costs would have increased net incomes of high-input farmers by more than 100% in two regions and by 28% in the third region (Howard *et al.*, 2000:25).

In Mozambique during the 1980s, Interquimica imported all agrochemicals, whereas large agricultural enterprises may now buy from their mother companies abroad or from local representatives (e.g. Agroquímicos, Tecap, Zeneca) of multinational chemical firms, e.g. BASF or Ciba-Geigy (Howard *et al.*, 1998:16). In a market so small, this fragmentation eliminates any possibility of achieving bulk-order discounts. Indeed, this is a general problem throughout sub-Saharan Africa.

> Although liberalisation has removed many of the restrictions on the type of fertiliser that may be imported, previous customs for specific formulations tend

[11] For example, in the Zambezi Valley, the concession companies hold a tight monopoly on agrochemicals. Of cotton growers who use pesticides, 96.6% get them from the concessionaires, and for tobacco, 93.9% do. Of the tobacco farmers who use fertilizers, 98.6% obtain them from the concessionaires (Benfica *et al.*, 2005:19). Also, in early 2006, the Sociedade Algodoeira de Namialo (Sanam) bought out the Sociedade de Desenvolvimento Algodoeiro de Manialo (Sodan), thus gaining a monopoly over cotton production in Nampula Province and, consequently, a monopsony over the supply of agrochemicals for the 50,000 farmers in the concession areas (*Notícias: Economia e Negócios,* 3/2/06:1).

> to be followed, which eliminate the possibility of bulk orders. Debrah (2000) gives the example of the minor differences in cotton fertiliser formulations across neighbouring West African states, leading to the necessity of small, individual import orders and consequent higher prices.
>
> (Tripp, 2003:10)

Recognizing the problem, a study in 1999 by the Economics Directorate in the Ministry of Agriculture and Fishing recommended:

> investigating the possibility of conglomerating regional orders for fertilizers when this would achieve economies of scale in transportation and distribution and, hence, reduce significantly the costs of fertilizers for farmers. Mozambique is strategically positioned to take advantage of economies of scale through a system that would combine regional fertilizer orders since joint orders with Malawi, Zimbabwe, and South Africa could enter through the Nacala, Beira and Maputo ports.
>
> (DAP, 1999:52)

Similarly, an International Fertilizer Centre study by Debrah (2000:30) recommended that 'restrictive product specifications can be simplified to international norms ... [and] regional cooperation through primary ports can provide the means to achieve economies of scale and on-shore bagging of bulk shipments'. More recently, Chianu *et al.* (2008:68–69) confirmed that:

> improvements in structural supply issues [ordering trucks in advance, actively negotiating, use of large ships that stop in few locations, ordering 'generic' mass-produced NPK product, ordering in bulk, sourcing from the lowest cost plants, selecting low cost (but 'right') fertilizers, etc.] have been found to lead to 11% [to] 18% reduction in fertilizer farm gate price (Kelly *et al.* 2003; Kumar 2007). Kumar (2007) noted that these could reduce fertilizer farm gate price in Malawi by 14.5%, from US$482/t to US$412/t.

Cutting fertilizer costs would encourage usage and increase yields and incomes. Depending on the price elasticity of fertilizer demand (and application), Chianu *et al.* (2008:65) estimated that:

> structural change in fertilizer procurement (reducing price by 15%) led to 6% additional income (US$ 125 million) under low [demand] elasticity (−0.38), 22% (US$ 472 million) under medium elasticity (−1.43), and 34% (US$ 730 million) under high elasticity (−2.24) compared with base. Switching from one scenario to another indicated the potential for 20% [to] 32% further increase in farm income.

An experiment by the International Fertilizer Development Centre's (IFDC) Maize Intensification Project in eight locations in four provinces in Mozambique found that, with 'traditional management and no fertilizer', saved seed achieved merely 1.84 t/ha on the test plots, whereas with the application of NPK fertilizer, yields rose 55% to 2.86 t/ha. Moreover, 'important increases in yield were obtained by moving from saved seed (2.86 mt/ha) to OPV seed (3.55 mt/ha) and then to hybrid seed (4.29 mt/ha) even though all three received identical fertilizer (12-24-12-0 + urea) and cultural management' (IFDC, 2009:14–15).

More than a decade after the initial recommendations, Norway's Yara International, with support from the Alliance for a Green Revolution in Africa, is investing in 'two ports in Dar es Salaam and Beira (Mozambique) with around \$30 million each ... to set up large bulk terminal facilities to handle imports for fertiliser which would reduce costs' (Roelf, 2009). The facility in Beira should be operational in December 2010 and, as part of the Beira Agricultural Growth Corridor, will supply both Mozambique and inland countries. Depending on how the investment is structured and owned, this should help to achieve these discounts on a regional scale. However, an evaluation is still needed of the practical and economic viability of mobilizing or requiring importers within Mozambique to form private buyers' associations to conglomerate purchases of the same chemicals going through the same ports (e.g. Nacala, Beira, Maputo) and thereby achieve discounts.[12]

Though the topic is little researched and the evidence scant, similar problems may exist with agrochemicals. For example:

- 'According to interviews for this study, industry norms for mark-ups [on pesticides for cotton] range from 15% [to] 20%' (GDS, 2005:36). That is the claim, but an analysis of data for three different insecticides imported by one agent revealed that the normal mark-up on the c.i.f. price range is, in fact, between 35% and 57% (GDS, 2005:37; Coughlin 2006:21). Moreover, 'during peak seasons and when there are supply shortages, dealer/distributors enjoy even higher margins, particularly for more expensive insecticides. According to interviews, margins may go as high as 65% [to] 100% of f.o.b. price' (GDS, 2005:36).
- 'Although cotton companies have a number of insecticides to choose from, prices between various insecticides do not vary widely and thus do not justify the wide discrepancy between the estimated cost of delivering sprays to farmers (151,819MT/ha) and the cost claimed by the joint venture concession companies (313,800MT) and deducted from the cotton farmer's revenue. No reasonable explanation could be found to rationalize this discrepancy, which suggests that further investigation might be required' (GDS, 2005:38).

The impact of taxes on agriculture

Under colonial rule and after independence during the period of central planning, the agricultural sector was heavily protected (Alfieri *et al.*, 2007:7). In 1992, with the peace accord and Mozambique's signature of the General Agreement on Trade and Tariffs, privatization and liberalization began with a series of structural adjustment programmes and a progressive reduction and simplification of import duties. By 2010, most grains, including maize, rice and

[12] The idea of a mandatory but privately owned joint purchasing body would probably meet resistance, albeit camouflaged, from any importers or concessionaires using transfer pricing to shift profits out of the country without paying taxes.

beans, were only subject to a 2.5% tariff, or 0% if from a Southern African Development Community country.

During the late 1970s and most of the 1980s, the National Rate of Assistance (NRA) coefficients[13] were 'highly negative', and small farmers endured strong discrimination. Between 1991 and 1997, the rates turned positive 'due to the introduction of taxes on imports and exports'. During the period 1995 to 2000, the 'NRA rates seem to oscillate . . . around the value of the import tariff, plus the VAT in some cases'. The 'lack of government intervention in the sector implies that the actual NRA values should theoretically converge to import tariffs and VAT rates', as has happened for maize, rice, beans and groundnuts (Alfieri *et al.*, 2007:15–17). Compared to the 1980s, the impact of fiscal policies and market liberalizations hugely increased the NRA values for import-competing crops (rice, maize, beans, groundnuts) and, though less, for export crops too, especially cashew nuts and cotton (Fig. 13.6). Especially targeting the most popular crops raised by smallholders, this assistance has continued to increase in the new millennium (Figs 13.7 and 13.8).

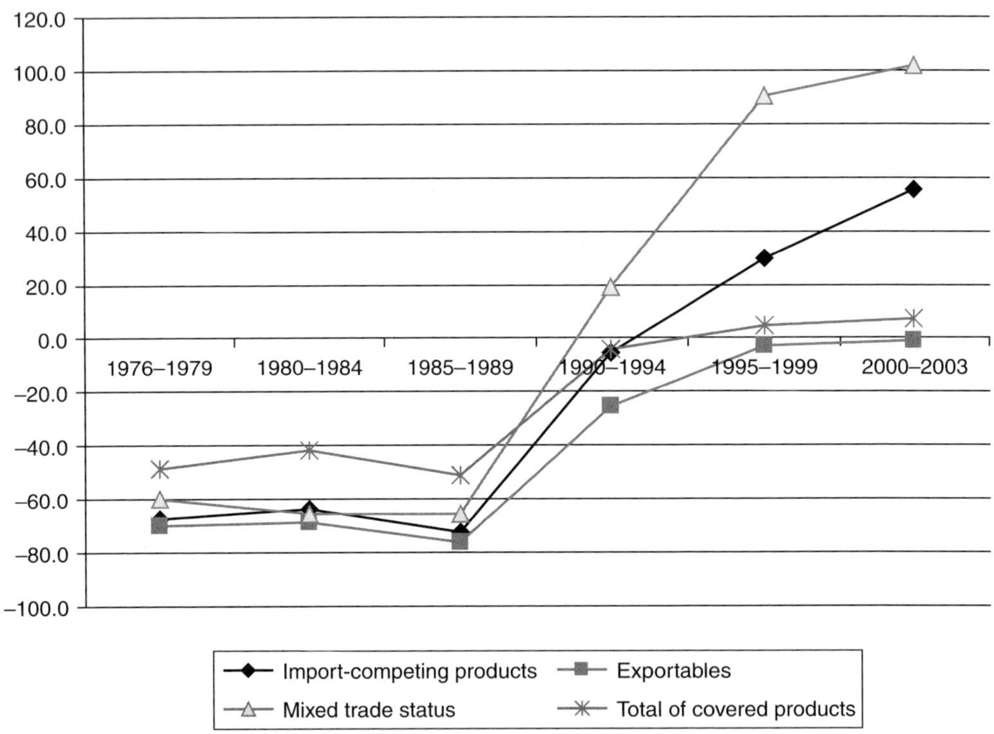

Fig. 13.6. Nominal rates of assistance to exportables, import-competing and all agricultural products, Mozambique, 1976–2003 (%). (Adapted from data in Alfieri *et al.*, 2007.)

[13] The NRA coefficients are 'the level of distortions induced by government policy interventions in the agricultural sector', as measured by the percentage 'gap between domestic prices and what they would be under free markets' (Alfieri *et al.*, 2007:11).

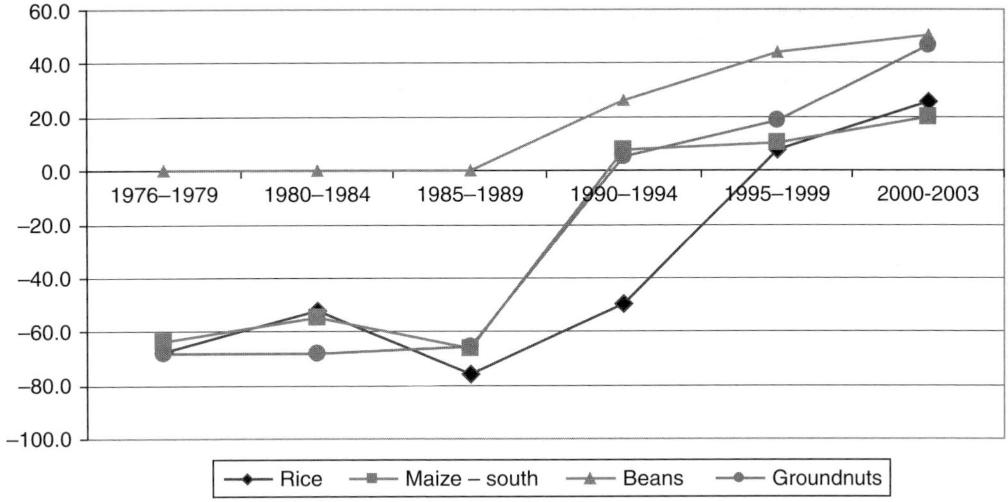

Fig. 13.7. Nominal rates of assistance to four import-competing crops, Mozambique, 1976–2003 (%). (Adapted from data in Alfieri *et al.*, 2007.)

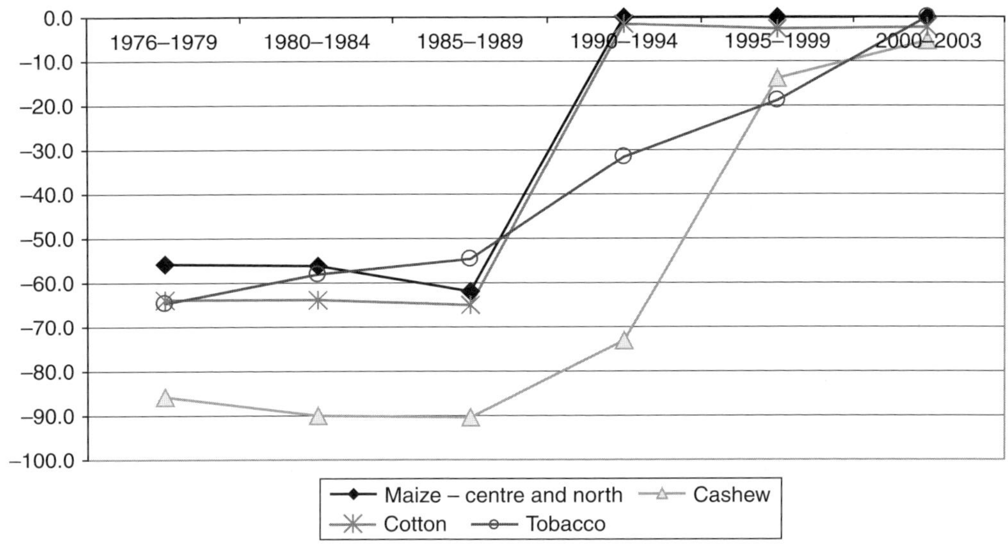

Fig. 13.8. Nominal rates of assistance to four exportable crops, Mozambique, 1976–2003 (%). (Adapted from data in Alfieri *et al.*, 2007.)

These tax measures and their impact on farm-gate prices, coupled with the elimination of controls over crop prices and movement, have been the principal way that government policy has benefitted small farmers, most of whom do not benefit from the concessionaire schemes. This shift in policy has enabled the country to become nearly self-sufficient and much less vulnerable for major grains, except rice, while largely eliminating the discrimination against export crops.

Institution Building

Peace in 1992 was the turning point. Refugees returned to farms en masse, and institutional building and rehabilitation occurred in all sectors, e.g. transportation, health, education, agriculture. More than 1200 largely loss-making state enterprises[14] were privatized, the big ones usually successfully, the small and medium ones, less so. Within agriculture, between 1992 and 2005, four major achievements occurred in institution building and reform:

- The agricultural extension system became operational[15] and grew into a pluralistic system involving extension workers from the private sector, non-governmental organizations (NGOs) and the government.
- To overcome the agricultural research system's lack of strategy and connectivity between its own organs and with the extension service and farmers, the Institute of Agronomic Research of Mozambique (IIAM) was created in 2005, amalgamating three research institutes and two centres. The new institute also includes economists and social scientists to improve the linkage with farmers and ensure that research results and the consequent changes in agricultural practices advocated by the extension service will consider market conditions and be profitable and not too risky for farmers.
- The curriculum for primary schools changed in early 2004. They now teach 'carpentry, sewing, and various skills related to agriculture and animal husbandry', de facto making the schools a dynamic component in agricultural education. With nearly four million 'kids learning improved agricultural techniques, ... the impact – with their parents and, eventually, when the students have their own farms – could be great' (MADER, 2004:115).[16]
- A significant institutional reform in 1999 created a 5-year sector-wide programme for agriculture (ProAgri I), whereby numerous donors pooled funds to support activities, build institutional capabilities and greatly reduce

[14] The state enterprises received subsidies amounting to 1% of GDP (Cramer, 2001:86).

[15] Though created in 1987, the National Directorate for Rural Development was hampered by war and insufficient resources and did not become truly functional till peace came (Gemo *et al.*, 2005:2, 22).

[16] Since primary schools must now teach farming and animal husbandry, the new curriculum creates scope for the extension services to assist the schools and, perhaps, the teacher training institutions. Despite the reform, the training institutes still grow little of their own food and, most commonly, greatly underutilize their model farms. Given the new curriculum, farming and animal husbandry could be part of the training while also supporting the institutes' budgets. In 2005, the National Directorate of Extension initiated contacts with the Ministry of Education to explore how the schools and extension service might cooperate but, with the change in the government, the initiative was put on hold. With financing by FAO and cooperation from the Ministries of Agriculture and Education, a pilot project, Projecto Hortas Escolares – functioning at 12 schools in Tete, 12 in Inhambane and 12 in Gaza – has nearly completed the preparation of a manual for teaching farming and animal husbandry in primary schools throughout the country. (Interviews with Hélder Gemo, National Director of Extension Services, Ministry of Agriculture, 13 February 2006; Abel Assis, Director, National Institute for the Development of Education, 6 March 2006; and Hassane Rachid, Ministry of Agriculture, 21 February 2006.)

reporting and other overhead expenses.[17] ProAgri II, endorsed by the ministry and originally planned to start in 2005, was delayed, though partial interim finance was provided. To start the full programme, the ministry had to comply with donor requests for institutional and workforce reform and develop clear and agreed-upon statements of priorities concerning environment, gender and HIV/AIDS.

As designed, ProAgri II would shift much power and more than three-quarters of its budget to the provinces and districts while also setting up Multi-Stakeholder Agricultural and Rural Development Councils ('comprising representatives of other government sectors, private agricultural companies, NGOs and smallholders') to introduce a demand-driven element into the preparation of the provincial annual activity plans and budgets (MADER, 2004:128).[18] Now, the concept is to avoid the creation of institutional redundancies and, instead, to utilize existing provincial forums after including additional stakeholders.

In 2000, the government also launched its Programme for the Reduction of Absolute Poverty (PARPA) as a strategic framework for sectoral work, including agriculture. As it evolved, PARPA shifted from a short- to a medium- and long-term focus promoting fast, widespread growth as the best way to benefit the poor (Mozambique, 2001:2). This has, at least, obliged the ministries to analyse systematically how their policies and programmes affect the poor and especially women. Though sometimes perfunctory, this analysis often inspires changes to their benefit.

Finally, despite changes elsewhere in the system, a major institutional deficiency continues to afflict most farmers: the extreme weakness of the credit and savings institutions in rural areas.

The following sections only focus on the agricultural extension and research systems and the availability of agricultural credit, while leaving aside the structural issues of the Ministry of Agriculture, ProAgri and the autonomous institutions such as the national cotton and cashew institutes. Nor do they focus on the primary school curriculum reform, since no evaluations exist yet from the perspective of agriculture.

[17] The reporting requirements for a plethora of uncoordinated donor programmes and projects can absorb huge amounts of professional time – both local and foreign – to produce disjointed, often redundant evaluations requiring diverse reporting procedures. For example, in Tanzania in 1999, donors sent '1,000 missions per year and the government was producing 2,400 quarterly reports annually to meet their requirements' (Gemo *et al.*, 2005:16, based on World Bank, 2002).

[18] The plan for ProAgri II also foresaw the creation of a Horizontal Management Board within the ministry's headquarters (MADER, 2004:129). As originally conceived, this would create demand pull to counterbalance the power and bureaucratic inertia in the vertically organized national directorates and prioritize the activities and resources across them more in accordance with clients' needs. But, as finally approved, the plan placed 'representatives of central and provincial MADER' on the board (chaired by the minister) but without any form of client representation, thus partly undermining the purpose for the distinction between vertical and horizontal organization, namely to distinguish between service suppliers and users. As proposed under ProAgri II, central power would remain vertical but with minimal modification.

The agricultural extension system: a brief assessment

Initiated in 1987, the National Directorate for Rural Extension expanded rapidly after peace came in 1992 but, due to scant resources and international pressure, was 'kept on hold' after 1999, capped off at a maximum of 800 extension workers, each normally serving 225 farm households (Gemo *et al.*, 2005:107). Nevertheless, since the system is pluralistic and 127 of the 128 districts have at least some extension workers from the National Directorate of Rural Extension Services (DNER), NGOs or private companies, it might seem that most districts are covered. In fact, the coverage is typically quite thin, with 1.3 extension workers per 10,000 rural inhabitants, i.e. about one-sixth of the ideal coverage of one extension worker per every 225 households (MADER, 2004). Moreover, during 2002–2003, only 14% of farmers had received advice from an extension worker.[19] Though only 9.4% of villages have an extension office or post, even in those villages 'only 20% of the households ... actually benefited from it'. Of all farm households, 32% acknowledge having 'access to extension services' in their village (Perumalpillai-Essex, 2005:8, 18). Access? A vague, inclusive concept! Ambiguities aside, most farmers get no extension services, directly or indirectly.

Why? Distance is a big factor. Though '20% of villages are within 30 km of an office, ... 43.5% have more than 200 km to travel to visit an office', an impractical distance for both the extension workers and poor farmers (Perumalpillai-Essex, 2005:8,11). DNER's extensionists use a modified train and visit methodology that is less top-down, more participatory and flexible about scheduling visits in tune with farmers' needs. This approach increasingly emphasizes working with farmers' associations as being both faster and more cost-effective (Gemo *et al.*, 2005:42). For example, during our field visit to Murrupula in September 2005, the District Agricultural Director informed us that he has seven extensionists using the modified train and visit model[20] and seven (paid by CARE) exclusively dedicated to promoting farmers' associations. In 1 year, the latter seven have set up 82 associations, which, in his opinion, render benefits far beyond those achieved using the standard approach. Though community leaders typically have limited

[19] TIA's estimate of 14.1% coverage corresponds remarkable well with the 17.2% estimated in MADER (2004) on the basis of the norm of 225 farm households per extension worker. On the other hand, the World Bank (2005b:17) uses a different and rather vague concept: access. Accordingly, in 2002, 32% of communities 'had access to extension services over the past 12 months ... [though] only 20% of the households in villages with an extension service, actually benefited from it'.

[20] Between 1975 and 1995, the World Bank promoted the train and visit model for extension organization in more than 70 countries. Despite being '25% to 40% more costly than the systems' it replaced, it was 'intended to deal with accountability by improving management's ability to monitor staff activities, taking advantage of the strict visit schedule, identifiable contact farmers, intensive hierarchy of supervisory staff, and other quantifiable measures. ... Several features of the design could not stand up to practical realities, however. The quality of extension services remained mostly not monitorable, and the lack of accountability to farmers was not resolved' (Anderson and Feder, 2004:49).

education and the associations need capacity building, especially by participatory methods, our study in the sample villages revealed that associations can help to facilitate market access, negotiate for better prices for inputs and crops, and serve as a vehicle for the promotion of new or improved technologies such as fish ponds or the use of pedal pumps to irrigate vegetables.

Given its resource constraints, however, the DNER has deliberately chosen to concentrate its efforts in high-potential areas while ignoring others. The strategy is justifiable since 'the global experience shows that there is a high payoff for concentrating extensionists in high-potential agro-ecologies and districts rather than sprinkling extensionists throughout the countryside' (Gemo *et al.*, 2005:95). Yet a problem exists. Even inside those high-potential areas, many farmers receive no advice, even indirectly, from extension workers. The extension workers are simply too few to go to all the communities and villages, even the nearby villages. But a choice has been made: to a great extent, even within the target districts, the extension system assists the same farmers in the same villages year after year while permanently ignoring others. Many farmers have no prospects of seeing an extension worker, even within a decade, while others have the service guaranteed year in, year out. DNER does not have a strategy to rotate some, though perhaps not all, extension workers every 3 or 4 years to previously uncovered villages.[21]

Is this justified? Some rationale exists to keep extension workers permanently focused on particular areas within a district, especially if the remaining areas have little agricultural potential. Moreover, agriculture is dynamic and the problems in a given zone are not the same every year. Even so, most extension messages do not change year to year. The villages within the extension worker's circuit get saturated with mostly the same messages, whose marginal utility declines, since, within the first years, the ready learners will have already adopted them. Though Gemo *et al.* (2005:45) argue that 'many of the simple technology messages are still relevant', their prolonged repetition to the same farmers has declining returns.

A rotational system might maximize the number of households and farmer associations that will have received assistance over, say, a decade.[22] However, both this and the present strategy of permanency of geographical focus have costs and benefits. Rotation might have a larger impact than that achieved by fixing extension workers in nearly permanent circuits but it would probably increase costs for housing and transportation. Over, say, a decade, which strategy would manifest the best cost–benefit ratio? Only a prospective cost–benefit analysis would suggest the answer, an answer that would need subsequent confirmation in practice. Nevertheless, so long as the extension system is so direly short of resources[23] that it assists but a fraction of the farmers in a given district, an evaluation of alternative strategies might reveal whether and under what

[21] Although DNER is currently finalizing a programme (financed by FAO) to expand to 93 the number of districts it serves, the problem of gaps in coverage inside the districts will persist.

[22] Wider geographical focus would also complement the need that rural primary schools have for extension advice about farming techniques specifically relevant for their agronomic conditions.

[23] DNER is currently amending its master plan and, with major assistance from FAO, plans to extend its coverage from 66 to more than 90 districts. Even in those districts, however, only a fraction of the farmers will receive assistance directly or even indirectly.

circumstances an alternative approach would be useful. Intermediate options could also be considered, for example the possibility of leaving some extension staff in the original circuits to work, albeit less frequently, through farmers' associations to deal with new problems.

Ostensibly, one of the advantages of a pluralist extension system is the ability to experiment with and learn from different approaches. In Mozambique, however, 'there are *few examples* of horizontal linkages and systematic exchange of substantive experience and financial information among Mozambique's three extension providers' (Gemo *et al.*, 2005:97, emphasis added).

The extension system also suffers from the inadequate preparation of many of its extension workers.

> The training of extensionists about high-value crops is a new challenge as most of them do not have the technical knowledge required for production of these crops; how to add value to the commodities and how to find information about prices, grades, and standard; WTO regulation; and access to regional and global markets.
>
> (Gemo *et al.*, 2005:74)

Another acute problem concerns the usefulness of the technologies and methods promoted in Mozambique by extension workers. The technologies advocated for small farmers have been both too risky and frequently unprofitable in view of market demand and farm-gate prices, e.g. Sasakawa's maize technology kit (Howard *et al.*, 1998, 1999, 2000, 2001). For example, Eicher (2004:26) reported that 'during our field visit a provincial agricultural officer reported that "we need new technical messages. We have preached the same messages such as planting on line for 10 years. We need messages on conservation farming, tobacco, animal husbandry, and fish farming."' All three extension providers – government, NGOs and private companies – suffer a 'general lack of technology that is profitable to small-scale farmers on a recurring basis and at an acceptable risk' (Gemo *et al.*, 2005:92). Moreover – as Mole (2000:8, 9) found regarding the technology for the chemical control of the powdery mildew disease that attacks cashew trees – to be successful, the strategies must focus not only on increasing yields but also on reducing costs, while simultaneously ensuring that the messages are, indeed, appropriate for the local soil and climatic conditions. Consistent with this, our field survey revealed that many of those who know a certain improved agricultural technique do not, in fact, use it (Coughlin and Givá, 2009: Table 16). For example, though 59% claim to know about crop rotation, only 27% actually did this, and 62% knew about the methods and benefits of using manure but only 5% used it (see Appendix, Table 13A.1). Surprisingly, however, the farmers claimed to have learned almost everything from their parents or neighbours, while only 1% of those interviewed claimed to have learned a technique from the extension officers (Table 13A.2).

Without capital, however, messages – even when good – are hard to implement. When extension agents work inside the context of a project or concessionaire scheme that furnishes inputs on credit and perhaps invests in infrastructure, then farmers implement their messages much more readily. This may be done through the promotion of inventory credit schemes coupled with other investments in irrigation or animal traction. Moreover, the use of hermetically sealed storage techniques would seem to make inventory credit schemes far more profitable. With such storage, sales during the 'hungry season' would, on average, yield 337% higher net income for maize and 129% more for beans compared to sales at harvest with minimal storage (Coughlin and Givá, 2009:37–39).

Supporting access to inputs and extension services may not suffice, however. Often, the government or even private companies bungle campaigns to encourage farmers to plant new crops because no ready markets exist. This creates losses and inspires scepticism about later advice. For example, in Naminhalo in Nampula Province, our field interviews revealed that:

> an agricultural marketing firm, CANAN, [had] encouraged farmers to plant tobacco but by harvest time the company had collapsed, leaving the farmers with nowhere to sell their crop. Later, extension workers encouraged them to plant jatropha but, again, it had no buyers. [Now,] an NGO is promoting resin harvesting but even the local extension worker and the village leaders in Naminhalo doubt it will have a market. Since the Ministry of Agriculture does not vet efforts by private firms and non-governmental organizations to promote new crops, it cannot inform farmers whether those campaigns merit confidence. Such coordination and oversight would speed the acceptance of new crops with real markets while helping to avert disasters too.
>
> (Coughlin and Givá, 2009:29)

The agricultural research network

Despite some successes, the apparently low overall impact of extension services in Mozambique and the persistent complaints about the lack of profitable, low-risk and location-specific technology ready for dissemination raise questions about the efficacy of the linkages between research, extension and the market (Gemo *et al.*, 2005:60). For example, Eicher (2002:14) reported that:

> in our May 2002 field visits to six districts, we found there was a lack of cost of production studies of present and improved technology for the family sector and a general lack of connectivity between research stations and extension programs. A number of research stations were inactive because of disbursement delays, lack of qualified staff and inadequate computer and support services.

Earlier, an evaluation by the Royal Tropical Institute found that:

> Agricultural research is largely planned and coordinated from headquarters in Maputo. ... A consequence of this strategy with centralized planning is isolation from the producers' reality and weak involvement of [farmers, extension workers and other] agents in setting priorities and planning research.
>
> (KIT, 2000:11)

The National Director of Rural Extension and his co-authors affirmed, in 2005, that 'unfortunately the linkages between extension, research, and marketing have not improved significantly over the 1999–2004 period'. They identified 'three main reasons for this impasse':

- Though highly complementary, research and extension services 'continue to work on their own agendas and priorities' due to the 'fragmented approach to decision-making and implementing decisions'.
- 'Both services have serious funding and human-capital constraints'.
- A transparent career ladder does not exist to provide 'training and incentives to work [for] a cadre of highly committed professionals in both extension and research, who are on a particular job for long enough to develop contacts and trust with professionals in other services' (Gemo *et al.*, 2005:57).

Indeed, Mozambique – in the same category as Rwanda – has less than one agricultural researcher per 50,000 people economically active in agriculture, whereas the ratio for Reunion, Mauritius, Libya, Egypt, Cape Verde, South Africa, Tunisia and the Seychelles is 1:2500 or better, and for developed countries it is roughly 1:400 (Roseboom *et al.*, 2003:68, 69).

In 2002, a further complication was that 'in spite of the directives from MADER/ProAgri toward the Farm–Systems–Research approach, the public research system' still had no social scientists.

> The few adaptive research interventions done in Nampula and Niassa provinces have been dominated by teams comprised by natural scientists (mainly agronomists). The lack of social scientists results in experimental programmes based on physical parameters while little attention is given to socioeconomic analyses and systems perspective as well as basic costs and benefits, and return-to-investment analysis of the technologies under development.
>
> (SANAGRI, 2002:8)

To deal with these problems, the government, in late 2004, consolidated a training centre and four research institutions,[24] including their regional research facilities, into the Institute of Agrarian Research of Mozambique. The Agronomy and Veterinary Science Faculties of Eduardo Mondlane University conduct mostly academically oriented research and remain independent. In late 2005, IIAM received technical assistance from Michigan State University, created a unit for socio-economic research and recruited new specialists in recognition of the urgent need to focus its research on economically profitable options for farmers in view of the exigencies and opportunities in national and international markets.[25] Furthermore, as now structured, IIAM allows the regional research centres considerable autonomy in setting research priorities in accordance with local needs. To ensure better connectivity between the research centres and farmers and other agents in the production chain, regional

[24] The Centre for Agrarian Education, the National Institute for Agronomic Research (INIA), the National Institute for Veterinary Research (INIVE), the Institute for Animal Production (IPA), and the Centre for Forest Research (CEF).

[25] Calesto Bias, IIAM, 2006.

forums have been set up, including representatives from government, NGOs, farmers' associations and other stakeholder groups.

Financial services

In rural areas, savings and credit facilities are largely absent, especially for small farmers, and handling or keeping cash is often risky. Since few rural districts have a bank branch, farmers, especially cash croppers, risk assaults and theft after receiving large cash payments for their harvests,[26] and pathetic anecdotes tell of hoarded notes being eaten by insects or consumed by flames.

Only 2.9% of rural households have access to credit, and this nearly always comes from the concessionaire companies managing large cash crop schemes. Mozambique's banks lend only 10% of their portfolio to agricultural activities and little or nothing to smallholders (Nathan Associates, 2004:9–10). 'A number of funds ... provide a specific financial product to a selected target group of rural entrepreneurs', and some micro-credit initiatives exist for the rural areas. Nationwide, 'about 30 micro-financial institutions are operating, ... [of] which World Relief International, CCCP, CARE, Tchuma, SOCREMO, and Novo Banco are considered the major ones', though only about 18% of their credit goes to agriculture (Varajidás, 2005:10).[27] The sector has, however, been growing dramatically. 'In 1998, ... there were approximately 9,000 loan beneficiaries of which about half received services from only two providers (World Relief's FCC programme and the UGC's poultry input credit programme). ... By 2005, the picture [had] changed considerably' (de Vletter, 2006:11, 13). De Vletter (2006:11, 12) listed at least 54 operators 'reaching 66,000 borrowers and 63,000 savers', although there was a 'high rate of dropout or institutional transition over the period 1997–2005'. Despite this growth, the absence of any national insurance fund to protect members' deposits against bankruptcy also hampers their expansion.[28]

Present Strategies for Agricultural and Agro-industrial Development

In 2007, the government launched its strategy for a Green Revolution in Mozambique based on five pillars:

1. Natural resources (land, water, forests and wildlife).
2. Improved technologies.

[26] Notícias: *Economia e Negócios* 7/4/06:4–5.

[27] In 2008, Rabobank and other partners set up Banco Terra in Maputo, with plans for the rapid establishment of branches in Nampula and Pemba.

[28] For several years now, Tchuma, Lda., and others have been lobbying the government to create such an insurance fund to facilitate the promotion of savings and loan cooperatives (Gildo Lucas, CEO, Tchuma, Sarl, 2006).

3. Markets and up-to-date information.
4. Financial services.
5. Formation of human and social capital (MAG, 2007a:8).

In discussing each pillar it merely lists the gamut of useful activities within each category or pillar. Real priorities are not discernible. Seemingly everything is valid and everything is a priority.

The revolution's vision is overly technical and silent about the need to plan, coordinate and integrate the efforts of diverse actors to provide the infrastructure and supporting linkages required for specific products and industries to start and grow (Castel-Branco, 2008a:11). It builds no mechanisms to guarantee that the efforts of different ministries, local governments, financial institutions, NGOs and entrepreneurs responsible for creating a particular value chain, cluster or region will, in fact, act in a timely and synchronized way. The document provides not a hint of how those efforts should be prioritized, coordinated and consciously, strategically financed while guaranteeing market linkages.

A coherent, well-integrated strategy must have adequate resources to motivate and propel investments all along various value chains, e.g. research, education, inputs, institutional and financial support, input suppliers, production techniques and standards, agricultural processing, transportation, power and communication infrastructure and provision, and local and international marketing. A proper strategy requires well-planned, pro-active, integrated policies and implementation, with strong complementarities and beneficial externalities able to inspire confidence and investment in specific value chains and economic clusters (Ernst & Young and EconPolicy Research Group, 2005). Dozens of disjointed, underfunded ministerial strategies (some ministries have four or five) replete with unranked 'priorities' hardly comprise a national development strategy (Castel-Branco, 2008b:20).

> Overall, the government must have 40 or 50 strategies at the national level, and that is without considering dozens of district and provincial strategies and many sectoral, sub-sectoral and sub-national strategies by donors. Should the country function with nearly 250 strategies? Is this necessary? Is it viable? How much does it cost in time and human, financial, institutional and informational capacity to manage all these strategies? Will this proliferation of strategies not fragment and weaken the state and the government?
>
> (Castel-Branco, 2008b:29 translated)

Sectoral, subsectoral and ministerial inertia, precedent and in-fighting over the relative size of each ministerial directorate's budget guide financial and manpower allocations. And behold! The budgetary line items that bubble up correspond approximately to the thousands of strategic 'priorities'. In truth, however, there is no overall guiding strategy and no high-level mechanism for interministerial coordination – with partial power to reallocate resources between ministries – to conceptualize and mount coherent, well-articulated and adequately funded campaigns involving the synchronized efforts of various ministries, NGOs and private sector institutions and companies to accomplish complex developmental programmes.[29] A strategy lacking a coherent, well-integrated conceptualization of activities and with no guiding organ and no power to reallocate resources to ensure results is but a showy illusion *para o Inglês ver*. Though it might inspire some activities in the right direction (e.g. animal traction), the task – rural and agricultural development – requires a far more robust, comprehensive instrument for planning and implementation.

Conclusions

After independence, Mozambique passed through more than a decade emphasizing state farms, large cooperatives, low farm-gate prices, a monopsony over agricultural inputs by Agricom and the regulation and policing of crop movements, all severely discriminating against small farmers, who also suffered from the capriciously brutal civil war led by Renamo and financed by Portuguese ex-colonialists, Rhodesia and, later, South Africa's Bureau of State Security. Hunger ensued and more than a quarter of the population became internal or external refugees. Poorly managed and crushed from without, the socialist experiment failed.

Taking an abruptly new tack, the government began, in 1987, through the Economic Rehabilitation Program, to liberalize market controls and, after the Peace Accord of 1992, started to privatize small, medium and large enterprises en masse while taking fiscal measures to protect local farmers. For most crops, 'the gap between domestic prices and what they would be under free markets' went from highly negative to very positive – a boon to small farmers (Alfieri *et al*., 2007:11). Peace, roads, other infrastructural projects and, in some areas, the revival or expansion of concessionaire schemes also helped many farmers.

For most, however, the benefits were small or slow in coming. Remote from markets and seldom, if ever, seeing an extension worker, farmers remained on their tiny plots, using traditional manual technologies: no chemicals, no improved seeds or farming techniques and no animal traction. No technological change, neither intensification nor extensification, and certainly no Green Revolution! Such farmers are the majority: impoverished and largely isolated from the benefits of policy, good advice and improved technologies.

Capital-poor and rarely receiving advice from extension workers, Mozambique's small farmers are ensnared in a low-technology, low-output trap. Although strategic infrastructural investments in roads and communication help them reach and benefit from markets, and agricultural research helps

[29] In 2005, the final report for the Reformulation of Industrial Policies and Strategies recommended a weak, narrow version of this for the Ministry of Industry but it was never implemented (Ernst & Young and EconPolicy Research Group, 2005:85).

them to confront threats or improve productivity, these efforts have, so far, been too gradual and insufficient to change their fundamental reality: low productivity, low incomes and dire poverty.

Receptivity alone is, however, insufficient. Escape from the low-level production trap requires large, synchronized infrastructural and industrial investments to facilitate commerce and create value chains. Villages also require capital investment: focused, moderately sized, short term and, preferably, rotational, so that the funds move on to other farmers and villages. In some circumstances, small irrigation schemes can be initiated, often coupled with very profitable inventory credit projects, preferably with a fast transition to saving instead of borrowing and to individual instead of village-level management, made more viable by use of small hermetic grain bins that require no chemicals to control pests and fungi. This strategy would allow farmers to avoid the most serious risks and vulnerabilities of such projects: (i) corrupt or incompetent management; and (ii) the occasional big fall in net income (after interest charges and other costs) in years when hungry-season prices fail to rise or even fall below harvest-time prices. In other situations, animal traction can help for cultivating crops and transporting harvests to nearby cities instead of merely selling to merchants who go to remote villages and offer farmers far from advantageous prices.

Investment may also enhance farmers' receptivity to and application of the messages promoted by extension workers. At least in the ten villages studied, extension workers reach few farmers, and the farmers themselves state that the vast majority of the agricultural techniques they know about or use come from family, friends and neighbours but very little and very rarely from the extension workers. There was, however, an exception: the farmers benefitted greatly when advice came in the context of significant investment, for example in irrigation or in fish ponds complemented by pedal pumps.

For villages like these – none inside of concessionaire zones – the ability to inject capital to boost output and incomes significantly may well be crucial to enhancing the relevance and productivity of extension workers. Without complementary capital investments, the extension workers are hamstrung, minimally effective and often ignored. For the majority to escape poverty requires the synergies of well-coordinated advice and significant capital investment in the villages. Without that, the lives and production systems of small farmers will change little. And, in 30 years, with nearly triple the current population and far less freely available cultivatable land, the majority of Mozambique's small farmers will – as in much of Asia and Latin America – still use traditional techniques on tiny plots, producing barely enough to survive.

Without capital, not only for marketing and processing infrastructures but also for improved inputs and techniques, draught animals, irrigation systems, better storage and extension services, neither yields nor plots sizes will increase significantly. Thus, for the majority, change requires large, wide-scale programmes to enable villages and individual farmers to invest to change the scale and intensity of their operations.

The vision and the effort must be big! And also at the level of the village! If not, the majority will be ignored and impoverished – for generations.

References

Abbot, P. (2009) *Development Dimensions of High Food Prices*. Consultancy report for the OECD.

Alfieri, A., Arndt, C. and Cirera, X. (2007) *Distortions in Agricultural Incentives in Mozambique*. Directorate of Studies and Policy Analysis, Ministry of Planning and Development, Mozambique, Working Paper 54.

Alston, J., Marra, M., Pardey, P. and Wyatt, T. (2000) *A Meta-analysis of Rates of Return to Agricultural R & D: Ex Pede Herculem?* International Food Policy Research Institute, Washington, DC.

Anderson, J. and Feder, G. 2004. Agricultural extension: good intentions and hard realities. *World Bank Research Observer* 19(1), 41–60.

Arndt, C., Benfica, R., Maximiano, N., Nucifora, A. and Thurlow, J. (2008) *Higher Fuel and Food Prices: Economic Impacts and Responses for Mozambique*. International Food Policy Research Institute, Discussion Paper 00836, Washington, DC.

Banda, S. Jr (2009) Mozambique rice production increases. Africa News Network, July 17.

Benfica, R., Miguel, A., Zandamela, J. and Sousa, N. (2005) The economics of smallholder households in tobacco and cotton growing areas of the Zambezi Valley of Mozambique. Directorate of Economics, Ministry of Agriculture and Rural Development, Mozambique, Research Report 59E.

Castel-Branco, C. (2008a) *Notas de Reflexão Sobre a 'Revolução Verde': Contributo para um Debate*. Instituto de Estudos Sociais and Económicos, Discussion Paper 02/2008, Mozambique.

Castel-Branco, C. (2008b) *Desafios do Desenvolvimento Rural em Moçambique: Contributo Crítico com Debate de Postulados Básicos*. Instituto de Estudos Sociais e Económicos. Discussion Paper 03/2008, Mozambique.

Chianu, J., Adesina, A., Sanginga, P., Bationo, A., Chianu, J. and Sanginga, N. (2008) Structural change in fertilizer procurement method: assessment of impact in sub-Saharan Africa. *Journal of Business Management* 2(3), 65–71.

Connolly, L. and Athenry, T. (2008) Rising costs reduce farm profit. *Irish Farmers' Journal*, 31 May.

Coughlin, P. (2006) Agricultural intensification in Mozambique infrastructure, policy and institutional framework: an overview. An extensive literature review prepared for the Afrint Project of Lund University, Sweden.

Coughlin, P. and Givá, N. (2009) *Agricultural Intensification in Mozambique: Lessons from 10 Villages—Afrint II*. Report for the Afrint II Project of Lund University, Sweden.

Cramer, C. (2001) Privatisation and adjustment in Mozambique: a 'hospital pass'? *Journal of Southern African Studies* 27, 79–103.

Cravinho, J. and George, E. (2007) Mozambique: economy. In: Frame, I. (ed.) *Africa South of the Sahara 2008*. Routledge, London.

Cunguara, B. (2008) Pathways out of poverty in rural Mozambique. MSc thesis, Michigan State University, East Lansing, Michigan.

DAP (1999) Avaliação dos impedimentos para o desenvolvimento do sector dos insumos agrícolas e veterinários. Draft report. Department of Policy Analysis, Ministry of Agriculture and Pescas (DAP), Mozambique.

Debrah, S. (2000) *From State Monopoly to Private Sector Oligopoly: African Agricultural Input Markets in Crisis*. International Fertilizer Development Centre, Lome, Togo.

de Vletter, F. (2006) *Microfinance in Mozambique: Achievements, Prospects & Challenges*. A report of the Mozambique Microfinance Facility.

Eicher, C. (2002) Mozambique: an analysis of the implementation of the Extension Master Plan. Department of Agricultural Economics, Michigan State University, East Lansing, Michigan, Staff Paper 2002-31.

Eicher, C. (2004) Mozambique: building African models of agricultural extension. In: Riveira, W. and Alex, G. (eds) *National Strategy and Reform Process: Case Studies of International Initiatives,* vol. 5 of *Extension Reform for Rural Development*. Discussion Paper 12, Agriculture and Rural Development Department, World Bank. Washington, DC. Available at: http://lnweb18.worldbank.org/ESSD/ardext.nsf/11ByDocName/ExtensionReformforRuralDevelopmentVolume5NationalStrategyandReformProcess/$FILE/Extension_Reform_V5_final.pdf (accessed 6 April 2010).

Ernst & Young and EconPolicy Research Group, Ltd (2005) Reformulação da política e estratégias industriais em Moçambique. Report prepared for the Ministry of Industry and Commerce, Mozambique.

Evenson, R. (2001) Economic impacts of agricultural research and extension. In: Gardner, B. and Rausser, G. (eds) *Handbook on Agricultural Economics*. Elsevier, Amsterdam, pp. 573–628.

Evenson, R. and Mwabu, G. (2001) The effect of agricultural extension on farm yields in Kenya. *African Development Review* 13, 1–24.

FAOSTAT data (2005) Available at http://faostat.fao.org/default.aspx.

Ferizli, A., Navarro, S., Donahaye, J.E., Rinder, M. and Azrieli, A. (2001) Airtight granary for use by subsistence farmers. In: Donahaye, E., Navarro, S. and Leesch, J. (eds) *Proceedings of International Conference on Controlled Atmosphere and Fumigation in Stored Products*, 29 October–3 Nov, Fresno, California. Executive Printing Services, Clovis, California, pp.37–43.

GDS (2005) Value chain analysis for strategic sectors in Mozambique. Report prepared for The Enterprise Development Project (PoDE). Global Development Solutions, LLC, Mozambique.

Gemo, H., Eicher, C. and Teclemanrian, S. (2005) *Mozambique's Experience in Building a National Extension System*. Michigan State University Press, East Lansing, Michigan.

Giovannucci, D., Varangis, P. and Larson, D. (2001) Warehouse receipts: facilitating credit and commodity markets. Available at: http://web.worldbank.org/wbsite/external/topics/extard/0,,contentMDK:20440946~pagePK:210058~piPK:210062~theSitePK:336682,00.html (accessed 7 April 2010).

Hanlon, J. (2010) Some notes on rural income in Mozambique. Unpublished summary based on preliminary data from the *Trabalho de Inquérito Agrícola ao Sector Familiar*, 2008–2009.

Howard, J., Jeje, J., Tschirley, D., Strasberg, P., Crawford, E. and Weber, M. (1998) What makes agricultural intensification profitable for Mozambican smallholders? An appraisal of the inputs subsector and the 1996/97 DNER/SG2000 Program. Department of Agricultural Economics, Michigan State University, East Lansing, Michigan, International Development Working Paper 69.

Howard, J., Kelley, V., Maredia, M., Stepanek, J. and Crawford, E.W. (1999) Progress and problems in promoting high-external-input technologies in sub-Saharan Africa: the Sasakawa-Global 2000 experience in Ethiopia and Mozambique. Paper for the Annual Meeting of the American Agricultural Economics Association, Nashville, Tennessee, 8–11 August.

Howard, J., Jeje, J.J., Kelly, V. and Boughton, D. (2000) Comparing yields and profitability in MADER's high- and low-input maize programs: 1997/98 survey results and analysis. Directorate of Economic, Ministry of Agriculture and Rural Development, Research Report no. 39E, Mozambique.

Howard, J., Low, J., Jeje, J.J., Boughton, D., Massingue, J. and Maredia, M. (2001) Constraints and strategies for the development of the seed system in Mozambique. Directorate of Economics, Ministry of Agriculture and Rural Development, Research Report no. 43E, USA.

Huang, W., McBride, W. and Vasavada, U. (2009) Recent volatility in U.S. fertilizer prices: causes and consequences. *Amber Waves* 7(1), 1–21.

IFDC (2009) Mozambique: maize intensification in Mozambique (MIM). Available at: www.fertilizer.org/ifa/content/download/21536/309188/version/2/file/AF_2009_durban_hammond_2.pdf (accessed 7 April 2010).

INE (2008) *Contas Nacionais de Moçambique*. INE, Maputo, Mozambique.

Issufo, A. (2003) Current legislative and policy changes and their effect on forestry and land use in Mozambique. Paper prepared for the First International Workshop on Promoting Common Property in Africa: Networks for Influencing Policy and Governance of Natural Resources (Co-Govern), South Africa.

Kelly, V., Adesina, A. and Gordon, A. (2003) Expanding access to agricultural inputs in Africa: a review of recent market development experience. *Food Policy* 28, 379–404.

KIT (2000) O ProAgri em Moçambique: reforma institutional do sistema de investigação agrária. KIT (Royal Tropical Institute). Report for the Ministry of Agriculture and Rural Development, Mozambique.

Kumar, S. (2007) Dynamics of the global fertilizer market. PowerPoint presentation, Institute of Agriculture and Environment Research, Oslo, Norway.

Langyintuo, A. and Mekuria, M. (2005) Accounting for neighborhood influence in estimating factors determining the adoption of improved agricultural technologies. International Maize and Wheat Improvement Center (CIMMYT), Mount Pleasant, Harare, Zimbabwe.

Langyintuo, A. and Mekuria, M. (2008) Assessing the influence of neighborhood effects on the adoption of improved agricultural technologies in developing agriculture. *African Journal of Agricultural and Resource Economics* 2(2), 151–169.

MADER (2003a) Alguns aspectos da comercialização agrícola: acesso de produtores e integração nos mercados. Contribuição ao *Relatório Conceitua,* nota de discussão para MADER(Ministério de Agricultura e Desenvolvimento Rural) ProAgri2.

MADER (2003b) *Trabalho de Inquérito Agrícola ao Sector* Familiar, 2002–2003. CD ROM.

MADER (2004) *Strategy Document: ProAgri II*. MADER, Maputo, Mozambique.

MADER (2006) *Trabalho de Inquérito Agrícola ao Sector Familiar*, 2004–2005.

MAG (Ministry of Agriculture) (2007a) Conceito, princípios e estratégia de Revolução Verde em Moçambique, Maputo, Mozambique.

MAG (2007b) *Trabalho de Inquérito Agrícola 2007 (TIA)*. Ministry of Agriculture, Maputo, Mozambique.

MAG (2008) Trabalho de Inquérito Agrícola 2007 (TIA): Dissemination Summary, Maputo, Mozambique.

Mole, P. (2000) Smallholder cashew development opportunities and linkages to food security in Nampula Province, Mozambique: summary of findings and implications for policy, research and extension efforts. Directorate of Economics, Ministry of Agriculture and Rural Development, Research Report no. 42E.

Mole, P. (2006) *Smallholder Agricultural Intensification in Africa: Mozambique Micro Study Report—Afrint*. Report for the Afrint Project of Lund University, Sweden.

Mozambique (2001) Plano de Acção para a Redução da Pobreza Absoluta, 2001–2005 (Parpa): Documento de Estratégia e Plano de Acção para a Redução da Pobreza e Promoção do Crescimento Económico. Conselho de Ministros do Governo de Moçambique, Maputo, Mozambique.

Nathan Associates (2004) *Mozambique: Diagnostic Trade Integration Study—Main Report*. Study for the Ministry of Industry and Commerce funded by the Trade Capacity Building Project, U.S. Agency for International Development, Mozambique.

OECD-FAO (2009) *OECD-FAO Agricultural Outlook 2009–2018*. Available at: www.oecd.org/dataoecd/2/31/43040036.pdf (accessed 7 April 2010).

Oehmke, J. and Crawford, E. (1993) The impact of agricultural technology in sub-Saharan Africa: a synthesis of symposium findings. Michigan State University, Department of Agricultural Economics, East Lansing, Michigan, International Development Paper 14.

Oehmke, J., Anandajayasekeram, P. and Masters, W. (1997) Agricultural technology development and transfer in Africa: impacts achieved and lessons learned. USAID, Office of Sustainable Development, Bureau for Africa, SD Publication Series. Available at: http://pdf.dec.org/pdf_docs/pnacb618.pdf (accessed 7 April 2010).

Owens, R., Hoddinott, J. and Kinsey, B. (2003) The impact of agricultural extension on farm production in resettlement areas of Zimbabwe. *Economic Development and Cultural Change* 51, 337–358.

Pender, J., Abdoulaye, T., Ndjeunga, J., Gerard, B. and Kato, E. (2008) Impacts of inventory credit, input supply shops, and fertilizer microdosing in the drylands of Niger. IFPRI Discussion Paper 00763, Washington D.C., USA.

Perumalpillai-Essex, J. (2005) Impacts of extension services in rural Mozambique. Report by ECON Analysis for the Environment, Rural and Social Development Department, Africa Region, World Bank, Washington, DC.

Roelf, W. (2009) Fertilizer prices expected to increase in 2010. *Africa News Network*. Available at: www.ask.com/bar?q=AGRA+Beira+fertilizer+plant&page=1&qsrc=2106&dm=all&ab=2&u=http%3A%2F%2Fwww.africanagricultureblog.com%2F2009%2F06%2Ffertilizer-prices-expected-to-increase.html&sg=Iam2w72yXNrSE1vVvdGMAIXjFVIL8nxZc7Vcw1POJxs%3D&tsp=126737830638 (accessed 7 April 2010).

Roseboom, J., Beintema, N. and Mitra, S. (2003) Building impact-oriented R&D institutions. Background Paper No. 3 commissioned by the InterAcademy Council (IAC) Study Panel on Science and Technology Strategies for Improving Agricultural Productivity and Food Security in Africa. Available at: www.interacademycouncil.net/Includes/DBLink.asp?ID=9063 (accessed 7 April 2010).

SANAGRI (2002) *Review of Danida-supported Extension and Research Activities within the Agricultural Sector Programme Support (ASPS): Final Report—Mozambique*. Report commissioned by the Mozambique Ministry of Foreign Affairs and Danida.

Skalsky, S. (2008) Impact of fuel and nitrogen prices on profitability of selected crops: a case study. *Agronomy Journal* 100(4), 1161–1165.

Tripp, R. (2003) Strengthening the enabling environment for agricultural technology development in sub-Saharan Africa. Overseas Development Institute, Working Paper 212. Available at: www.odi.org.uk/publications/working_papers/wp212.pdf (accessed 6 February 2006).

Uaiene, F. (2006) Introduction of new agricultural technologies and marketing strategies in central Mozambique. Institute of Agricultural Research of Mozambique, Directorate of Training, Documentation, and Technology Transfer, Mozambique, Research Report No. 2E.

Uaiene, R., Arndt, C. and Masters, W. (2009) Determinants of agricultural technology adoption in Mozambique. National Directorate of Studies and Policy Analysis, Ministry of Planning and Development, Mozambique, Discussion paper no. 67E.

USDA (2009) *Agricultural Projections to 2018*. Long-term Projections Report OCE-2009-1. USDA (Interagency Agricultural Projections Committee, US Department of Agriculture) Washington, DC.

Varajidás, B. (2005) Contract farming's credit schemes as an alternative credit source for the smallholders: a case study from Mozambique. MSc Thesis, University of Oslo, Oslo. Available at: http://wo.uio.no/as/WebObjects/theses.woa/wa/ these?WORKID=28342 (accessed 7 April 2010).

Villiers, P., deBruin, T. and Navarro, S. (2006) Development and applications of the hermetic storage technology. *Proceedings of the 9th International Working Conference on Stored Products Protection (IWCSPP)*, Sao Paulo, Brazil.

Vitek, F. (2009) An assessment of external price competitiveness for Mozambique. IMF Working Paper 09/165. Available at: www.imf.org/external/pubs/ft/wp/2009/wp09165.pdf (accessed 7 April 2010).

Walker, T., Tschirley, D., Low, J., Tanque, M.P., Boughton, D., Payongayong, E. and Weber, M. (2004) Determinants of rural income, poverty, and perceived well-being in Mozambique in 2001–2002. Economics Directorate, Ministry of Agriculture and Rural Development, Mozambique, Research Report 57E.

World Bank (1996) Mozambique agricultural sector memorandum, AFTA1. Washington, DC.

World Bank (2002) African Development Indicators 2002: drawn from World Bank Africa database. World Bank, Washington, DC.

World Bank (2003) Mozambique country assistance strategy. Memorandum of the president of the International Development Association, International Finance Corporation and the Multilateral Investment Guarantee Agency to the Executive Directors on a Country Assistance Strategy for the Republic of Mozambique, Report. World Bank, Washington, DC.

World Bank (2005a) *Mozambique Country Economic Memorandum: Sustaining Growth and Reducing Poverty*. Report 32615-MZ, Washington, DC, pp. 1–139.

World Bank (2005b) *Mozambique Agricultural Development Strategy: Stimulating Smallholder Agricultural Growth*. Report No. 32416-MZ, Washington, DC, draft of 19 September.

Zavale, H., Mabaya, E. and Christy, R. (2005) Adoption of improved maize seed by smallholder farmers in Mozambique. Department of Applied Economics and Management, Cornell University, Ithaca, New York.

Appendix

Table 13A.1. Percentage of respondents knowing or using improved agricultural techniques, by sex and region. (From: Coughlin and Givá, 2009:16.)

	% of all respondents within each region								% by sex of farm manager			
	North		Centre		South		All three regions		Male		Female	
Technique	Knows	Uses	Knows	Uses	Knows	Uses	Knows	Uses	Knows	Uses	Knows	Uses
Crop rotation	81	56	37	7	61	9	59	27	62	31	53	18
Intercropping	99	94	89	82	58	56	87	82	93	89	73	65
Intercropping with nitrogen-fixing crops (beans, etc.)	91	83	60	42	96	85	79	67	80	66	77	69
Fallowing	80	57	36	4	44	8	55	26	60	29	44	19
Improved fallowing	3	1	7	1	5	3	5	1	5	1	6	2
Animal manure	21	0	45	1	62	5	39	2	37	1	42	3
Zero or minimum tillage	1	0	23	23	63	61	22	21	14	13	43	41
Breaking the hard pan	1	0	12	4	18	5	9	3	7	3	14	3
Green manure/compost/residue incorporation	6	0	24	19	25	1	17	8	16	8	20	7
Chemical fertilizer	22	0	33	0	39	0	30	0	30	0	31	0
Soil and water conservation (level bunds, grass strips, terracing, etc.)	21	7	13	1	27	1	19	3	18	4	20	2
Improved planting practices	24	6	48	31	38	16	36	18	35	17	39	19
Integrated (soil) nutrient management (INM)	1	0	4	0	5	0	3	0	3	0	2	0
Integrated pest management (IPM)	1	0	6	0	5	1	3	0	4	0	3	0
Agro-forestry	3	0	7	1	3	0	4	0	5	0	3	0
Pesticides/herbicides	18	1	25	4	4	0	18	2	20	1	12	2
Rainwater harvesting	4	0	11	1	37	3	13	1	11	1	20	1
Irrigation	54	0	47	3	76	1	55	1	56	1	55	2
Average number of respondents per question	160	161	163	160	79	78	401	398	280	277	121	121

Table 13A2. Agricultural techniques applied and how they were learned. (From: Coughlin and Givá, 2009:16.)

		If used, from where did you learn the technique? (%)								
	Not practising this technique (%)	My parents or a family member	A fellow farmer or a neighbour	An extension agent, an NGO or other formal organization	The radio, newspaper or TV	An NGO	An input supplier or private trader	Another source	Total %	Total respondents
Crop rotation	68	25	6	1				1	100	344
Intercropping	13	73	12	1				1	100	373
Intercropping with nitrogen-fixing crops (beans, etc.)	28	59	11	1	1				100	377
Fallowing	68	23	8	1				1	100	346
Improved fallowing	98	1	1						100	335
Animal manure	99		1						100	333
Zero or minimum tillage	76	22	2	1					100	355
Breaking the hard pan	96	3	1						100	336
Green manure/compost/residue incorporation	91	7	1						100	339
Chemical fertilizer	99								100	334
Soil and water conservation (level bunds, grass strips, terracing, etc.)	96	1	3						100	334
Improved planting practices	81	10	8	1		1		1	100	345
Integrated (soil) nutrient management (INM)	100								100	335
Integrated pest management (IPM)	100								100	336
Agro-forestry	100								100	336
Pesticides/herbicides	98		1	1					100	336
Rainwater harvesting	99								100	335
Irrigation	97	1	1						100	336

Note: A blank means zero. Rows may not sum exactly due to rounding.

14 Conclusions: What Direction for the Future of African Agriculture?

Ernest Aryeetey,[1] Göran Djurfeldt[2] and Aida C. Isinika[3]

[1]*The Brookings Institution, Washington, DC, USA;* [2]*Department of Sociology, Lund University, Lund, Sweden;* [3]*Institute of Continuing Education, Sokoine University of Agriculture, Morogoro, Tanzania*

The story of African agriculture has always been a mixed one. Agriculture in the region has often performed less well than expected, and it has not always been obvious what the way forward should be. The first publication of the *Afrint* research project (Djurfeldt *et al.*, 2005), to which the current volume is a sequel, concluded that while the Asian Green Revolution could not be transplanted into Africa, there were several lessons to be drawn from the experience. These included the fact that there was scope for a state-driven, market-mediated and smallholder-based Green Revolution. This lesson was important for a smallholder-driven agricultural transformation to meet the Millennium Development Goals in Africa. It was also recognized that while many of the obstacles to progress in African agriculture were basically related to national policies and of a structural nature, one could not ignore the international setting for trade in goods and services and in the exchange of knowledge and how these impacted African agriculture. It would be difficult for African farmers to compete in a globalizing environment that still has significant protection and unfair trading practices. In light of the declining trend in production and productivity for all major food crops in the countries represented in the previous study (*Afrint I*), a number of challenging questions were posed for further reflection. Some of those questions were: Can sub-Saharan Africa handle agricultural development? Can sub-Saharan Africa export itself out of the agricultural crisis? These questions remain relevant today, even if there are still no clear answers.

The current volume presents examples of how selected African governments have played or not played their mediation role to influence the market and entice effective smallholder participation in the staple food subsector. As stated in the introductory chapter (Djurfeldt *et al.*, this volume), African governments have been playing a more active role in terms of policy and institutional reforms, in order to achieve food security and even use agriculture to drive economic growth. There are reports here of subsidies (Malawi, Kenya and Tanzania), government-led agricultural development programmes and initiatives (Zambia, Tanzania, Ghana) and more inclusion of the private sector within government plans and programmes in most countries. These efforts seem to have slowed down somewhat the downward spiral of production and productivity, which was evident throughout the nineties. The evidence presented in this volume indicates that, despite some marginal improvements, the productivity gap remains wide, hence more room for improvement.

The various chapters in this book provide significant new insights into how the poor performance of African agriculture can best be explained. They also show how different types of farmers have responded to policy initiatives and public programmes in different countries, while providing explanations for the variations. It is important to keep in mind that the role that the state can or should play in agriculture is far from settled and the different country experiences show how those countries approach this issue and what the responses from their farmers and others have been.

Understanding the Poor Performance of Agriculture

Many reasons have been assigned for the poor performance of African agriculture, and indeed the reasons do not vary much from country to country. The list is endless and covers such issues as inappropriate agricultural/farming practices (including indiscriminate burning to clear land and shorter fallow periods), lack of appropriate soil information and the mining of soil nutrients, inappropriate land-use planning and practices, insecure land tenure systems and hence a need for land tenure reforms, lack of efficient and effective extension delivery services, widespread illiteracy, non-use of new technologies, lack of adequate policy measures on sustainable agriculture and diversification, lack of policy harmonization among the various sectors of the economy, lack of other employment opportunities besides agriculture, lack of choice on the part of people to be committed to maintaining natural resources, poor human settlements planning, lack of rural infrastructure, including irrigation, storage, roads and energy, lack of agricultural credit, lack of adequate opportunity for non-farm activities and poor organization and definition of market structures. This list is not by any means exhaustive. There are some that are clearly symptomatic of other deeper problems. The various chapters of this book have explored the current situation in a number of countries in light of national and global policy changes and contribute here to the discussion of solutions to these challenges.

Holmén and Hydén (Chapter 2, this volume) analyse the failure of a Green Revolution in Africa over the last three decades and consider the prospects for one in the future. They note that the levels of both expectations and emphases on agriculture have shifted over the years in the donor and research community, often for the wrong reasons. In discussing the failure of past efforts, the authors point to the 1960s and 1970s, i.e. before structural adjustment programmes (SAPs). They note that state interventions, however well intentioned, were inefficient and often circumvented by smallholders. They were often very poorly managed by the state institutions responsible for them. They also observe, however, that the removal of these programmes and their replacement with SAPs from the early 1980s may have made things even worse, as essential markets remained missing at the same time that government support was being removed. Interestingly, while the situation was unfolding, donor support for agriculture also fell in the 1980s and 1990s. The food crisis in 2008 only exposed the vulnerability of a system where nothing seems to work well. In particular, it revealed that smallholders, rather than benefitting from increasing prices, were actually more likely to withdraw from markets, as they had become net buyers of food. There are concerns that price volatility, combined with the global financial crisis, will further reduce investments in agriculture. Clearly, the issue of what role governments in Africa choose to play will remain critical in explaining the problems of agriculture. Considering, however, that many African governments do not have adequate resources for supporting agriculture, a lot will also depend on the relationship that development partners have with African governments. If the development partners choose not to support agriculture, it will be very difficult for governments in the region to find the means for making the development of agriculture meaningful.

Government agricultural policies and agricultural outcomes

All countries in Africa have their own experiences of how they envisioned agricultural development and how these led to different policy choices. The policy choices have been driven by the nature of the domestic social and economic debates and by the nature of the national politics. Whether the policies worked or not in delivering expected outcomes depended on how all of these factors were accounted for in policy design. A good example of how policies have evolved in this context is provided by the study of Mozambique (Chapter 13, this volume). Mozambique's experience is not by any means unique and reflects the challenges that confront agriculture in the region.

Coughlin (Chapter 13, this volume) discusses agricultural developments in Mozambique within the context of the government's policies. He begins by tracing the shift from socialist agriculture to liberalization through the Economic Rehabilitation Programme. After this structural adjustment programme, driven by the state and donors, he contends that there are great possibilities for increased investment. Coughlin notes that Mozambique is notable for its sparse use of agricultural inputs and low use of extension services. He indeed pays a lot of attention to poor extension services as a reflection of poor institutional development. The extension system has not expanded, and thus there are only 1.3 agents for every 10,000 rural inhabitants. Most farmers thus receive no services, and a deliberate strategy of focusing on 'high potential' areas has further reduced coverage. Coughlin argues that better rotation of these workers could increase the marginal impact of their services and that they should promote strategies that both increase yields and reduce costs, so that they are actually implemented. Despite poor extension services, infrastructure is improving, and competition in the input sector is growing. Thus, the author argues that a paradigm shift is called for in order to change the way that extension services are used, such that alternative inputs are encouraged and better value chains are developed.

Although growth in Mozambique has been high and agriculture is expanding, yields are low and stagnant, as in most places. The macroeconomic environment has encouraged agriculture in the short run, in part by suspending taxes on petroleum to blunt the impact of fuel prices on farmers. However, this policy may not be sustainable. Low profitability, stemming from these conditions, may be one cause for low yields, but credit access is also severely lacking, hampering the ability to access new technologies. Projects aimed to improve productivity, often inspired by non-governmental organizations (NGOs), have been unsustainable. Coughlin instead proposes focusing on simple things, like better storage technology, to improve farm incomes in the short term. He also argues that because seed and fertilizer constitute the vast majority of non-labour production costs, organic fertilizers and reduced costs for inorganic fertilizers could have a positive effect. He does not mention the possibility of fertilizer subsidies, which have been common in recent years in sub-Saharan Africa. He is, however, positive about the government's support for agriculture through other tax measures that have assisted farm product prices, like the National Rate of Assistance.

Coughlin is also interested in how research knowledge is created, as indeed in most countries. The agricultural research network, which was previously very fragmented, was consolidated in 2005 under the Institute of Agrarian Research of Mozambique. This institute has significant autonomy, although it is unclear how much more effective it is than the previous institutions.

Finally, Coughlin argues that Mozambique's strategy for agriculture is vague and lacking in specific priorities. This is clearly a sentiment that may be shared in many countries. Adequate resources do not exist to implement the strategy, and resources are spread thinly and incoherently. Despite this, Coughlin argues for larger investments in infrastructure and value chains, arguing that advice from extension workers is more effective when coupled with these investments. It remains unclear, however, how the 'big vision' he calls for will avoid the unfocused pitfalls of the development strategy that he critiques, but that is the nature of African agriculture and its policies.

How far governments will go in terms of policy is again reflected by the Nigerian study (Chapter 11, this volume). The wide array of policies can sometimes be confusing. For example, in pursuit of the goal of modernizing agriculture the government has increased import duties on rice and soybeans, subsidized export of crops in general and banned cassava imports and maize exports. There have been modest input policies with partial subsidies for fertilizer, although these have failed to reach smallholders. The government has made credit available directly and has supported research activities aimed at both new seed varieties and human capacity development. Nigeria has also partnered with donors on irrigation projects. Clearly, government is very much at the centre of agricultural development in Nigeria, and its failures mean failure for Nigerian agriculture.

The inability of governments to organize markets properly is also the main theme of the work on Tanzania (Chapter 12, this volume). The authors focus on three key constraints to agricultural development. First, the subsistence nature of markets is still prevalent. Sale of maize has increased slightly, but only 38% sell their maize. Maize sales are extremely local, and prices received by farmers are low, which the authors argue is a result of poor bargaining power from high need for cash. Secondly, transaction costs are high because of difficult infrastructure issues and lack of information. Competition is low, as middlemen collude to offer lower prices. Thirdly markets are very thin, both for inputs and for credit, evidenced by high prices of both and particularly low access to credit. The researchers argue that market development is crucial for the agricultural reform agenda and will result in increased investment and bargaining power for farmers, as well as lower transaction costs. They argue that extension services must be expanded and should help to create linkages throughout the value chain in order to increase market participation. These types of complementary policies are needed to strengthen the weak trends of improvement and bring about real transformation.

Inadequate spending on agriculture and the role of the state

One of the commonest summaries of the problems facing agriculture is the fact that not enough is spent on it, as seen throughout this volume. Spending on agriculture by both the state[1] and by individuals through investment is simply inadequate. Most African countries devote less than 5% of total expenditure to agriculture, despite the fact that they have pledged among themselves under the Comprehensive Africa Agricultural Development Programme (CAADP) to devote as much as 10% of national budgets to the sector. As a result, most of the other challenges that agriculture faces can be associated with this difficult situation. Again this is a fact that is linked to how governments perceive their role in agriculture.

There is ample evidence that many governments spent a little more on agriculture in the past than they do now, even though the efficiency of such spending has always been questionable. Figure 14.1 shows that even though in a number of the *Afrint* study countries the share of agriculture in the national budget went up (especially in Ethiopia, Ghana, Malawi, Mozambique and Nigeria), it was almost always from a very low base. The only exception here is Ethiopia. In Tanzania, Uganda and Zambia agricultural spending went down as a share of the national budget.

Indeed, apart from the low level of spending by the state, there have been several questions about the effectiveness and efficiency of public spending (World Bank, 2008). In many countries the objectives for public spending on

[1] 'State' is used throughout this chapter to refer to government in the broadest sense.

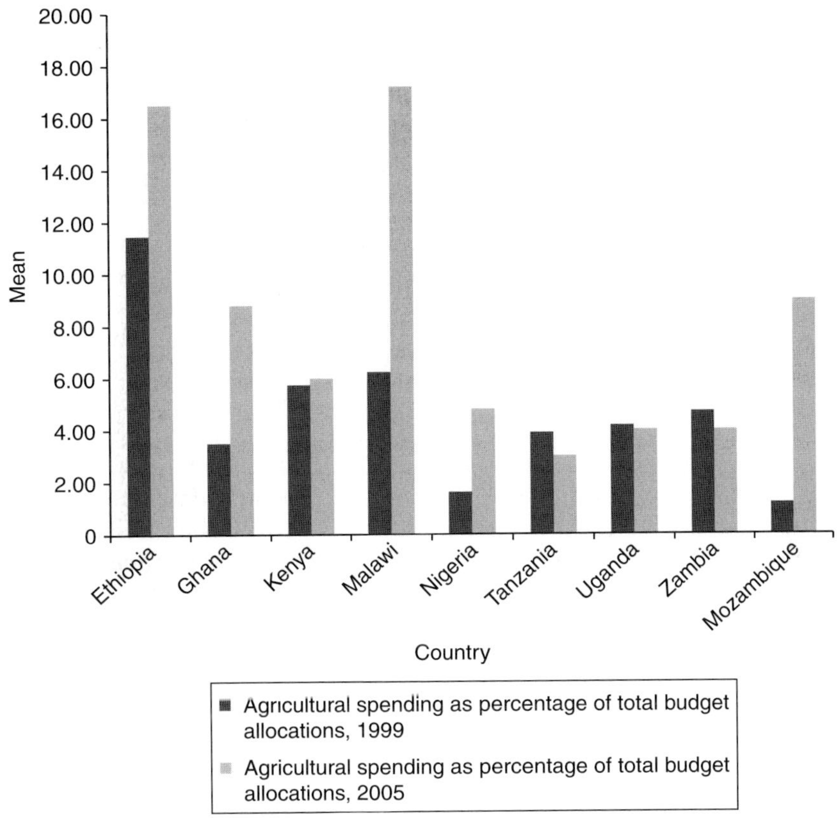

Fig. 14.1. Agricultural spending as a percentage of total budget allocations by country and year. (Adapted from: data from various sources.) Note: the figures for public expenditure refer to the situation in 1999 and 2005, respectively.

agriculture are not well defined. Many have argued that the states' spending is poorly structured and seldom crowds in any private investment (World Bank, 2008). In the study of Zambia (Chapter 10, this volume) the researchers note that, given that the aim is to enhance the private sector's viability, the programmes have not been successful on that score. Crowding-out is evident on both the input and output sides, and uncertainty regarding the timing and magnitude of government interventions has exacerbated this effect. Subsidies and food reserve programmes have come at the expense of other investments in agricultural public goods, such as roads and extension programmes. The programmes have also failed to expand access to agricultural credit, as poor targeting has limited the usefulness of the programmes for poorer farmers. Thus, the authors conclude that the Zambian government must strike a better balance with its agricultural spending to encourage real development in the sector.

In Kenya, the *Afrint* study (Chapter 9, this volume) shows that between the first post-independence period and the more recent period, there have been changes in the role of the state with regard to funding. The more recent period after ending structural adjustment and the immediate post-independence period are considered to be similar because of the heavier involvement of the government in the agricultural sector. In the second period in between, the authors argue that budgetary support for agriculture dropped because of structural adjustment programmes, and gains in performance slowed down as well. The authors argue that government support is necessary for all of the six 'I's[2] to be adequately emphasized. Regarding the most recent period, the authors particularly point to increased credit access as a main driver of productivity gains through technology and input channels. In particular, government support for savings and credit cooperatives was essential to the flow of credit. Additionally, renewed subsidies for inputs, better extension service provision and more money for infrastructure and irrigation are described as key factors for increasing productivity.

In their paper on the drivers of maize production (Chapter 5, this volume), Andersson *et al.* show that increased budget allocations to agriculture had no traceable effect on production between 2000 and 2008 in the eight countries studied. This ties in with the issues of efficiency and effectiveness of government spending. Still, there is no doubt that government spending on agriculture will remain crucial to the direction that developments in the sector will follow. What is important is for governments to not only raise their commitments to agriculture but also structure expenditures in a manner that allows such expenditures to draw in private investments and enhance productivity. Governments need to be strategic in their spending choices. Strategies must aim at removing bottlenecks to private investment, such as the provision of essential infrastructure (including roads, irrigation, energy and research) and the development of market institutions, including for agricultural finance. Governments should thus seek to minimize the risks associated with agriculture.

Increasing population pressure and land scarcity

Africa remains the fastest-growing region in the world in terms of population. It is projected that Africa's population will more than double between 2010 and 2050, while the population in Asia (excluding China) will only grow by 36%. The Americas will see their population increase by only 25% in the same period and Europe's population will be reduced by 5% (UN, 2009). With this rapid increase in population in Africa, a trend that has been present over most of the last century, it is not surprising that larger numbers of people are often shown to be scrambling for the more productive lands and poorer and less powerful households or groups get pushed to less productive or marginal lands. It is generally known that the extent of land scarcity varies by country.

[2] Innovation, Inputs, Infrastructure, Institutions, Information and Incentives.

The *Afrint* studies provide some striking results in this regard, as shown in the paper by Jirström *et al.* (Chapter 4, this volume). First, they find that farm sizes have been decreasing or remained stagnant in almost all the countries studied, and that per capita land can be very small in some cases, such that the bottom quartile in some countries are nearly without land. Inequality in land distribution has increased in some countries but not in others. As a result, crop composition has remained pretty stable. The Tanzania study (Chapter 12, this volume) confirms the situation when it is noted that more land was brought under cultivation between the two *Afrint* studies but yields remained stagnant in the same period. The situation is not any different in the other countries, as increased production is very strongly correlated with land under cultivation, as is shown in the paper by Andersson *et al.* (Chapter 5, this volume) on the drivers of maize production.

The most challenging question arising from the land and population issues is whether there is a need for land tenure reform in African countries. Obviously the answer to that question has to be driven by individual country contexts, but it is also obvious that all countries have to find ways of encouraging both equitable and efficient uses of land. There are not many proposals for achieving this dual objective but the proposal by Aryeetey and Udry (2010) for Ghana offers a fresh new look at land tenure reform. It is important that land tenure reform leads to a more equitable as well as efficient use of land. It should not increase social and political tensions unduly, and there are opportunities for doing that through the land bank concept (Aryeetey and Udry, 2010).

Technology use and innovations in African agriculture and the role of the state

The low spending on agriculture is amply reflected by the limited use of new inputs that could lead to higher yields. Indeed, while the rest of the world has moved much faster, relying on new technologies and innovations that are generally well known, this has not happened in Africa. Biotechnology, improved fertilizer use, improvements in seed quality and storage, etc. have helped to expand the production of food and other agricultural products in many dimensions in other parts of the world. Jirström *et al.* (Chapter 4, this volume) apply the concept of yield gaps to show the potential for smallholder productivity. Instead of relying on experimental yields, they set the target as the yield in the top 5% of farm households. The data shows a significant variation in yields at country, region and village levels over time. The breakdown of the production specifics for a number of crops shows that for maize, for example, total production and yields have fallen for most countries. Expansion in production, even in countries like Malawi, has only occurred through significant land expansion, not yield growth, and this contrasts with the national data. Sorghum yields have also fallen in most countries. Rice production has expanded in some countries and contracted in others. However, across the different staple crops, yield gaps have consistently been between 54% and 66%, and although the authors do not explain the yield gaps in full, they point to technological adoption as a key driver. The use of fertilizer and better seeds has been stagnant at best.

Akande *et al.*, in the chapter on Nigeria (Chapter 11, this volume), argue that the commercial incentive to adopt more expensive inputs is missing. Furthermore, much of the harvest never makes it to market, even for those who do sell their crops at all. Although they can only draw conclusions for maize, the authors conclude that Nigeria's Green Revolution is yet to get started and that stronger incentives are needed at the farm level. The government must ensure that fertilizer availability is not stunted by its own decreased procurement and that technological adoption increases.

While it is possible to simply attribute the low level of new technology usage in Africa to the low level of education of farmers, it is also important to understand that there are many other institutional factors that explain the low application of new technologies in African agricultural production. Admittedly, low education levels make the transfer of technology more difficult. But low education levels can be contained with appropriate extension services and proper adaptation of new technologies by public agencies, and this is often not available to farmers. The need for new technologies is compounded by the growing challenges of climate change, which require new measures for mitigation and adaptation, developments that have not yet been factored into national policies on agriculture. Clearly the institutional arrangements for passing on new technologies and new knowledge to farmers are very much challenged in most countries.

In many countries, the use of new technologies is generally influenced by the role that the state chooses to play. In Zambia, for example, the authors argue that the Fertilizer Support Programme (FSP) has done more harm than good, through a number of channels. In making this suggestion, they analyse the effects of the Food Reserve Agency (FRA) on the private market, offer a macro-analysis of the programmes and raise a number of issues about FSPs. They argue that, first, because of poor targeting, the subsidies and direct provision of fertilizer have displaced private spending on the same goods and crowded-out private sellers of fertilizer from the business. Secondly, uncertainty regarding the timing and volume has exacerbated this effect. They also note that FSPs have been enabled by a removal of donor conditionality and debt relief, but that provision levels have been inconsistent and are potentially unsustainable (Haantuba *et al.*, Chapter 10, this volume).

The way in which the state can influence new technology application and how this affects output is well reflected by the Zambian study (Chapter 10, this volume). For their micro-analysis, using *Afrint* data, the authors provide a comparison of production before and after implementation of FSPs, which shows that the subsidies provided by the state have indeed increased the planting of maize at the expense of other crops, including sorghum. They show that where sorghum is grown, usage of hybrid seeds is much lower than in maize crops, signalling that advancements in other crops have been stunted by the targeting of maize. They find that the FRA has crowded-out private traders, as the FRA became a much more important marketing channel for surveyed households by 2007, primarily at the expense of private traders. Their regression analysis also demonstrates that fertilizer expenditure and the use of state agencies for selling crops are key determinants of maize production.

There is no doubt today that fertilizers, properly applied, can make a major difference to production and productivity. Similarly, newly developed seeds hold significant promise in many areas. What is not yet well established is the exact role that the state should play in order to get them properly introduced to farmers. The recent Malawian experience with fertilizer subsidies has often been used as a major argument for subsidies, even though the jury is still out. The one conclusion we draw from all of these experiences is that improved technologies and innovations can have large positive effects on agriculture in Africa, but it is unlikely that significant infusions of new approaches will take place simply in response to market conditions. The market institutions have not been properly developed and this therefore requires the state to be more engaged than necessary in the supply of new technologies. What states in Africa can do is to develop functioning market institutions which will therefore not require them to engage in the actual supply of technology. As will be discussed below in our conclusion, the African Green Revolution may see another role for the private sector in the supply of technology and other inputs than the Asian Green Revolution did.

Food security, food self-sufficiency and African agriculture

One of the biggest challenges to policy makers in Africa is whether agriculture can provide the means to attain food security and food self-sufficiency. The challenge remains one of the most daunting, largely in view of the fast-growing population. The study by Dzanku and Sarpong (Chapter 8, this volume) provides some insights into how issues of food security and self-sufficiency are shaping up in Ghana. Their chapter uses a micro-analysis to determine the relationship between agricultural diversification, food self-sufficiency and food security in Ghana, as well as the role that infrastructure and institutions play in this relationship. They define diversification as the diversion of resources to increased production of non-staples relative to staple foods. While comparative advantage theory may posit one optimal allocation of resources, high transaction costs and other market failures may lead to an increased emphasis on food security and thus a desire for food self-sufficiency at the expense of diversification. The researchers use as their policy context the Ghana government's desire to modernize agriculture as one way of promoting broad growth and transformation. They show that Ghana's smallholder farms use limited technology and are rainfall dependent, leading to low and volatile productivity. The country as a whole has a high self-sufficiency ratio for many staples but has an overall deficit because it largely imports several items, including rice, wheat, sugar and meat.

Dzanku and Sarpong (Chapter 8, this volume) model the diversification decision at the household level as a function of a desire for self-sufficiency and a desire for the benefits of diversification. They investigate their model using the *Afrint* data from 2002 and 2008. They find that households that are self-sufficient in staple foods are more likely to diversify, although this effect is stronger in the Eastern Region but absent in the Upper-East Region and significant in the overall sample. They find that proximity to a market and membership of a farmer organization is linked to diversification in the Eastern Region, while contact with extension agents is an important factor in the Upper-East. Also, a non-farm income and an index of physical asset holdings is positively correlated with diversification in the Eastern Region, implying complementarities between diversification and non-agricultural activity.

The authors also investigate whether diversification hurts food security. The results here are inconclusive in the Upper East but find a positive relationship in the East. They speculate that the better market conditions in the Eastern Region, which is closer to cities, may explain the discrepancy. Dzanku and Sarpong conclude that geography is an important factor in determining these relationships and that transaction costs are also quite likely a major factor. Institutions, as proxied by organization membership and extension contact, quite likely reduce the need for self-sufficiency. They do not find convincing evidence that diversification helps or hurts food security, except in the Eastern Region. Thus, they conclude that enhancing productivity in staple crops is necessary to improve food security and to encourage diversification.

What this study on Ghana shows is that the issues of food security and self-sufficiency are generally country and area specific, and can be influenced to a very large extent by the nature of investments made in the area. When the geography of an area is more difficult than in other places, it pays for the state to make the necessary investment with an eye on efficiency enhancement. Building on the comparative and competitive advantages of different geographic areas is key in determining what investments to make.

Poorly functioning credit markets

One of the most frequently recurring themes in agriculture is the difficult access to finance, including credit, by farmers. This has often led to governments intervening in the markets, often without much success. The last two decades have seen many studies on the functioning of rural financial markets, with diverse views on what role the state should play. While the state was seen to have a necessary direct role to play in credit delivery in the early post-independence period in most of Africa, that view changed in the midst of structural adjustment programmes. It is today not very clear what role is expected of the state. Many take the position, however, that the state needs to be very pragmatic and support the development of the institutions that will facilitate exchanges. The state must build the market institutions and support them and not replace them (Nissanke and Aryeetey, 1998). This view has led to the evolution of many different types of institutions and arrangements in different countries, the effectiveness of which for agricultural finance is not very clear.

The case study from Ethiopia (Chapter 7, this volume) discusses agricultural credit with state involvement in that country. In analysing financial institutions and results of household surveys, Wolday Amha considers the prospects for the provision of financial and credit services among smallholders in Ethiopia. He first notes that credit is necessary to move smallholders away from subsistence agriculture but that loan delivery has been unsustainable because of poorly structured subsidies and poor repayment, clearly a failure of government. A number of other key issues need to be addressed, including high transaction costs from small loans and a dispersed population, weak institutional capacity and weak infrastructure, both physical/IT and contractual. At the macro level, the Ethiopian government has supported banks, microfinance institutions (MFIs) and cooperatives to provide rural finance. While the regulatory environment has made operation for MFIs easier and has clarified which other institutions and actors can and cannot provide financial services, there are still a number of weaknesses. For instance, multipurpose cooperatives have been emphasized over more appropriate financial cooperatives and savings and credit cooperatives (SACCOs). The role of the latter institutions is not fully fleshed out in current regulations, and a separate law for cooperatives may offer better protection for members and reduce risks of fraud and other illicit behaviours.

The overview of the financial system is supplemented by analysis of the *Afrint* surveys from 2002 and 2008. The results show that about half of respondents had access to agricultural input credit and even more were able to save. Ability to repay is high, at 77.5%, and most report increased access relative to 2002. The surveys also show that cooperatives and MFIs are the primary providers of credit, and multipurpose cooperatives in particular. Loans are primarily for farm inputs and generally occur on 8–12 months terms. The regression analysis provides some interesting results, including the following: (i) land holdings and education are positively correlated with probability of taking loans, while income is negatively correlated; (ii) loan size is positively correlated with land, savings and marketable surplus; (iii) ability to repay is positively correlated with land size and access to extension, and negatively correlated with family size; and (iv) likelihood of saving is correlated positively with education, cash income and land, and negatively with household size.

The author argues that finance must shift to provide products specifically tailored to smallholders. He maintains that, even though an enabling regulatory and policy framework does exist in Ethiopia, loans are not reaching the poorest farmers. Some issues flagged to be addressed include the need for a registry system for financial assets, financial literacy, technological capacity and more training for financial specialists.

One conclusion that we draw from the study of agricultural finance in Ethiopia is that agricultural finance is extremely difficult, regardless of which players are involved in the delivery of credit. This provides an opportunity for the different actors to consider specialized roles. It is our view that the state should specialize in removing the risks associated with agricultural and rural finance. Once the state invests adequately in rural infrastructure, most of the risks associated with agricultural credit would disappear. With improved irrigation farmers would depend less on rainfall and the risk associated with rainfed agriculture would be reduced. Similarly, providing roads takes care of the risks associated with transportation. But for the financial markets, one significant contribution that the state can make is to help put together the information base on potential borrowers, possibly through the development of credit reference bureaus that are locally based. This can be best done by arrangements that involve local government bodies and not the central government. It calls for financial institutions working together with the local government agencies within decentralized structures. The state can also provide fiscal incentives for financial institutions that provide rural and agricultural finance to expand their rural activities. Tax rebates are a good example of what can be provided. Partnering with the private sector to provide agricultural credit is another option that might be considered by assisting properly structured microfinance institutions that are interested in agriculture. The state must concentrate on removing the obstacles to the effective functioning of rural financial markets.

Market functionality and social networks

There is a growing body of literature that suggests that in view of the poorly functioning nature of various markets in Africa, networks provide a valuable source for exchanges and for providing what markets fail to deliver (Fafchamps, 2003). What it is not very well understood, however, is how the networks operate under varied circumstances and what limitations are placed on their operations. The *Afrint* study also considered some of the related issues by looking at in-kind transfers between households.

The study by Andersson (Chapter 6, this volume) uses the *Afrint* surveys of 4000 households in nine sub-Saharan African countries to analyse the causes and effects of maize remittances among smallholders. These maize remittances occur in-kind and bypass market channels. The paper seeks to determine whether remittance patterns are a result of market failures or whether they supplement market activity. From the survey, 2857 smallholders produce maize, of which 1206 are maize remitters. Simple comparisons of the group demonstrate that remittances are not indicative of poor market access, as remitters both produce more maize and sell a greater percentage of it in the market than non-remitters. This finding is primarily driven by low amounts sold in the bottom two quartiles of maize production among non-remitters, as the upper two quartiles are similar in both groups. The author thus concludes that remittances appear to be part of a multi-spatial support network, perhaps reflecting a culture of gift-giving and reciprocity. The remittances do not appear to be a sign of a largely in-kind economy, as in-kind labour payments are minimal in the sample, and the remittances are more likely given with some sort of implicit or latent reciprocity.

The recipients of remittances are another key focus of the paper by Andersson (Chapter 6, this volume), and the data show that they are largely rural, some nearby and others not, and are probably food insecure. Remitters actually seem to forfeit their own consumption needs, as their consumption is similar to non-remitters. Nevertheless, market limitations causing price volatility and regular shortages may make avoiding the market quite valuable for receiving households. Cash reciprocity is limited, thus underscoring the likely importance of cultural obligations. Considering all of these aspects, the author notes that household size is effectively larger in terms of consumption needs, as higher-producing smallholders inevitably spread their surplus among a large group of receivers. The author also notes that staple foods constitute a surprisingly small percentage of cash income. Thus, to ease food security burdens among remitters and receivers, market failures may need to be addressed to incentivize higher productivity.

One conclusion to draw from this study of remittances is that social networks can be quite significant under varied circumstances and that different actors involved in them may have different motives for joining them. There is every reason to believe with greater economic development and greater diversity of the economy the networks and their members will have changing incentives for continuing membership. For the time being, social networks help to satisfy some of the needs of persons that cannot be satisfied by markets.

Moving Forwards on African Agriculture

Currently many African governments have expressed a strong desire to modernize their agricultural sectors and this is reflected by their acceptance of the Comprehensive Africa Agriculture Development Program (CAADP). The CAADP principles are:

- Agriculture-led growth as a main strategy to achieve the Millennium Development Goal of halving poverty and hunger by 2015.
- The pursuit of a 6% average annual growth rate for the agricultural sector at the national level.
- The allocation of 10% of the national budget to the agricultural sector.
- The exploitation of regional complementarities and cooperation to boost growth.
- The principles of policy efficiency, dialogue, review and accountability, shared by all New Partnership for Africa's Development (NEPAD) programmes.
- The principles of partnerships and alliances to include farmers, agribusiness and civil society communities.

These are ambitious principles but certainly achievable with greater commitment by African governments and their development partners.

There are a number of options to consider in terms of moving African agriculture forwards. These will have to incorporate what farmers and their governments in African countries can learn from other regions, especially Asia, as well as from their own history. Indeed there have been several studies of how African countries can learn from the Asian experiences in this regard. Common themes in what can be learned from Asia (Gabre-Mahdin and Johnson, 1999; Djurfeldt and Jirström, 2005) have included the following:

1. Technological progress based on increases in farm labour productivity supported by an effective agricultural research programme oriented to the needs of small farmers.
2. A broad-based agricultural development strategy supported by rural infrastructure and institutions to foster widespread increases in farm cash incomes, rather than a dualistic structure favouring large-scale units.
3. Rural industrialization based on rising rural demand that provides a source of non-farm, rural employment growth.
4. Rapid commercialization of the agricultural sector based on both exports and growth of the internal market in response to an expanding non-farm population dependent on purchased food.
5. Commitment to education and the strengthening of human capital, including a priority for maternal and child health programmes to improve child survival prospects, in order to provide a basis for industrial growth and to promote changes in fertility behaviour.

If Africa seeks to pursue this path, the role of governments cannot be overemphasized. A major lesson coming out of the country studies is that governments in the region have been important for the development of agriculture, even if there have been problems with the nature of their involvement. Their investments in agriculture had been expected to crowd in additional private investment, and when they failed to do so the private sector has been incapacitated. The chapter by Holmén (Chapter 3, this volume) summarizes the very mixed picture of the role of government in the agricultural sector. In looking at what governments have generally done, he suggests that there has been some coherence in terms of policy priorities. For instance, poverty reduction is now almost uniformly targeted as the ultimate goal, and attention to women is often a secondary goal, given their role in the sector. While budgetary commitments to agriculture have not been met in recent years, they are increasing in a number of places, thus reversing the trend from SAPs. Support from outside is still lagging, as G8 countries are falling short of their aid targets.

Holmén (Chapter 3, this volume) also notes that the private sector has been emphasized differentially across different countries and that crowding-out, whether intended or not, has been common. But even when governments have abandoned their direct activities in the agricultural sector, many specific crops and prices are still targeted by regulatory boards, and subsidies have made a comeback in many places. Despite these trends, the private sector does not appear to be ready to take over the government's role as a major investor in agriculture, as this is still very limited, especially in output markets. Granted that infrastructural and regulatory burdens are barriers to entry, Holmén is convinced that governments cannot yet exit from the process. Decentralization of government is considered crucial to the process of strengthening government involvement, but limitations at the local level have constrained their advancement. More capacity building is undoubtedly necessary to shift control to local areas. In sum, there is here a clear rejection of the approach reflected by the structural adjustment programmes, i.e. that government involvement should be almost entirely removed.

Moving forward, and despite the dire conditions at hand, Holmén and Hydén (Chapter 2, this volume) point out that there are three new factors in African agriculture, which hold some promise. First, there is a renewed commitment to agriculture as a driver of exports and growth, and the effort is led by African governments and institutions. There is now also broad recognition, especially among the donor community, that agricultural advancement is a multilateral process in which a wide range of stakeholders must be involved. The commitment has taken an official form through the CAADP, as earlier mentioned, and the commitment of governments to dedicate 10% of their budgets to agriculture. There is also promise that change will come from the bottom, as farmers begin to demand services from extension officers and become more integrated into the broader framework.

Also important for the future of African agriculture are scientific advances, which are taking place both in multinational corporations and, increasingly, in African research institutions. Research is much more targeted to local conditions and focused on productivity, and this research is backed by an array of donor-funded networks. Finally, policies that are more directly targeted at farmers, like input subsidies, are returning to the centre stage and are increasingly moulded to country conditions by African governments.

While identifying the opportunities at hand, Holmén and Hydén (Chapter 2, this volume) also identify key challenges that need to be overcome. External factors, such as the financial crisis and climate change, are a growing threat to agriculture. Domestic limitations may also affect the effectiveness of agricultural programmes, as governance gains have been modest. Increasingly, bottom-up processes must be put in place so that groups of smallholders can press for good management of programme funds at the local level. Finally, there are agricultural debates that have been around for a while, such as whether countries should focus on food crops or export crops, big farms or small farms, and what to do about land tenure reform. The studies here do not offer definitive answers to these questions, but in each case, they caution balance in addressing them. They conclude by noting that a smallholder orientation is necessary for agricultural development in Africa and that while a Green Revolution is possible, it is not guaranteed. They put the focus on African institutions and stakeholders to realize the possibilities.

There are some specific policy orientations countries might pursue, depending on their own circumstances. Holmén (Chapter 3, this volume) has observed that fertilizer subsidies, as mentioned above, are becoming common once again and appear to have worked in countries like Malawi, although they are expensive. Producer organizations have also been encouraged but, again, capacity is lacking for them to be a driver of progress. Commodity exchanges and warehouse receipt systems are types of market information systems and are being implemented in a number of countries. These offer producers promise of better returns and increased access to credit. Progress in infrastructure has been mixed, as budgetary allocations have increased in some countries and stagnated in others. Extension programmes, previously a target of SAP cuts, are increasingly recognized as important, especially because private sector provision of those services is weak. Contract farming and out-growing schemes are presented as a partial solution to improve smallholder agriculture but not as a substitute for public services. A conclusion to be drawn from Holmén (Chapter 3, this volume) is that, even though the state, sometimes against donor wishes, is again becoming more involved in agriculture, while the private sector remains weak, the emphasis on smallholders is still lacking, and this omission may put other key goals, including poverty reduction, in jeopardy.

Summary

Reflecting, at last, over the contrasts between the African Green Revolution, yet to take shape, and the well-known case of Asia, it is appropriate to ask how the African Green Revolution might differ from the Asian one. At one level, the answer is simple: it would involve other crops, be less focused on rice and wheat and be adapted to other water and climate regimes. More fundamentally, the African Green Revolution might not show the three characteristics observed in Asia, i.e. state-driven, market-mediated and smallholder-based, which were the focus in the earlier *Afrint* study (Djurfeldt and Jirström, 2005).

One of Karl Marx's best known aphorisms is that '…all great world-historic facts and personages appear, so to speak, twice… the first time as tragedy, the second time as farce' (Marx and Engels, 2001). Attempts to replicate the Asian Green Revolution in contemporary sub-Saharan Africa may come to look as tragic farces. Examples of these failures are already seen in Nigeria (Akande *et al.*, Chapter 11, this volume) and in the first attempts in Ethiopia in the early years of the new millennium to replicate the Asian Green Revolution (Wolday Amha, Chapter 7, this volume). In Nigeria, since the restoration of democracy, after the switch from military dictatorship, governments have pursued agricultural policies that have looked almost like textbook examples of Asian policies but with rather limited outcomes in terms of production and, above all, on the area productivity of food crops. Similarly, during the early years of the Meles regime in Ethiopia, the government adopted a credit-fuelled extension of high-yielding varieties, with the tragic effect of busting maize markets, making it impossible for farmers to sell their crops and repay their debts.

One comes to think of Marx's dictum when reflecting on recent crop biotechnology developments. The International Service for the Acquisition of Agri-Biotech Applications (ISAAA) claims that 'a second wave of biotech growth and development (has begun)' and that new biotech applications are spreading fast and benefitting millions of smallholders all over the world, with one prominent exception: sub-Saharan food crops are not yet drawn into the process (ISAAA, 2009). With the exception of South Africa, the first example in the region that reminds one of a potential revolution is the rapid multiplication of Bt cotton in Burkina Faso, which promises to help in reclaiming the country's position as a major supplier of cotton to the world market (after it had been nearly knocked out by the US dumping of Alabama cotton).

If this example is foreboding a future, crop technologies patented by the private sector will have to play quite a different role from what they did in Asia from the late 1960s onwards. Efforts to create partnerships with the multinational private sector will be critical and such organizations as the Alliance for a Green Revolution in Africa (AGRA) and others may provide some clear examples of how to stimulate public–private partnerships (PPP) in promoting a smallholder-based Green Revolution in sub-Saharan Africa. The PPPs that they promote foresee close cooperation with the private sector, including with multinationals like Monsanto, Syngenta, Yara and others. A redefined division of labour and responsibility between the public and the private sector might thus come to characterize the African Green Revolution when it comes of age and gains pace. The coming decade will show if such a recast of classical Green Revolution strategies will be potent enough to take the edge off the African food crisis. If so, capital may come to play another role in pro-poor agricultural growth than Marx had envisaged.

References

Aryeetey, E. and Udry, C. (2010) Creating property rights: land banks in Ghana. *American Economic Review, Papers and Proceedings* 100(2), 1–9.

Djurfeldt, G. and Jirström, M. (2005) The puzzle of the policy shift – the early green revolution in India, Indonesia and the Philippines. In: Djurfeldt, G., Holmén, H., Jirström, M. and Larsson, R. (eds) *The African Food Crisis: Lessons from the Asian Green Revolution*. CAB International, Wallingford, UK, pp. 43–63.

Djurfeldt, G., Holmén, H., Jirström, M. and Larsson, R. (eds) (2005) *The African Food Crisis: Lessons from the Asian Green Revolution*. CAB International, Wallingford, UK.

Fafchamps, M. (2003) Ethnicity and networks in African trade. *Contributions to Economic Analysis and Policy* 2:1, Article 14. Available at: http://www.bepress.com/bejeap/contributions/vol2/iss1/art14 (accessed 23 May 2010).

Gabre-Mahdin, E. and Johnson, B.F. (1999) *Accelerating Africa's Structural Transformation: Lessons from East Asia*. MSSD Discussion Paper 34, International Food Policy Research Institute, Washington, DC.

ISAAA (2009) *Global Status of Commercialized Biotech/GM Crops: 2009. The First Fourteen Years, 1996 to 2009*. ISAAA brief: Crop Biotech Update Special Edition, International Service for the Acquisition of Agri-Biotech Applications (ISAAA).

Marx, K. and Engels, F. (2001) *The 18th Brumaire of Louis Bonaparte* [electronic resource]. Electric Book Co., London.

Nissanke, M. and Aryeetey, E. (1998) *Financial Integration and Development: Liberalization and Reform in Sub-Saharan Africa*. Routledge, London.

United Nations (2009) World population prospects: the 2008 revision population database. Available at http://esa.un.org/unpp/ (accessed 23 May 2010).

World Bank (2008) *World Development Report 2008*. Oxford University Press, Oxford.

Index

Page numbers in **bold** type refer to figures and tables.

AEMFI (Association of Ethiopian Microfinance Institutions) 162, 171
African Union (AU)
  Abuja (2006) Declaration 288
  Maputo (2004) Declaration 3, 28, 237
  targets for agricultural budget allocation 53–54, 109, 223, 290
Afrint study projects
  Afrint I
    household survey questionnaire 16–18
    timing and scope 3–4
    village questionnaire 18
  Afrint II
    questionnaire and panel subset 18–19
    timing and objectives 4–5
  aims and findings 74–75, 78–81, 103–104
  attrition, factors responsible 19–20
  country-level methodology 10–11
  Ethiopian financial services
    kebele (peasant association) surveys 160
    modelling results 182–186
  fertilizer support programme, Zambia
    data collection and analysis 240, 248–250
    sampling frame choice 239–240
  maize-growers
    comparison of Nigeria with other countries 271–272
    panel data, use in modelling 110–114
    standardized remittance flow questionnaire 139–140
  Mozambique, sampling frame coverage 319
  sampling procedure
    bias, errors and corrections 15–16, 78, 111
    gender distribution, farm managers **99**
    household sampling 14–15
    region/site purposive sampling 13–14, 77–78
    sample subdivisions and sizes 13

Afrint study projects (*continued*)
selection of sampling frame 11–13, **12**
village definition and selection 14
AGRA (Alliance for a Green Revolution in Africa) 28–29, 76, 371
agriculture, in Sub-Saharan Africa (SSA)
crops and productivity 2, 23, 223, **224**
climate and soil fertility limitations 51, 191, 227
contract farming 65–67, 229, 304–305, 317
performance measurement 46
trends for staple crops 47–50, **48**, **50**
types grown and cropping systems 141
governmental role, variation between countries 354–355
budget allocations 28, 53–54, 109, 358–360, **359**
global comparisons 289, **289**, 290
government priorities 24–25, 50, 223
strategic policy choices 356–358
hunger/poverty alleviation prospects 1–2, 37, 47, 214, 219–220
policy challenges, current
agricultural sector issues 36–40, 52–54, 190–192
climate change 34, 294, 362
credit and financial service access 364–366
domestic capacity 34–36, 57–59, 286
equity and food security 282–283
extension service provision 64–65, 226, 286, 317, 318–319
food security and self-sufficiency 363–364
foreign investment 33–34, 290, 293, 336–337
political disturbances 230, 233, 316
population pressure and land scarcity 360–361
rural infrastructure development 50–51, 63–64, 193–194, 293–294
technology use and innovations 361–363
weak market structures 55–57, 138–139, 302–303, 366–367
progress, 21st century
balanced policy management 282
institutional commitment 27–29, 52–53, 310, 344–345, 367–369
pro-farmer policies 31–32, 59–63, 369–370
scientific advances 30–31, 369
reasons for marginalization/stagnation 214, 355–356
global financial and food price crises 26–27, 258, 309, 356
market interventions and reforms 25, 46–47
Structural Adjustment Programmes 25–26, 46, 221–223, 309

banks
credit for smallholders and small traders 57
formal, impact in Pakistan and India 163
loan applicant screening criteria 164–165
guarantee funds for microfinance 172–173
mergers and state capital grants 262
National Bank of Ethiopia (NBE) 162, 167–168, 169–170
private and state-owned, lending patterns 173–174, 295, **296**
biofuels
competition for agricultural land 4, 291–292

global demand, effect on commodity prices 329–330
government encouragement, Mozambique 318

CAADP (Comprehensive Africa Agricultural Development Programme)
emergence and adoption 3
growth rate targets 258, 288, 367
monitoring role 28
cash (export) crops
concessionaires, control of inputs 331, 333, 343
contribution to farmers' income 95–96, 142–143, 189, 195
household motivations for diversification 192–193
promotion, for economic development 36–37, 56, 317, 318
impact of tax measures **334**, 334–335, **335**
out-grower schemes (contract farming) 65–67, 229
Swynnerton Plan, Kenya 221
cassava
breeding for improved varieties 30
drought tolerance 243, 245, 253
government promotion 266
market growth and production 86, **87**, 269
export challenges 267
production data, survey problems 17, 86
yield increases 48, 299, **300**
climate change
need for irrigation development 294
research funding policies 34, 362
resistance of sorghum, breeding programme 86
contract farming 65–67, 229, 304–305, 331
cooperatives
benefits and limitations 38–39, 251–252, 307
federal and regional agencies 171
savings and credit unions 159, 168, 171–172, 228
rural sustainability 175, 307–308
state support and involvement 61–62, 159, 168, 170
see also farmer-based organizations
credit access *see* loans
credit guarantee schemes 172–173
cultivation, area of 9, 39
average, family farms 77
government extensification policies 47–48
specific staple crops, Nigeria **266**, 266–267

development strategies
African Union programmes 3, 53–54
financial services development 160–161, 185, 365–366
international aid 25, 33–34, 368
investment in public goods 243, 359
land tenure regulation 39–40, 291–292, 361
research on sustainable agriculture 31
six 'I's framework 215–216, **231–232**, 360
staple food and export (cash) crops balance 36–37, 56, 68, 192–193
strengthening policy-making capacity 32, 35, 59
support for smallholders 37–39, 59–63, 75–77, 108–109, 369–370
see also fertilizers, subsidies
diversification, land use 95–96
factors influencing decisions 195–196, 200–205, **203–204**
measurement methods 199–200
relationship with food security and self-sufficiency 189–190, 192–193, 205–207, **206**, 363–364

economic performance
effects of farming subsidies 32, 60–61, 228–229, 238–239
Ethiopia, recent growth rate 156
exchange rate fluctuations 327–328, **328**
GDP and agricultural sector growth rates
association with maize production 122–124, **123**
Ghana 191
Kenya 215, **215**, 216, **233**
Nigeria 257–258, 264, **265**
resource reallocation 223–224, **225**, 253–254
Tanzania 306, **314**
impact of global recession 27, 54, 107–108, 317, 318–319
inflation rate changes 287, 296, 316, 317
investment incentives and disincentives 290, 344–345
role of smallholder sector 75–77, 109, 257
tax regimes 262, 295, 333–335, **334**, **335**
endogeneity, statistical 111–112, 129–131
Ethiopia
agricultural sector performance 156–157
Derg era 157, 168
economic growth, 21st century 156
financial needs of smallholders 157–158
levels of provision 160–162, **161**
for technology investment 165
macro-level policy environment 162, 166–167, 187
regulatory frameworks 159, 169–170
rural development strategies 167–169
meso-level support institutions
infrastructural scope 162, 170–171, 173, 187
key players 171–173
micro (grass roots) level finance providers 162
formal 173–177
informal channels 177–178
semi-formal lending institutions (iqqub, iddir) 177
smallholder *Afrint I* and *II* surveys, analysis
access to finance 178, **179**
finance providers 178–180, **179**
loan access determinants 182–185
loan period 181–182, **182**
purpose of loans 180–181, **181**
study of financial service provision
comparable studies, worldwide 163–166
conclusions from *Afrint* surveys 185–187
data collection and analysis 160
objectives 159–160
export crops *see* cash crops
extension services
education initiatives 336
government investment 220–221, 357
implementation at local level 29, 294–295
improved farm management practices, training 247, 253, 317, 319
local usefulness of advice 340
reasons for low productivity impact 322–325, 356
linkages with research networks 341–343
providers and funding 64–65, 183, 247, **248**
public–private partnerships 226
related to gender and FBO membership 191, 200, **202**, **352**
staffing issues 65, 286, 295
geographical coverage 338–340
knowledge of market conditions 341

farm size
average, smallholders 77
changes, in *Afrint* study period 78, **79**
per capita distribution 78, **80**, 81
inequalities and variation 81, 98–103
policy direction, big or small farms 37–39, 76–77
farmer-based organizations (FBOs) 200, **202**
effect of membership on diversification 204
membership for subsidised fertilizer access 251–252
quality standard control 307–308
sustainability and funding resources 326
fertilizers
adoption and disuse by farmers 115, 270, 285, 296–297, **297**
access and dose rates 229, **229**, 242, **242**
maize-growing, expenditure 141–142, 241, 250
microdosing 320
prices
influence of energy costs **330**, 330–331
supply control and farmer profitability 331–333
subsidies 60–61, 131, 228–229, 287–288
beneficiary targeting 241
effect on price and supply 240–241, 261
support programme (FSP), Zambia 238–239, 240–243, 362–363
use related to staple crop productivity 94, **94**, 119–121, 252–253, 332
financial service provision
demand and delivery, Ethiopia 6–7, 365
cooperatives 171–172, 175
credit guarantee and training schemes 172–173
delivery approaches 158–159, 160–162, **161**
informal and semi-formal 177–178
microfinance institutions (MFIs) 171, 174–175
smallholder demand challenges 157–158
loan availability and uses 164–166, 180–181, **181**, 227–228
outreach to small farmers 163–164, 295, 343
small trader participation problems 57, 286–287
see also banks; loans
food security
causes of crisis in sub-Saharan Africa 23, 24–27
determinants in rural households 205–207, **212–213**
government and donor support programmes 176–177, 238–239
crop-specific targets 265–266, **266**
regional targeting 364
measures and estimates
household estimate 199
production per consumption unit (PCU) 99, 150
regional variation (Tanzania) **314**, **315**
production stability and food emergencies **49**, 49–50
related to food self-sufficiency 189–190, 192–193, 301
role of in-kind remittances 140, 149–151, **150**, 153, 365–366
see also self-sufficiency, staple foods

gender
related to productivity and income 98–103, **100–101**, **102**, 124, 207
status and support for women as producers 52–53
extension service access, Ghana 191, 200

Ghana
  agricultural production performance **191**, 191–192
  cash and subsistence crops 192–193
  development policies and targets 190–191, 194
  farmers and food, descriptive statistics 200, **202**
  infrastructure development 193–194
  modelling study of production choices, methods
    econometric models 196–197
    equations and hypothesis framework 195–196
    household decision assumptions 194
    study villages 197–198
    variables and their measurement 198–200, **199**, **201**
  study results
    food security, impacts of diversification 205–207, **206**, 363–364
    geographic regional variation 208
    impact of infrastructure and resources **203–204**, 204–205
    land allocation to non-staple crops 200, **202**, 202–204, **203–204**
governments, sub-Saharan Africa
  accountability 34–35
  agricultural policy priorities 4, 5, 27–28, 217, 356–358
    budget allocation commitment 53–54, 121, 156, 220–221, 223–224
    cultivation area expansion 47–48, 278
    farming subsidies 32, 60–61, 131, 228–229, 240–243
    food security programmes 176–177, 190–192, 258
    foreign influence 23–24, 46, 281, 336–337
    gender issues 52–53
    market support 61–67, 168–169
    relationships with private sector 55–57, 168, 226, 238–239, 253
  legislative and regulatory environment 159, 169–170
  macroeconomic stability 166–167, 233–234, 327–330
  political consolidation 24–25, 50, 58–59
    threats to stability 230
  relationships with financial donors 25, 290, 356
    Chinese investment 33–34, 63
    land ownership issues 39, 291–292
    local demand-driven funding 35–36, 57–58, 65, 230, 337
    rural loans promotion policies 167–169
    Western (overseas) aid 28, 30, 54, 281
  see also development strategies; economic performance; Structural Adjustment Programmes
Green Revolution
  African
    access to research and technology 41–42, 279
    policy aims and effectiveness 4, 124, 258–259, 343–345
    public–private partnerships 371
  Asian, relevance to Africa 67–68
    differences 35, 41, 309, 370
    recommended priorities 40–41, 45, 367–368
  budget investment levels 3, 290
    cost-effectiveness of interventions 278–279

income, household
  non-agricultural share 96–98, **96**, **98**, 142

cash remittances from family members 153
effect on diversification 205
small-scale micro-business 102
percentage generated from staple crops 95, **96**, 142
effects of cash crop production 193, 195
related to remittances 147, **147**
related to land assets and gender 98–103, **100–101**, **102**
rural and urban linkages 138–139, 147–148, 219–220
infrastructure, rural
consequences of neglect 222–223, 303
development costs 50–51, 226–227
financial and communication 162, 164, 194, 293–294
'hard' and 'soft,' definition 190
local identification of priorities 230
measurement indicator 200
road and transport systems 63–64, 193, 293
intensification, agricultural
maize production, Kenya 7–8, 220–221, 223–230
mathematical modelling 114, 275
potential for, regional variation 77, 221
international financial institutions (IFIs)
development aid for SSA states 25, 263–264
exclusion from loan delivery, Ethiopia 170
impact on economic policies 31–32, 37, 47
attitude to subsidies 60
debt relief and donor support 242
pressure for decentralization/liberalization 58, 59, 281
International Fund for Agricultural Development (IFAD) 172
irrigation
adoption by smallholders 227, 294
need for loan capital 324–325
climatic requirement 51, 204
effect on crop yields 88, 270
public investment 169, 234, 264, 365–366
as risk reduction strategy 194
see also rainfall dependence

Kenya
agricultural status and performance 214–215, **215**, 216
conditions for intensification 215–216
maize production trends
area cultivated **218**, 218–219
food and income importance 217–218
total production, tonnes 218, **218**
yield (productivity) 219, **219**
poverty statistics 216–217
productivity related to credit access 164, 222, 227–228, 360
socioeconomic history of agricultural policies 216, 233–234
performance revitalization, 2003–2007 223–230, **225**
post-election (2007) prospects 230–233, **233**
rapid productivity increase, 1963–1985 220–221
structural adjustment programmes impact, 1986–2002 221–223

land tenure
customary ownership transactions 39–40
distribution inequality 361
land leases to foreign companies 39, 292
statutory regulation 40, 169, 291–292, 361
transfer from European to African ownership 220, 234
land use
allocation to staple crops 81, **82**, 291, **291**
area of cultivation 9, 39, 47–48, 361

land use (*continued*)
competition, food and biofuels 4, 291–292
diversification 7, 95–96, 192–193, 200, 202–205
intensification 7, 215, 275
livestock farmers, uses of credit 164–166
loans (credit access)
donor capital funding schemes 172–173
effect on extension service effectiveness 324–325
exclusion of poorer households 174–175, 248, 343
interest rate control 262
legislative/government support for microfinance 159, 167–168, 227–228
loan period 181–182, **182**
semi-formal and informal borrowing 177–178, 305
sustainability 178–180, 186
variable features 365
borrowing probability 182–183, **183**
loan size 183, **184**
repayment performance 183–184, **184**, 222
warehouse receipt systems 62–63, 305

maize production
countries, differences between
analysis of drivers of change 272–279
maize-growing households 141, **141**
production levels 126, **127–128**, 141, **142**
remittances to relatives 143, **143**
data collection and analysis 110
household questionnaire survey 139–140
variables and indicators 114–116, **118–119**, **133–135**
historical role and crop qualities 139
influencing factors
drought and floods 119, 218, 223
farm household age (Chayanov effect) 116, 117
farm inheritance by descendants 117
farm size increase 117, 119
gender and elite (wealth group) membership 116, 124, **125–126**
geographical factors 285, 298–299
macroeconomic environment 115–116, 121, 124, **129**
market participation 115, 121, **122**, **136–137**, 143–149
recent production trends
commercial incentives and state involvement 131–132, 360
productivity variation, causes 83
total production and cultivation area 48, **49**, **84**, 86, 115
yield 47–50, 81, 83, **84**
subsistence production
consumption needs 149–151
payment for agricultural labour 148–149
support for urban relatives 144, 151–153
technology use 119–121, **120**, **135–136**
fertilizer inputs 115, 120–121, 141–142, 320
hybrid/improved varieties 142
ploughing, adoption of 115, 121, 128
see also Kenya, maize production trends; Zambia, maize productivity determinants
Malawi
agricultural budget allocation 53
AISP (Agricultural Input Support Program) 60–61, 108
credit access and financial infrastructure 164

staple crops, productivity increases 48, 83
markets
control by ethnic and traditional leaders 57, 59
government intervention 8, 25, 109, 287–288
exchanges and information systems 62–63, 224, 293–294, 366
liberalization policies 222, 238–239, 286–287, 333–335
producer organizations, promotion of 61–62, 319
transitional measures to market-led development 169, 306–307
interaction with in-kind remittances 139, 143–144, 366–367
commodities as labour payment 148–149
response to market shortages 146–148
marketing boards, benefits and limitations 304
missing markets
agent shortage 56–57
as barrier to smallholder commercialization 66, 189–190, 192, 304–306
role in agricultural development 2–3, 55–56, 131, 138–139
staple and high-value crops compared 94–96, **95**
transaction costs 68, 108, 190, 303–304
see also trade
microfinance institutions (MFIs)
equity and credit funds 172–173
government support 159, 167–168, 365
regulation and supervision 169–170, 171
target clients and risks 174–175, 343
Millennium Development Goals (MDGs) 3, 45–46, 216
on hunger and poverty, progress 47, 282
role and commitment of governments 5, 52, 54
incorporation in development plans 237, 239, 288–289
political instability, consequences of 233
modelling
credit access in Ethiopia
logit estimation models 182–185
multiple regression model 183, **184**
primary and secondary data sources 160
econometric models of crop production, Ghana
measurement of key variables 198–200
pooled tobit and probit model estimations 196–197
probit model hypotheses and results 205–207, **212–213**
sampling techniques 197–198
tobit model hypotheses and results 200, 202–205, **211–212**
two-sample t-test (non-staple land allocation) 200, **202**
fertilizer support programme impact, Zambia
multiple linear regression model specification 249–250
quantitative statistical methods 240
variables and hypothesized relationships 249, **255–256**
maize production, drivers of change
data characteristics 110–112
endogeneity 111–112, 129–131
model robustness 126, 128–129
reduced form modelling strategy 112–114, 117
monopsony, effect on small farmers 303–304, 318, 331

Mozambique
history
national economic growth 316–317
post-colonial political events 316, 345
small farmers' status and prospects 317–319
household and village survey methods 319
institution building
agricultural privatization and strategic reforms 336–337, 345
agricultural research network 336, 341–343, 357
financial services 343
rural extension system 338–341, **352–353**, 356–357
land and resource assets 320
production and productivity factors
crop yields **321**, 322
farm-level performance 320–325, **322**
fertilizer and agrochemical costs 331–333
impact of taxes 333–335
macroeconomic/international environment 327–330
sustainability 325–327
smallholder poverty 318, 320, 345–346
strategies for development 343–345, 346, 357

NEPAD (New Partnership for Africa's Development) 3, 28, 76, 237
NERICA (New Rice for Africa) project 30–31, 262, 298
Nigeria
agricultural sector performance
crops and livestock production 257–258
productivity growth rates 258–259
economy and government 257
food and agriculture policies and programmes 357–358
agricultural research support 262–263
fertilizer subsidies 261–262
fiscal (credit support) policies 262
individual commodity initiatives 259–260, 261
international cooperation 260, 263–264
outcomes and recent agricultural performance 264–265, **265**
price stabilization and strategic food storage 263
state-level policy framework (NEEDS) 260
trade tariffs 260–261
maize production, comparative analysis
commercialization, smallholder **276**, 276–277
cultivated area intensification 275
macroeconomic variables **277**, 277–278
ploughing impacts 275–276
production model results 272, **273–274**, 278–279
scope and study methods 271–272
seed fertilizer technology adoption 272, **275**
productivity trends, food crops
area under cultivation, specific crops 266–267, **267**
farm gate and market prices 268–269, **270**
irrigation development 270
production strategies and total output 267, **268**
targets 265, **266**
technology input use 270
variability, year-to-year 266
yield and yield gaps 268, **269**
non-governmental organizations (NGOs)
as channel for development aid 25
government regulation, Ethiopia 170, 173
continuing financial service provision 175–176

promotion of drought-resistant crops 245, 253
provision of farmers' extension services 64, 65, 226, 295

OECD (Organisation for Economic Co-operation and Development)
commodity price predictions **329**, 329–330
producer and export subsidies 4, 132

PAFP (Pan African Farmers Platform) 29
participatory rural appraisal (PRA) ranking 15
PASDEP (Plan for Accelerated and Sustainable Development to End Poverty, Ethiopia) 167–169
policies *see* development strategies; governments, sub-Saharan Africa
population
growth rate 23, 360
low rural density, consequences 50–51, 76
urban/rural balance 24–25, 75, **151**, 151–152
poverty
extent in sub-Saharan Africa 45–46, 216–217, 288–289
contributing factors, Kenya 217
regional variation, Tanzania **305**
impact on smallholder productivity 51–52, 98–103
financial access and farm investment 157, 160, 305, 318–319
pathways out of poverty trap 214, 345–346
remittances and reciprocity 148–149
prices
fertilizer and seed 240–241, 242, 252, 261
food
policies for stabilization 263
recent global increases 4, 26–27, 107, 328–329, **329**
related to domestic demand 51–52, 144
input/output ratio 288
profitability and crop storage 51, 324, 326, **326**, 341
pesticides 333
petroleum, effect on food and agriculture 26, 318, 327
impact on fertilizer prices **330**, 330–331
price information, and commodity exchanges 62–63, 303–304
private sector
agricultural market agents 56–57, 286
coordination challenges 226
participation constraints 57, 238–239, 290, 368
relationship with public sector 55–56, 245–246, 253
role in biotechnology supply 363, 371
productivity
drivers of change 108–109, 131–132, 271–279, **273–274**, 298–300
effects of farm size 37–39, 76, 99
improvement through research 30–31
staple crops
cassava 86, **87**
maize 81, 83, **83–84**, 86
rice 88, **91–92**
sorghum 86, 88, **88–89**
trends related to population/labour 285, **285**
variation between countries 48, **48**, 223, **224**, 361
see also maize production; yield

rainfall dependence
diversification as insurance strategy 193
due to undeveloped irrigation potential 270
small scale water management 227
see also irrigation

remittances, in-kind
amounts, by country 143, **143**
gift and reciprocal support culture 148–149, 366
rural kinship networks 152
by household, compared with sales 144, **144**
correlation analysis 144–146, **145**
related to productivity 147
related to market performance 139, 143–144, 146–148, 367
urban food security impacts 144, 151–153
research, agricultural
focus on African food crops 30–31
government support 220, 262–263
networks and coordination 30, 341–343
products as public goods 41–42
sustainable development 31
socio-economic research 342–343
rice
land management
mechanization 292
multinational investment, Mozambique 318
NERICA cultivar 30–31, 262, 298
production and yield data 88, **91–92**, 299, **300**
RUFIP (Rural Financial Intermediation Program) 172

SACCOs (Savings and Credit Cooperatives) 159, 168, 171–172, 175, 228
seed varieties, improved
adoption rates by farmers 93–94, **94**, 297–298, **298**
dependence on output price 327
maize 142, 228, **246**, 246–247
rice 30–31, 298
sorghum 246, **246**
effect on productivity 252
self-sufficiency, staple foods
effect of import tariffs and tax regimes 333–335
as food security priority 189–190, 283
maize equivalent ratios 198, **199**
political motivation 107–108
trends in recent decades (Tanzania) 300–302, **301**
urban and rural family links 140, 148, 153
Simpson Index 199–200, 207
smallholders
access to agricultural extension services 64–65, 226, 294–295, 317
sustainability of project impacts 325–327
age profile 292
commercialization and market integration 65–66, 94–96, **95**
maize growers 115, 121, 131
marketing outlets, staple crops 229–230, **230**, 245–246, **246**, 302–303
poverty traps 51–52, 140
urban subsistence linkages 140, 144
farm size 77, 78–81, **79**, **80**
land assets and gender 98–103, 124, **125–126**
related to loan repayment 185–186
related to subsidy targeting 241
financial services access 157–158, 178–180, **179**
influencing factors 182–185, **183**, **184**, 247–248
opportunities and uses of money 157, 160, 164–165, 180–181
income 6
farm gate and market prices 268–269, **270**, 303, 331–332
influences of trade middlemen 303–304
non-farm sources 96–98, **96**, **98**
related to savings 184–185, **185**, 186
staple and cash crop sales 142–143, 229–230, 302
productivity, staple crops 81–88, 147, 204, 208–209

training in farm management practices 247, 253
yield gaps 88, 93, **93**, 108
as proportion of farming sector 51, 215
protection of, in contract farming 66–67, 304–305
role in African development 75–77, 138–139
women's roles 52–53, 103, 152, 251
see also technology adoption
social networks *see* remittances, in-kind
soil fertility 51, 191, 285
sorghum
cultivation area expansion 47–48
high-yield hybrids
adoption by farmers 246, **246**
breeding 86
impact of maize seed and fertilizer subsidies 243, **244**, 245, 253
productivity data 86, 88, **88–89**, 299–300
state intervention *see* governments, sub-Saharan Africa
Structural Adjustment Programmes (SAPs)
administrative decentralization 57–59, 281–282, 316–317
effects on poverty 3, 46, 282
impact on agricultural sector 25–26, 47, 49–50, 221–223, 309
subsistence farming
consumption requirements, modelling 195–196
proportion of smallholder households 95, 104
variation between crops, regions and years 302
prospects for diversification 76–77, 189–190, 195
as response to unstable input prices 285–286, 304
responsibility for in-kind remittances 139, 149–151

Tanzania
government policy structures
ASDS (Agricultural Structure Development Strategy) 283–284, **284**, 289
global macroeconomic environment 281
history and aims of market reforms 281–283
institutional reform and decentralization 283–285
Kilimo Kwanza (Agriculture First) programme 299, 301
productivity decline from 1986 **285**, 285–287
recommendations 310, 358
related to 21st century international targets 288–291
restoration of subsidies 287–288
transformative policies, objectives for future 306–308
Kilimo Kwanza (Agriculture First) programme 9, 28
production and food security **314**, **315**
crop production 298–300, **299**, **300**
farming transaction costs 303–304
food self-sufficiency 300–302, **301**
missing markets 304–306
subsistence markets (maize, rice, cassava) 302–303
resources, access and use
extension and financial services 286, 294–296, **296**
fertilizer and other inputs 285, 296–298, **297**, **298**
infrastructure (transport and communication) 293–294
irrigation 294
land use and land tenure **291**, 291–292
tools and implements 292–293
taxes 262, 295, 327, 333–335
technology adoption 119–121, **321**, 361–363
agrochemicals 297, **297**
efficiency of use 320
financial constraints 222, 286
dairy farmers 165–166

technology adoption (*continued*)
free access to public goods 41–42
risks for smallholders, due to poverty 51–52, 94, 98, 318
survey data on adoption rates 93–94, **94**, 228, **228**
tools and implements
hand hoe use, Tanzania 292–293
oxen and mechanical tools 250–251, 275–276
ploughing benefits 115, 121, 128
use of improved seed and fertilizer **246**, 246–247, 252, 272, **275**
regional and crop comparisons 296–298, **297**, **298**
trade
global negotiations, Doha development round 121, 132
globalization 76, 354
tariffs and import controls 260–261, 333–335
see also markets
transport *see* infrastructure, rural

urbanization 24–25, 75, **151**, 151–152

World Development Report 2008 (World Bank) 4, 31, 75

yield
annual fluctuations **49**, 49–50, 266
gaps
compared to experimental potential 268, **269**
compared to world standards 47, **50**, 88, 332, **332**
between farmers 88, 93, **93**, 108, 361
see also productivity

Zambia
agricultural budget resource allocation 237–239, **238**, **239**, 359
need for diversification 253–254
FRA (Food Reserve Agency) 238–239, 245, 253, 362–363
FSP (fertilizer support programme) elements
50/50 seed and fertilizer subsidy 238, 241
Food Security Pack (free fertilizer distribution) 238, 245
public fertilizer production 238
impact of FSP, analytical study
analysis methods and survey scope 239–240, 248–250
long-term sustainability and value 242–243, 253, 362
macro-level analysis 240–243
micro-level analysis 243–248
policy inconsistency and implementation 242
policy recommendations 252–254
maize productivity determinants
non-significant 251–252
statistically significant 250–251
study results
beneficiary targeting and distributions 241, 242, **242**, 250
credit access 247–248, **249**
crop diversification (drought-tolerant staples) 241, 243, **244**, 245, 253
extension services access 247, **248**, 252
farmers' output marketing channels 245–246, **246**, 251
fertilizer delivery delays and mis-timing 241
gender and productivity trends **244**, 244–245, 251, 252
improved seed varieties, use of **246**, 246–247, 252
private sector participation and market distortion 240–241, 253
regional differences 252